...duction to
Fuzzy Control

Springer
Berlin
Heidelberg
New York
Barcelona
Budapest
Hong Kong
London
Milan
Paris
Santa Clara
Singapore
Tokyo

Dimiter Driankov Hans Hellendoorn
Michael Reinfrank

An Introduction to Fuzzy Control

With cooperation from Rainer Palm,
Bruce Graham and Anibal Ollero

Foreword by Lennart Ljung

Second, Revised Edition

Springer

Prof. Dr. Dimiter Driankov
Dr. Hans Hellendoorn
Dr. Michael Reinfrank
Siemens AG, ZFE T SN 4
Otto-Hahn-Ring 6
D-81730 München
Germany

With 152 Figures

Cover picture by Karl Sims, Thinking Machines Corporation, Cambridge, MA, generated on the Connection Machine®

ISBN 978-3-642-08234-4

Library of Congress Cataloging-in-Publication Data applied for

Die Deutsche Bibliothek - CIP-Einheitsaufnahme
Driankov, Dimiter: An introduction to fuzzy control/Dimiter Driankov; Hans Hellendoorn; Michael Reinfrank. With cooperation from Rainer Palm ... Foreword by Lennart Ljung. 2.ed. - Berlin; Heidelberg; New York; Barcelona; Budapest; Hong Kong; London; Milan; Paris; Santa Clara; Singapore; Tokyo: Springer: 1996

NE: Hellendoorn, Hans; Reinfrank, Michael

Printed in the United States of America

Cover Design: Künkel+Lopka, Ilvesheim.
Printed on acid-free paper

Foreword

Fuzzy control has met a tremendous interest in applications over the past few years, and also among manufacturers of control equipment. Researchers in automatic control, however, have shown less interest – and sometimes also suspicion – for the area.

There are a number of reasons for the "popularity" of fuzzy control. First, real life control objects are nonlinear: their dynamics change with the operating point and there may be other essential nonlinearities in the process. This calls for regulators with nonlinearities. Traditionally, this has been dealt with by various *ad hoc* tricks, such as max selectors and so on. A more advanced version is called gain scheduling – i.e., changing the controller parameters with the operating point. Also these more traditional techniques have been met with relatively little interest in the automatic control research community, perhaps primarily since it is very difficult to analyze the performance of such regulators. Instead, changing control modes, depending on various signal levels has been created on a case by case basis by the control designer, reflecting his or her insight into the control object's properties.

In my mind (which is a conventional control theorist's) what fuzzy control offers is a better user interface to this process of translating system insight into controller nonlinearities. This is useful and valuable, since performance improving nonlinearities are often underutilized features in our regulators.

However, the fact remains that the performance of the resulting regulators is difficult to analyze due to their nonlinear effects. It is even difficult to establish general and non-conservative criteria for such an essential, qualitative property as closed loop stability. Sometimes these fundamental features of feedback have not been properly taken account of in the fuzzy control approach and that may be a main reason for the control community's hesitation. It is therefore a great pleasure to see this book by Driankov, Hellendoorn and Reinfrank that gives a mathematical description of fuzzy control in itself, and links it with classical control tools for analyzing closed loop stability. At the same time, the book gives an overview of the techniques and the potentials of fuzzy control. I would like to see this as a good starting point for bringing the techniques of fuzzy control into closer contact with traditional control theory.

Linköping *Lennart Ljung*

Preface to the Second Edition

The growing interest in fuzzy techniques and in their application to difficult control problems and the encouraging reception of our book have called for a new edition. After publication of the previous edition, many reviewers sent us their comments. We are grateful for their valuable suggestions and have incorporated them in the new edition.

The main differences from the first edition are that Chapter 4, "Nonlinear Fuzzy Control," has partly been rewritten and brought up to date, and some of the material in Chapter 2, "The Mathematics of Fuzzy Control," has been reorganized.

We would like to thank Springer-Verlag, in particular Hans Wössner, for publishing this book and for the excellent collaboration.

Munich, November 1995

Dimiter Driankov
Hans Hellendoorn
Michael Reinfrank

Preface to the First Edition

There are a few bottle-necks hindering industry from a broader exploitation of the application potential of fuzzy control. In the first place, a better and more systematic design and analysis methodology for fuzzy control applications is needed. However, it should be clear that such a universal theory does not exist for conventional control engineering either, so we have to proceed from the few isolated spots where we already know exactly how to design a fuzzy control algorithm to clusters of problems and related design methodologies. The absence of such a coherent and systematic methodology makes the second main problem for fuzzy control even more serious: the lack of well-trained and experienced "fuzzy control engineers." It means that engineers who want to or have to get into this new technology have only limited support and guidance and must by and large rely on their own experience gained through experimental work. This book has been written to "widen" exactly these bottle-necks. Of course, there are many books on fuzzy control available. Most of them fall into one of the following three categories:

- text books on fuzzy logic and fuzzy set theory, requiring a high degree of mathematical sophistication of the reader,
- collections of papers describing applications of fuzzy control, usually not providing the necessary theoretical background,
- user manuals and guidelines, focusing on a particular fuzzy control tool.

What is still missing, in our opinion, is an introductory text book for control engineers, covering just the relevant part of the theory, and focusing on the principles of fuzzy control rather than particular applications or tools. Therefore, we tried to structure this book along the lines of a standard control engineering introduction: starting with the relevant mathematics, in this case fuzzy logic rather than Z-transforms and differential equations, etc., going through design parameters and choices and discussing fuzzy control in control engineering terms such as linear, nonlinear, and adaptive control, and stability criteria.

The book is explicitly aimed at readers from the control engineering community, in particular at engineers in an industrial environment who want to learn about fuzzy control, and at university students who need a set of lecture notes and a reference for their studies. However, Chapter 2 can be read without any knowledge of control theory and Chapters 1 and 3 require just some basic control theoretic knowledge; thus they can be of interest to members of the fuzzy and non-fuzzy AI community too.

In **Chapter 1**, Section 1.1, we discuss the relevance of the use of fuzzy control in an industrial environment. We show the benefits of fuzzy control, discuss the common prejudices against and mistaken ideas about it, and deal with some requirements that are necessary to bring the theory into action. Furthermore in this chapter, we introduce two major classes of knowledge based systems for closed-loop control: one class where the knowledge based system is involved in the supervision of the closed-loop operation, and another class where the knowledge based system directly realizes the closed-loop operation.

In Section 1.2 we consider the above two particular classes in the perspective of a variety of other uses of knowledge based systems for problems in process control.

In Section 1.3 we introduce the notion of a knowledge based controller by identifying it as a knowledge based system used for the specific task of on-line, closed-loop control. We discuss two principally different uses of a knowledge based controller: (1) as a direct expert control system, and (2) as a supervisory expert control system. Furthermore, we introduce the notion of a fuzzy knowledge based controller (FKBC) as a particular instance of a direct expert control system realized with the help of fuzzy logic knowledge representation and inference formalisms.

In Section 1.4 we discuss the general knowledge representation (modelling) issues involved in the design of knowledge based systems for supervisory and direct expert control. With the help of examples we illustrate the basic knowledge representation constructs and their levels of abstraction and resolution.

In **Chapter 2** we introduce those parts of fuzzy set theory and fuzzy logic which are relevant for the design of FKBCs.

In Section 2.1 we introduce the basics of fuzzy set theory in parallel with the corresponding notions from classical set theory.

In Section 2.2 we introduce the concept of a fuzzy relation and a set of operations on such relations which are used in the inference engine of a FKBC.

In Section 2.3 we consider in detail the knowledge representation constructs of approximate reasoning such as linguistic variable, fuzzy propositions and fuzzy if-then statements. We emphasize on the formal treatment (symbolic and meaning representation) of fuzzy if-then statements and inference with such statements.

In Section 2.4 we consider inference with a rule base consisting of a set of fuzzy if-then statements. We only consider those types of inference which are relevant in the design of FKBCs. We study the properties of a rule base such as completeness, consistency, continuity, and interaction and show their relevance for the performance of a FKBC.

In **Chapter 3** we introduce the principal design parameters of a FKBC and discuss their relevance with respect to its performance. Different design options for particular design parameters are presented; choice of membership functions, defuzzification methods, form and meaning of rules, and the inference engine are considered in detail. Other design parameters like scaling factors, derivation of rules, tuning of the FKBC, and stability analysis, are only informally discussed. Thus, this chapter gives a general view of the FKBC design problem and prepares the reader for the formal treatment and/or presentation of systematic techniques for rule derivation and scaling factors determination (Chapter 4), FKBC tuning and adaptation (Chapter 5), and stability analysis (Chapter 6).

In Section 3.1 we introduce the principal structure of a FKBC in terms of fuzzification model, knowledge base, inference engine, fuzzification and defuzzification model, and rule base. For each of these components we identify the corresponding design parameters.

In Section 3.2 we consider in detail the rule base of a FKBC. We emphasize the design of rules for different kinds of FKBC and the interpretation of these rules in the phase plane.

In Section 3.3 we describe the design parameters of the data base of a FKBC, such as membership functions and scaling factors, and their relevance for the performance of the controller.

In Section 3.4 we describe the inference engine of a FKBC. We introduce the representation formalism for single if-then rules and a set of such rules, and for the controller input and output. Furthermore we consider formal inference with if-then rules, which is of two types: composition based inference and individual-rule based inference.

In Section 3.5 we present the formal treatment of FKBC input fuzzification in the context of composition based inference and individual-rule based inference.

In Section 3.6 we present the six most-used defuzzification methods and introduce some criteria for their comparison and evaluation.

In **Chapter 4** a FKBC is considered as a particular type of nonlinear controller. From this point of view, the rule base, consisting of a set of if-then rules, is transformed into a nonlinear transfer element. Thus, if there is a good enough analytic model of the process, we can adopt design and analysis methods from nonlinear control theory. One such method, presented in detail in this chapter, is the so-called sliding mode control.

In Section 4.1 we introduce the reader to different types of linear, nonlinear and fuzzy knowledge based controllers.

In Section 4.2 we present the computational structure of a FKBC and within it identify sources of nonlinearity which make the FKBC a nonlinear transfer element. We also consider a rule-based representation for conventional transfer elements aimed at making the relationship between FKBCs and conventional controllers more transparent.

In Sections 4.3 and 4.4 we consider three basic types of fuzzy controllers: (1) PID-like FKBCs with variants PD- and PI-like FKBCs; (2) Sliding mode FKBCs with variants involving boundary layer, compensation term and higher order control; (3) Sugeno FKBCs, i.e., linear filters with variable coefficients which are nonlinear functions of the state vector. For the first type of FKBC, we illustrate the design procedure for a PD-like FKBC using a phase plane approach. For the second type of FKBC we show the similarity between sliding mode control with boundary layer and FKBCs in which the data base has been derived from the phase plane. Furthermore, based on this similarity, we propose design rules for a FKBC with respect to a normalized phase plane. For the third type of FKBC we present the controller structure and describe local and global stability analysis techniques for this type of controller. These techniques are exemplified in the case of control of a one-link robot manipulator.

In **Chapter 5** we consider tuning and adaptation of FKBCs. FKBCs are nonlinear and so they can be designed to cope with a certain amount of process nonlinearity. However, such design is difficult, especially if the controller must cope with nonlinearity over a significant portion of the operating range of the process. Also, the rules of the FKBC do not, in general, contain a temporal component, so they cannot cope with process changes over time. So there is a need for adaptive FKBCs as well. A FKBC contains a number of parameters that can be altered to modify the controller performance, i.e., the scaling factors for each variable, the fuzzy set definitions, and the if-then rules. A self-tuning FKBC modifies the fuzzy set definitions and the scaling factors on- or off-line. A self-organizing FKBC modifies the rules on- or off-line.

In Sections 5.1 and 5.2 we introduce self-tuning and self-organizing FKBCs and consider their adaptive components such as the process monitor and the adaptation mechanism. We also consider the use of a fuzzy process model in on-line identification.

In Section 5.3 some of the main approaches to the design of adaptive FKBC are described in detail. These include (1) membership function tuning using gradient descent and performance criteria, (2) rule modification using self-organizing FKBC with a performance monitor, and (3) rule modification using a fuzzy model-based FKBC.

In **Chapter 6** we consider several approaches for stability analysis of FKBCs. In the fuzzy control literature this type of analysis is usually done in the context of the following two views of the system under control: first, classical nonlinear dynamic systems theory, where the system under control is a "non-fuzzy" system and the FKBC is a particular class of nonlinear controller; second, dynamic fuzzy systems. The work presented in this chapter corresponds to the first view.

In Section 6.1 we introduce the basic approaches to the stability analysis of FKBCs.

In Section 6.2 we consider the so-called state space approach, based on the relationship between the phase plane of the system under control and the FKBC rules. In combination with a geometric interpretation of the contributions of the vector fields of the plant and the controller, this approach can be used to predict stability as well as other dynamic phenomena like limit cycles, isolated areas, oscillations, etc.

In Section 6.3 we formalize the problem of stability analysis by using concepts extracted from the qualitative theory of nonlinear dynamic systems. These concepts are used to interpret instabilities and bifurcations in terms of so-called stability and robustness indices. These indices are not only used to determine the stability and robustness of the closed-loop system, but are also used as a description of its dynamic behavior.

The above methods for stability analysis are based on the internal representation of dynamic systems. In the framework of the general stability theory there are two main directions: (1) stability in the Lyapunov sense, which refers to internal representation, and (2) input-output stability, which refers to the external representation. Section 6.4 is devoted to general concepts in input-output stability.

In Section 6.5 we cover the application of input-output stability in FKBC analysis by the use of the circle criterion. Single-input single-output as well as multi-input multi-output cases are considered.

In Section 6.6 we cover the application of input-output stability in FKBC analysis by the use of the conicity criterion.

Acknowledgements

It was clear from the start that our task of producing an urgently needed textbook on fuzzy control could only be fulfilled by involving external help in a few places. We asked several well known specialists in sub-areas of fuzzy control theory whose work could be clearly distinguished as bringing together the fuzzy control approach with conventional control theory to help us with contributions of their own. Rainer Palm from Siemens R&D, Germany, wrote Chapter 4 "Nonlinear Fuzzy Control." Bruce Graham from the John Curtin School of Medical Research in Canberra, Australia, wrote Chapter 5 "Adaptive Fuzzy Control." Prof. Anibal Ollero, Alfonso Garcia-Cerezo, Javier Aracil and Antonio Barreiro from the Universities of Malaga, Sevilla and Vigo, Spain, wrote Chapter 6 "Stability of Fuzzy Control Systems." Prof. Lennart Ljung from the Linköping University in Sweden wrote the foreword.

We want to thank all of these for their contributions and the complete freedom they gave us in changing their texts wherever we deemed it necessary. Thanks to the high quality of each of these contributions we mainly used this freedom to achieve unified notation and terminology, escape unnecessary repetitions, improve the quality of graphs and drawings, and restructure the individual contributions so that the links to conventional control theory were made clear.

Finally, we are grateful to Hans Wössner from Springer-Verlag for his advice in the production of this book.

Munich, March 1993

Dimiter Driankov
Hans Hellendoorn
Michael Reinfrank

Table of Contents

1. Introduction

A very general definition, which encompasses the majority of knowledge based systems (KBS) applications to problems in closed-loop control, may be as follows:

> A KBS for closed-loop control is a control system which enhances the performance, reliability, and robustness of control by incorporating knowledge which cannot be accommodated in the analytic model upon which the design of a control algorithm is based, and that is usually taken care of by manual modes of operation, or by other safety and ancillary logic mechanisms.

In the context of this general definition one can distinguish between two major classes of KBS for closed-loop control: (1) one class where the KBS is involved in the supervision of the closed loop operation, thus complementing and extending the conventional control algorithm, and (2) another class where the KBS directly realizes the closed loop operation, thus completely replacing the conventional control algorithm. These two classes of KBSs for closed loop control are the subject of this chapter.

But before we deal with KBSs and closed loop operations, we will show the use of fuzzy control from an industrial perspective. It is well-known that fuzzy control has been implemented in home appliances, cars, chemical processes, etc., but why has fuzzy control been so successful in areas where classical control theory has been dominant for so many years? This question, and other questions concerning the benefits of fuzzy control, are discussed in Section 1.1.

In Section 1.2 we consider the two particular classes mentioned above in the perspective of a variety of other uses of KBSs to problems in process control.

In Section 1.3 we introduce the notion of a knowledge based controller (KBC) by identifying it as a KBS used for the specific task of on-line closed-loop control. We discuss two principally different uses of a KBC: (1) a KBC as a direct expert control system (DECS), and (2) a KBC as a supervisory expert control system (SECS). Furthermore, we introduce the notion of a fuzzy knowledge based controller (FKBC) as a particular instance of DECS realized with the help of fuzzy logic knowledge representation and inference formalisms.

In Section 1.4 we discuss the general knowledge representation (modelling) issues involved in the design of KBS for supervisory and direct expert closed-loop control. We consider two examples, one of a typical KBS for supervisory expert control, and another one of a FKBC. These examples are used to illustrate the basic knowledge representation constructs and their levels of abstraction and resolution of description with the aim of identifying the conceptually different nature of the two types of KBC.

1.1 Fuzzy Control from an Industrial Perspective

1.1.1 A Lot of Fuzz About Fuzzy Control

The "fuzzy wave" has reached Europe. Boosted by success stories in Japanese white goods and home electronics, popular but also industrial interest in fuzzy control, which hitherto had not been recognized as a serious discipline, has been dramatically increasing since 1990. There are still, however, two predominant, extreme positions as to the benefits of fuzzy control. On one hand, many proponents of this technology claim that fuzzy control will revolutionize control engineering, promises major breakthroughs, and will be able to solve complex engineering problems with very little effort. On the other hand, many representatives of the control engineering community still proclaim the philosophy that "everything that can be done in fuzzy control can be done conventionally as well" and announce a breakdown of the "fuzzy hype" in the near future.

The insight that neither of the two positions accounts for the real potential of fuzzy control is only gradually increasing. In this preface, we will try and explain that, of course, many of the expectations triggered mainly by popular press articles are exaggerated but that fuzzy control has its advantages in many respects, and that it is here to stay: in most cases as an add-on to conventional technology, in some cases as an enabling technology. The main reason why it will become an integral part of control engineering is that fuzzy control is a useful technology from an industrial, business-oriented point of view, which is the one we are taking here.

In order to be able to explain what the benefits of fuzzy control are, however, we first will reconsider it from two different technological perspectives.

1.1.2 Fuzzy Control: Real-time Expert Systems or Nonlinear Control Systems?

Out of the many possible ways of looking at fuzzy control, we consider the following two as most appropriate. A fuzzy control system is a real-time expert system, implementing a part of a human operator's or process engineer's expertise which does not lend itself to being easily expressed in PID-parameters or differential equations but rather in situation/action rules. This view is introduced in the next sections of this chapter. However, fuzzy control differs from main-stream expert system technology in several aspects. One main feature of fuzzy control systems is their existence at two distinct levels: first, there are symbolic if-then rules and qualitative, fuzzy variables and values such as *if* pressure is high *and* slightly increasing *then* energy supply is medium negative.

These fuzzy values such as "slightly increasing" and fuzzy operators such as "and" are compiled into very elementary numerical objects and algorithms: function tables, interpolation, comparators, etc. The existence of this compiled level is the basis for fast, real-time implementations, as well as for embedding fuzzy control into the essentially numerical environment of conventional control.

In artificial intelligence, the field of qualitative physics follows a similar approach, using qualitative variables and so-called "qualitative differential equations," too. A main technical difference is their usage of crisp intervals – corresponding to rectangular fuzzy membership functions – for representing qualitative values. The major difference between fuzzy control and qualitative physics, however, can be found in their relationship to classical engineering disciplines: while the "Qualitative Physics Manifesto" aims at replacing differential equation-based techniques and solving the whole problem with artificial intelligence methods, fuzzy control has right from the beginning been considered as an extension to existing technology, seeking hybrid solutions by enhancing control engineering where it is needed and where it makes sense. In fact, most of the inventors of fuzzy control have a strong control engineering or systems theory background.

From their perspective, fuzzy control can be seen as a heuristic and modular way for defining nonlinear, table-based control systems. Reconsider the rule above: it is nothing but an informal "nonlinear PD-element." A collection of such rules can be used and, in fact, results in the definition of a nonlinear transition function, without the need for defining each entry of the table individually, and without necessarily knowing the closed form representation of that function. One way to combine fuzzy and PID-control then is to use a linear PID-system around the setpoint, where it does its job, and to "delinearize" the system in other areas by describing the desired behavior or control strategy with fuzzy rules.

A representation theorem, mainly due to Kosko [120], states that any continuous nonlinear function can be approximated as exactly as needed with a finite set of fuzzy variables, values, and rules. This theorem describes the representational power of fuzzy control in principle, but it does not answer the questions, how many rules are needed and how they can be found, which are of course essential to real-world problems and solutions. In many cases, relatively small and simple systems will do, and that is why already several hundreds of real, industrial applications of fuzzy control exist. What exactly are the benefits of using fuzzy control?

1.1.3 The Benefits of Fuzzy Control

Considering the existing applications of fuzzy control, which range from very small, micro-controller based systems in home appliances to large-scale process control systems, the advantages of using fuzzy control usually fall into one of the following four categories.

I. Implementing expert knowledge for a higher degree of automation

In many cases of industrial process control, e.g., in chemical industries, the degree of automation is quite low. There is a variety of basic, conventional control loops, but a human operator is needed during the starting or closing phase, for parametrizing the controllers, or for switching between different control modules. The knowledge of this operator is usually based on experience and does not lend itself to being expressed in differential equations. It is often rather of the type "if the situation is such and such, I should do the following." In this case, fuzzy control offers a method for representing and implementing the expert's knowledge. As an example, consider digester management in paper industries. A Portuguese company has implemented a fuzzy based digester management system on top of its process control software. Some twenty-five rules are used for expressing the core of the operator's control strategy. The main advantage of the higher degree of automation that has been achieved this way is the resulting consistent control strategy around the clock, yielding a substantial reduction (up to 60%) of the variation of product quality. Furthermore, subsequent optimizations of the software system led to a significant reduction of energy and base material consumption. All in all, a return on the investment needed for the fuzzy control software package and the knowledge base of the actual control system was achieved after only a few months.

II. Robust nonlinear control

Consider the following problem: a robot arm with several links has to move objects with different masses along a predefined path. Since there are good and exact models of this system available, it is not too big a problem to realize a PID-controller that works pretty well for known masses within a narrow range. Substantial parameter changes, however, or major external disturbances, lead to a sharp decrease in performance. In the presence of such disturbances, PID-systems usually are faced with a trade-off between fast reactions with significant overshoot or smooth but slow reactions, or they even run into problems in stabilizing the system at all. In this case, fuzzy control offers ways to implement simple but robust solutions that cover a wide range of system parameters and that can cope with major disturbances. In this particular case, a so-called fuzzy sliding mode controller (see Chapter 4) was implemented, that exhibits similar performance for a given mass with only slight variations, but that outperforms the PID-solution as soon as major variations are imposed.

There are two other advantages which can be achieved by using fuzzy or rather hybrid solutions which do not relate to the performance of the control system but rather to other aspects -- that are nevertheless important from a business standpoint.

III. Reduction of development and maintenance time

In some cases, two different groups of experts are involved in the development of a control system. Field experts know the application problem and how to design appropriate control strategies. Often, however, they are not electronics experts who know the bits and bytes of numerical algorithms and micro-controller implementations. The actual system realization then is carried out by electronics and systems programming engineers, who on the other hand are not familiar with the application problem. This often results in communication problems between these two groups of people, and hence in delays and extended development times. Fuzzy control, which – as pointed out above – "lives" at two levels of abstraction, offers languages at both levels of expertise: the symbolic level is appropriate for describing the application engineers' strategies, while the compiled level is well understood by the electronic engineers. Since there is a well-defined, formal translation between those levels, a fuzzy based approach can help reducing communication problems.

As an example, consider idle-speed control in automotive electronics, where sophisticated control strategies have to be implemented and run in time-critical situations on simple 8-bit micro-controllers. Besides major variations in system parameters due to mass series production and to aging problems, which require extensive system tuning, communication problems between engine and micro-controller specialists lead to very long development times. We recently had the chance to drive an experimental car with two idle-speed controllers, a conventional PID-system, and a fuzzy control system. The major control goal was to keep a constant idle-speed of 800 rpm, independently of disturbances imposed by different road conditions or additional power consumers such as power steering and air conditioning. Practically no differences could be observed concerning the system's behavior, but while almost two man-years went into the development of the conventional solution, development time for the fuzzy solution was around six months only.

IV. Marketing and patents

In Japan, "fuzzy" has become one of the most popular words. In particular in the realm of home appliances, the label "fuzzy-controlled" raises positive associations in the sense of modern, high quality, or user friendly. As a matter of fact, the use of fuzzy control has contributed heavily to the functionality of quite a few home appliances. The "one-touch-button" washing machine, for example, constitutes a major advance with respect to the large control panels of the older washing machines. In this machine, an additional infrared sensor is used to assess the quality and the quantity of dirt in the laundry, and a fuzzy evaluation procedure for these sensor data and subsequent fuzzy rule based inference is used to determine the details of the washing program. In 1990, Japanese manufacturers reported sales of fuzzy-controlled home appliances in the range of several billion US dollars.

This and other successful applications of fuzzy control, however, have led to a state where merely using fuzzy control – whether or not it really makes sense from a technological point of view – has become a marketing argument. Therefore, the majority of newly introduced home appliances are labelled fuzzy or "neuro-fuzzy." We do not believe that the mere use of fuzzy control, independently of its actual merits, will play a role as a marketing argument for industrial applications. In a few cases, we expect fuzzy control to be used in industrial applications even when it is not actually needed, just to enable companies to demonstrate their technological competence in fuzzy control.

There is yet another reason, however, for using fuzzy control even if it does not improve system performance or reduce development costs. In some business areas, the patent situation is such that it is hard to come up with new solutions using conventional technology without violating or at least interfering with a competitor's patent. Even if the patent turns out to be untenable, this usually leads to significant delays in introducing new products. In some cases, using fuzzy control for a qualitatively equivalent solution will – and actually does – help to by-pass existing patents.

So far, we have listed several success stories of fuzzy control, highlighting the benefits. It should be clear that in most cases, these are no pure fuzzy solutions but rather hybrid solutions, using fuzzy control to augment conventional technology. We should, however, address the question: where are the limits?

1.1.4 The Limits of Fuzzy Control

In some articles in the popular press, examples such as those above are overly generalized to statements about fuzzy logic as a really fabulous technique. Some of these statements are definitely misleading, for some others we simply need a much broader basis for induction in terms of many more applications in order to verify or deny such general statements. So let us reconsider the examples above and their erroneous generalizations:

Fuzzy control leads to a higher degree of automation for complex, ill-structured processes.
This is true in many cases, but only if there is relevant knowledge about the process and its control available that can be well expressed in terms of fuzzy logic. There are processes for which that kind of knowledge simply is not at all or not to the necessary extent available.

Fuzzy controllers are more robust than conventional controllers.
Again, there are many applications, where the use of fuzzy control – pure or in combination with, say, PID-control – resulted in simple yet highly robust control systems. There are other cases, too. We know of two attempts to control air conditioning systems with fuzzy logic, with only minor differences in structure and knowledge base: one turned out to be highly robust even in the presence of major disturbances, the other one was unstable. It is not yet fully understood for which kinds of control engineering problems fuzzy control really leads to

improved robustness and stability, and which are the relevant design choices that affect these properties.

It is not true, however, as opponents of fuzzy control often argue, that there are no stability criteria available for fuzzy control systems. Such criteria are discussed in several different places in this book. We also have to realize that in this respect, fuzzy control competes with nonlinear conventional control, where stability issues are not as easy to handle as for simple linear systems.

Fuzzy control reduces development time.

Again, yes in many cases it does, and the example above in automotive electronics is a real one. It can not be expected, however, as claimed in a number of publications, that using fuzzy control will help to solve really complex problems within a couple of weeks. The reduction of development time results from the fact that there is often knowledge about a process or about control strategies available which can be naturally expressed in fuzzy control but which is hard to encode in differential equations or conventional control algorithms. This knowledge remains to be acquired, encoded, tested, and debugged, and for all but the simplest problems, a strong control engineering background will be needed to realize good solutions.

In many ongoing projects in European industries, using fuzzy instead of conventional control will actually increase development time, just because the application engineers working on those projects are not yet familiar with the new technology. The shortage of well-trained personnel in fuzzy control is one of the major bottlenecks that prevents a broader exploitation of this new technology at this time. Fuzzy control is only beginning to be integrated into the academic curricula, and it will take several years until a sufficient number of engineers familiar with fuzzy control will be available from the universities. Since industry cannot wait for this to happen – in particular because the situation in Japan is completely different – an increasing number of special seminars and in-house training programs have been started.

Products using fuzzy control sell better.

While this is true for home appliances in Japan, we do not expect that this phenomenon will carry over to Europe. Marketing arguments will have to focus on user-friendliness, additional functionality, and environmental aspects such as energy savings or, say, reduction of fresh water consumption of washing machines. Since fuzzy logic can in many cases contribute to these points, we expect the technology to become positively associated with these advantages, but merely labelling a product "fuzzy-controlled" will not become a major sales argument. In some cases, in particular in home electronics and video equipment, the use of fuzzy control is not being mentioned at all. A fuzzy auto-focus system, for example, does not seem to be a quality label. Therefore, most suppliers of such systems use the term digital or intelligent auto-focus, even if it is implemented in fuzzy logic.

In the realm of industrial process control and related fields, marketing strategies differ from those in mass markets such as home appliances. While it is important to companies to demonstrate their technological competence in fuzzy control, the actual decisions for using fuzzy control or any other technique will be made in terms of its contribution to a higher degree of automation, better product quality, cost reduction, or other business-relevant aspects.

1.1.5 On When to Use Fuzzy Control

Having gone through a few examples where fuzzy control is being successfully used, we immediately cautioned about unjustified, too high expectations. But when, then, should fuzzy control be considered?

A satisfactory answer to this question cannot yet be formulated. What we would like to have at our disposal is a systematic decision procedure for analyzing a given problem and inferring whether and if so which variant of fuzzy control should be used based on problem characteristics. At this time, the statistical basis in terms of number of successful applications is not yet broad enough to generate useful generalizations. This should not be a surprise because the industrial and commercial use of fuzzy control did not really start until the late 1980s, compared to the immense number of existing conventional control systems with an accumulated experience of many decades. At this point, we can only formulate a few guidelines on whether to use fuzzy control or not.

First, it is clear that if there exists already a successful fuzzy control based solution to a problem similar to the one at hand, you are on the safe side. In terms of business advantages, however, these are the less interesting cases since somebody else might already commercially exploit that solution.

Next, if you have a good PID-solution, where system performance, development and maintenance costs as well as marketing policy are really satisfactory, stick to it.

Finally, if you are not satisfied with your existing solution with respect to any one of these criteria, or if you have to deal with a problem that could not be solved so far, it is important to analyze the reason why. If there is knowledge available about the system or process that could be used to improve your solution but that is hard to encode in terms of conventional control such as differential equations and PID-parameters, give fuzzy control a try. In many cases, this knowledge is of a more heuristic and qualitative nature and is easily expressible in terms of fuzzy if-then rules. We know of many cases of exactly this kind.

Just to name one, there is no exact and complete mathematical understanding of the processes going on in cooling systems in power generation devices. Hence, algorithmic solutions to infer diagnoses such as system leakage from sensor data exist only to a very limited extent, and the expertise of system operators and engineers is always needed to monitor such cooling systems. Both the evaluation of data such as pressure and change of pressure and the combination of symptoms to a proper situation identification, however, can be naturally expressed in fuzzy terms such as "normal pressure, but quickly increasing" and corresponding if-then rules. In this particular case, it took only a matter of months to design and implement a fuzzy logic based monitoring system for cooling system leakage that outperforms the existing classical solutions. It is important to note, however, that the fuzzy operators and membership functions being used in this system differ from the very simple min/max-inference and triangular membership functions that are predominant in most of the elementary control applications. We will come back to that issue later on.

1.1.6 The Market Place

So far, public interest in fuzzy control has been raised mainly by the immense number of fuzzy-controlled Japanese home appliances and the corresponding number of sales, ranging far above one billion US dollars. We expect fuzzy control to penetrate into European and US-American white good markets in 1993, presumably to a lower degree than that which can be observed in Japan. Another main area of current commercial activities in fuzzy control is automotive electronics. Here again, mainly Japanese suppliers have already brought fuzzy controlled anti-skid modules and engine management systems into the stage of series production. We expect this trend to continue, and major European and American car manufacturers have significant ongoing activities in fuzzy control, though we do not know of any product originating from Europe or the US that has already reached the market place.

Focusing on such spectacular products as "one-touch-button washing machines," however, it is often overlooked that the first real applications of fuzzy control were realized in the realm of industrial process control, such as cement kiln operation. An increasing number of applications is being reported in this area, and several major suppliers of process control hardware and software are now offering fuzzy control capabilities, for example, in programmable logic controllers or higher level process control systems. In the long run, we see the major application potential in this area, while in terms of market share and commercial benefits, home appliances, automotive electronics, and other microcontroller based applications will dominate the fuzzy control business for the next couple of years. As the technology matures, however, we foresee an increasing number of applications in complex systems engineering, the focus of fuzzy control moving from elementary control problems to higher levels in the system hierarchy such as supervisory control, monitoring and diagnosis, and logistics problems. It is important to note that also one of the major future industries, telecommunications, has started investigating fuzzy control for communication systems, and that several pilot projects have been started concerning, for example, routing and overload handling problems.

An increasing number of market growth studies are trying to estimate and forecast the economic relevance of fuzzy control. While these studies are worthwhile to carry out, it is very hard to delineate exactly the contribution of fuzzy control to the monetary value of a product. Consider the example of a washing machine, where a major part of the electronic control is realized in fuzzy logic. It is certainly not correct to take the price of the whole washing machine, but it also would be misleading to consider just the – usually standard – micro-controller or the number of lines of code realizing the fuzzy part of the system. Our assessment of the contemporary and future market relevance of fuzzy control is the following: we estimate that about 10-15% of all electric and electronic engineering applications will benefit from the use of fuzzy control, being roughly evenly split into fuzzy control as an enhancing and as an enabling technology. Considering the huge market volume of this area, and also its expected further growth, it becomes evident that fuzzy control is a major technology from a business point of view. In addition to that, there is a still small but increasing number of applications of fuzzy logic in other, purely information technology oriented fields such as databases, expert systems in the financial sector, and so on.

We should not, however, commit the same error as was done with "classical" artificial intelligence systems, where after some initial success in relatively narrow application areas it was believed that artificial intelligence could be used as a universal stand-alone technology to solve just about every problem in information processing. We expect that for a long time the main application potential of fuzzy logic will stay within the realm of fuzzy control, in an increasingly broader sense of "control." This is not to say that the relevance of fuzzy logic will not expand to other areas, where there are already a few spots such as data compression, but this is still mainly research-oriented work, whereas fuzzy control has already gained a significant role in daily industrial practice and its full potential in this area is by no means yet exploited or even known.

So far, we have discussed the relevance of fuzzy control applications as a business. There is yet another business area in fuzzy control, relating to the technology itself as a product, in terms of special hardware and software. So far, the majority of all existing applications is purely software-based. Although there is a large number of special fuzzy logic processors available and offered by semiconductor firms or small specialized companies, we expect that this situation will not change within the next years. Using special hardware in the sense of general-purpose fuzzy logic processors or coprocessors is and will only be justified in the following situations: the application is extremely time-critical, and the extra costs for an additional fuzzy processor can be recovered either from the sheer number of items (such as in automotive electronics) or from the fact that the system as a whole is large and expensive enough that the costs for a special processor are negligible (this might be the case when using a fuzzy processor for, say, a pattern recognition task in a complex plant automation system). Therefore, most suppliers of micro-controllers, PLCs, and other control equipment focus on developing specialized fuzzy control software for standard devices, and on integrating these packages into the existing development environments.

Besides major electric/electronic and industrial automation companies, an increasing number of small and specialized software companies is offering fuzzy control development tool kits for a broad range of standard processors. Our advice in general is to use as long as possible pure software solutions, even in higher programming languages such as C. The reason for this is that software offers a much higher degree of flexibility and, in most cases, the solutions are hybrid, i.e., there is always a need for a standard processor to implement the "non-fuzzy" parts of the system. Experience so far has shown also that the elementary fuzzy operations such as function table look-up and interpolation, comparators, and basic arithmetic functions are simple enough that fuzzy control algorithms can be optimized to run even very time-critical applications without specialized hardware. It may be the case that the role of hardware will become more important in the future, when more complex operators will be needed to adequately express and implement the expertise for control strategies. This might trigger the development of more sophisticated fuzzy processors and also semi- and full-custom integrated hardware designs. The point is, however, that it is not yet obvious which operators will be needed and that existing special hardware does not support any but the simplest fuzzy algorithms.

1.1.7 What Needs to Be Done?

Reconsider the example of the cooling system monitoring problem discussed above. During the system design and knowledge acquisition phase, it turned out that the human expert's usage of, for example, the AND-operation when combining symptoms could not be adequately expressed with the usual fuzzy logic interpretation of AND as a minimum operation. Also the fuzzy membership functions representing values such as "normal pressure" turned out to be more sophisticated than the simple triangular functions predominant in elementary control applications.

Several conclusions can be inferred from this example: first of all, it was extremely helpful to have a software tool available which supports experimental system implementation, evaluation, and tuning, and which offers the user ways to combine predefined standard operations with his or her own special code. There is an increasing number of such software tools available which – though not yet completely matured – provide valuable support. They are being steadily improved and are being integrated into standard software development packages for automatic control.

Secondly, considering the major part of existing applications of fuzzy control, it can be realized that the "application-proven" part, let us say the "first generation" of fuzzy control, exploits only a very small fragment of fuzzy logic theory. In many cases of more complex, ill-structured problems, this first generation technology is not sufficiently strong to represent and implement the knowledge needed for powerful solutions. It is often possible, though, to improve the technology additively by exploiting a larger part of the readily available theory.

The example above and many others make clear that the development of a mature, more powerful second generation technology still requires basic and experimental research. This research will have to focus on the questions of which parts of the still unused theory will eventually gain practical relevance, where does the theory itself need extensions, and which are useful ways of integrating fuzzy control with conventional and also other modern control algorithms such as neural network based approaches. We are convinced that only application-driven, experimental research can be sufficiently focused and guided in the right directions to make good progress. It simply does not make any sense to define a new operator "Fuzzy-And, Version 137" and to discuss its mathematical properties without being triggered by the fact that none of the existing operators can be used for a particular application.

Therefore, we believe that industrial research – as opposed to pure university-based research – will play a major role. Only industrial research labs have direct access to real-world problems, which is essential for proceeding in this field. This is not to say that the universities are not needed: most importantly, fuzzy control has to become an integral part of educational curricula in electronic and control engineering. Furthermore, academic people can and must contribute to the long-term, fundamental issues in fuzzy control theory. But the major push in fuzzy control within the next years will be application-driven and will necessarily have to be carried out in industrial research labs.

Also considering first generation technology, there are still a few bottle-necks hindering industry from a broader exploitation of the application potential of fuzzy control. In the first place, we need a better and more systematic design and analysis methodology for fuzzy control applications, supporting the whole life-cycle from initial decision making all the way through to deployment and maintenance, focusing on the following issue: which are the proper design choices given an analysis of the problem, and how do variations of system parameters affect the system's performance. It should be clear that we cannot expect a universal design and optimization strategy for fuzzy control which is of any practical use. Such a universal theory does not exist for conventional control engineering either. Instead, we have to proceed from the few isolated spots where we already know exactly how to design a fuzzy control algorithm to clusters of problems and related design methodologies.

Besides the necessary paper and experimental work which is needed to advance this methodology, we need a much broader basis of experience in terms of successful and of unsuccessful applications. We expect this methodology to gradually emerge as experience is being accumulated.

The absence of such a coherent and systematic methodology makes another main problem for first-generation fuzzy control even more serious: the lack of well-trained and experienced "fuzzy control engineers." It means that engineers who want to or have to get into this new technology have only limited support and guidance and must by and large rely on their own experience gained through experimental work.

1.2 Knowledge Based Systems for Process Control

A survey paper by Taunton [212] introduces a number of experimental applications of real-time (on-line) knowledge based systems which cover the following four major areas of process control.

1.2.1 Process Monitoring

A KBS used for monitoring deals with real-time data, operates continuously, and supports the process operator in choosing the "right" information which provides the "best" overview of the process in a certain temporal perspective, e.g., past, present, and future process outputs and combinations of these. In conventional monitoring systems the major effort has been devoted to collecting and displaying process data while very little has been done to aid the process operator in better understanding of the overall process output. Thus, the process operator is forced to focus on searching through large amounts of interrelated process data to identify causes for alarms, loss of product quality, deviations from normal operating conditions, etc. A KBS which provides a continuous identification of the overall process can be used to assist the operator in a number of tasks:

- Detect changes in the process output, thus revealing deviations from normal operating conditions and producing early warnings of evolving failures.
- Analyze causes for alarms in such a way that the cause of the primary alarm is not obscured by secondary alarms.
- Analyze the importance of alarms with respect to the current process output in order to avoid dealing with "nuisance" alarms that stem from harmless process output deviations or changes in external conditions.
- Provide assistance in avoiding or recovering from undesirable process output deviations by giving advice on recovery actions, indicating the degree of urgency of these actions and explaining their rationale.
- Store normal and abnormal operating conditions that have occurred in the past and use this type of process history to guide the process operator in predicting the process output, recognizing error situations which have occurred before, and giving suggestions on how to improve the current situation.

In order to be able to assist the process operator in these tasks the KBS for monitoring has to provide means for:

- Integration of extremely large amounts of numerical measurements with symbolic data which requires representation of numerical measurements at different levels of abstraction.

- Reasoning about events sequenced in time, time validity of measurements, and remeasurement strategies.
- Prescribed time decision making, i.e., making optimal decisions in prespecified time limits.
- On-line validation of the KBS and its inference engine.

Examples of KBS for monitoring include: GEN AID, for on-line identification of errors in power generators, developed by Westinghouse Corp. [154]; AFS, for alarm filtering, developed by Idaho National Engineering Lab. [40]; COOKER, for on-line monitoring of batch manufacturing processes, developed by Honeywell [1].

1.2.2 Fault Diagnosis

A KBS for fault diagnosis is used to handle problems like intermittent faults and faults which do not cause alarms. It is used as an on-line support tool for trouble shooting. The first generation of KBS for fault diagnosis are based on the process operator's heuristic knowledge about faults occurring "usually" and the causes for these faults. The drawback of such an approach is that it is process specific and can easily be outdated when the process is updated. Furthermore, it is impossible to detect unanticipated faults for which the process operator has no previous knowledge. The second generation of KBS for fault diagnosis are based on first principles knowledge (deep knowledge) which stands for knowledge that is generic and model based. Such systems model the correct, intended behavior of the process. During diagnosis, measurements of the process are compared against the process model and mismatches are interpreted as indicators of faults. The advantage of this type of KBS for fault diagnosis is that the KBS is generic and can detect, at least in theory, all possible faults. To be able to perform its on-line functions the KBS for fault diagnosis has to be able to represent and reason with and/or about:

- Real-time measurements which have tolerance ranges; time intervals representing the decay in validity of a measurement; cost of measurement information; trade-offs about how long one would like to keep a value versus how often one has to remeasure it.
- Deep model representations at different levels of abstraction and resolution; communication between models at different levels of abstraction; the interpretation of measurements at the abstraction level of each deep model.
- Different levels of time criticality which regulate the speed with which diagnostic decisions have to be made. Low criticality could be that of control failures which are not safety related, and high criticality could be that of safety related problems.

Examples of KBS for fault diagnosis include: FALCON, for fault diagnosis in an adipic acid reactor, developed by The Foxboro Corp., du Pont, and University of Delaware [123];YES/MVS, for the operation of IBM mainframe computers [141]. In [20] and [140] one can find a survey of typical problems in the design of KBSs for on-line fault diagnosis.

1.2.3 Planning and Scheduling

The task of planning and scheduling in a process control system is predominantly a management function ensuring that variable demand is met via effective utilization of production resources. This requires representation of combined knowledge about technical and economic constraints and can hardly be expressed as a purely analytic modelling task. There are two major aspects of planning in process control. Long term planning involves extension and reconfiguration of the process control system. Short term planning involves the planning of tasks which have to be performed in a limited time horizon to assure proper process and control system functions and optimization. Both types are concerned with the planning of a sequence of actions and the corresponding time scheduling of these actions. In process industry the typical application domains are batch processes and the operation of a continuous process during start up, shut down, and changes in production. Analytic techniques such as linear and dynamic programming are the tools normally used for planning and scheduling. However, the modelling assumptions on which these formal techniques are based do not often match the real planning situation or result in prohibitive amounts of computation. On the other hand, heuristics and qualitative preferences cannot be easily included in the analytic models. Thus KBS for planning and scheduling in process control can be seen as complementary to the existing analytic tools. Such a KBS normally incorporates knowledge of different types:

- Physical knowledge.
- Process configuration knowledge.
- Knowledge of pre-planned actions, e.g., work schedules, stops, repairs, etc.
- Experiential knowledge about process operation in similar planning tasks.
- Statistical knowledge about uncertainties in process parameters and measurements.
- Simulation facilities to predict process behavior as a result of intended actions.

Thus the KBS has to be able to represent and reason with conceptually different types of knowledge which involves the use of causal models, temporal and spatial data and relationships, taxonomic representations, etc. However, most of the difficulties in the design of KBSs for planning and scheduling stem from the real-time nature of the process control environment, namely its dynamics. A dynamic process output has two implications on the KBS plans. First, the sequence of actions in a plan are state dependent so that the KBS always needs to monitor the execution of actions: (1) to make sure that actions do occur, and (2) to watch for the effects of an action in order to move to the next action in the sequence. Each element of a plan, therefore, has implicit expectations about future states (typically the goal or subgoal behind an action) that the system needs to understand and be able to compare with its state description. Second, a plan needs to be robust in the face of change. As the process output is changing while the plan is being executed, this plan can be outdated before it is completed. There is no point in just discarding it and starting from scratch, it must be quickly adaptable to the new conditions. Other factors, such as limited time response, limited resources, and uncertainty with respect to the validity of measurements or actual occurrence of effects of actions, complicate further the design of the KBS. Applications involving planning for a ship positioning controller and a fuel system manager are described in [20]. In [196, 68, 69] constraint-based reasoning and its applications to job-shop scheduling are considered in detail. Applications to power generation and distribution are presented in [182]. The KBS ADAPOS, for load-flow planning, is the subject of [70], and a KBS application to flow-load forecasting, as an alternative to analytical methods, is described in [172].

1.2.4 Supervisory Control

Here the main control problems are related to tuning and adaptation so that maximum product yield is achieved. Usually, supervisory control is highly dependent on the experience of the process operators, which in turn varies due to different individual experiences and skills. In this area of process control applications, real-time KBSs are known as supervisory expert control systems (SECS). In a SECS the primary control functions are provided by conventional automatic control techniques and the KBS is employed in a supervisory mode to extend the range of application of these techniques. Examples include the work done in [6] where relay autotuning is used to build up process knowledge, and the performance adaptive controller described in [31, 121].

A very different approach to the use of real-time KBSs in process control is based on the fact that there may be many physical properties of the process that are not part of the analytical model used to design the controller. This may result from the complexity of the process or from lack of understanding of the physics involved. On the other hand, experienced process operators may have developed a number of heuristic control rules which allow them to control such a process in a satisfactory manner. While it may be extremely difficult to develop techniques which model a heuristic control policy in analytic terms, there are other modelling techniques which circumvent the use of analytic models. These techniques employ symbolic (qualitative) models which allow one to explicitly represent and reason with the heuristic knowledge available. In this case the

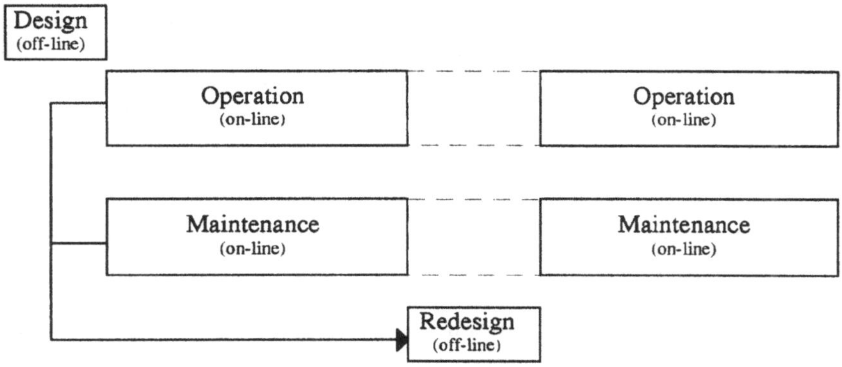

Fig. 1.1. The life-cycle of a process control system. The operation phase is the implementation environment for KBC. (Redrawn from [179].)

use of a KBS replaces completely the conventional control element of a process control system and is known as a direct expert control system (DECS). One particular instance of the class of DECSs is the so-called fuzzy knowledge based controller (FKBC) which employs a knowledge representation technique and an inference engine based on fuzzy logic.

1.3 Knowledge Based Controllers

A knowledge based controller (KBC) can be identified as a highly specialized KBS designed for performing a specific task during a particular phase of the life-cycle of a process control system. The life-cycle of a process control system is described by [179] as consisting of the following three phases: design, operation, and maintenance. The latter two phases run in parallel and are executed on-line as illustrated in Fig. 1.1. The operation phase constitutes the particular implementation environment of a KBC.

Within the operation phase researchers from ABB [179] distinguish between three specific tasks: monitoring, control. and planning. The first two, as well as short term planning of actions which have to be performed in order to assure proper process and control system function/optimization, inherit the on-line nature of the operation phase. Furthermore, each task can be characterized [179] in terms of its scope and nature as follows:

- Scope: single-loop operation – plant-wide operation
- Nature: operator assisted operation – closed loop operation

The first pair of scope characteristics defines whether the process control system operates on a single-loop level or is intended for a plant-wide, global level of operation. The second pair of nature characteristics defines whether the process control system is used mainly to provide information display and/or advice giving for the process operator, or whether it automatically, without any operator intervention, determines and executes appropriate control actions.

The KBS which can be employed during the particular phase of operation, for the specific task of control, can be classified, according to [179], into four main categories. The classification is done according to the intended use of the KBS in the control task and the scope and nature characteristics of the control task itself.

KBS for manual control assistance
The KBS is used to assist the process operator in the manual mode of control by providing a what-if type of support. In other words, it determines changes in the process output if the process operator attempts to execute certain control actions. The scope of such a KBS can be anywhere between single-loop operation and plant-wide operation. Its nature is operator assisted operation.

KBS for plant-wide tuning
The use of the KBS is to monitor and adjust the operation of the control loops, e.g., the tuning of PID-controllers in cascaded loops, on a plant-wide level. Thus the scope is plant-wide and the nature of such a KBS can be anywhere between operator assisted and closed loop operation.

KBS for quality control
The use of the KBS is to close the outermost control loop which determines the quality of the end product(s), by providing advice on how to achieve and maintain quality as close as possible to specifications. In general, this outermost control loop is characterized by the existence of unmeasurable control variables, variations in feedstock, a wide range of specifications for different end products, and unreliable instrumentation. Thus, its overall performance is exposed to operator intervention based on incomplete experiential knowledge and faulty instantiation of general heuristic rules. The scope of this type of KBS is closer to the plant-wide dimension of the scope characteristics. Its nature is operator assisted operation.

KBS for expert control
There are two principally different uses of the KBS: (1) A KBS for direct expert control (DECS, Fig. 1.2) and (2) A KBS for supervisory expert control (SECS, Fig. 1.3).

In DECS the KBS is used in a closed loop, thus replacing completely the conventional control element. The KBS replicates the process operator's manual control strategy by employing a rule base which determines the control output signal, given information about process output variables, e.g., error,

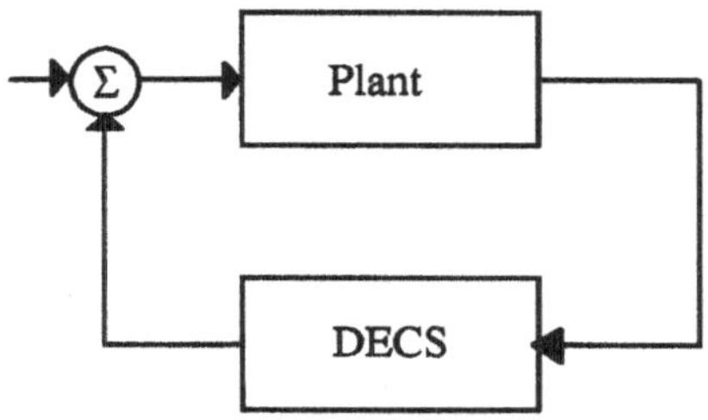

Fig. 1.2. The structure of a DECS.

change-of-error, etc. The need for such a KBS is motivated if the nature of the process under control is such that appropriate analytic models do not exist or are inadequate, but the process operator can manually control the process to a satisfactory degree. The scope of the KBS can be anywhere between single-loop and plant-wide operation. However, most of the work reported in the literature is closer to the single-loop dimension of the scope characteristics. The nature of the KBS is closed loop operation.

In SECSs the KBS is used out of the control loop acting as a supervisor of a conventional control element, thus complementing rather than replacing a conventional controller. The latter continuously implements the current control law while the KBS determines when and how this control law should be changed. In other words, the supervisory role of a KBS is to extend the range of application of a conventional controller by using explicit representation of general control knowledge and auto-tuning and adaptation heuristics. Thus the KBS replicates the knowledge and skills of the control engineer rather than that of the process operator. The KBS is built using a combination of knowledge representation techniques, e.g., object hierarchies, frames, causal models, production rules, etc. Thus the inference engine of such a KBS has to support different types of reasoning normally associated with each of the above knowledge representation techniques. This is in contrast to the KBS of a DECS where the predominant knowledge representation technique is production rule based and the inference engine is realized as a data driven, forward-chaining mechanism. The need for a SECS is motivated when the bulk of knowledge required to build a conventional controller is not associated with the analytical control algorithm, but with the heuristic logic required for its proper functioning. This heuristic logic deals with varying dead-time, effects of nonlinear actuators, wind-up of the integral term, transients due to parameter changes, etc. It should be noted here that well designed industrial PI- and PID-controllers already have encoded the heuristic logic which allows them to cope with the above mentioned problematic operating conditions. As discussed in [9], heuristics are even more important in multi-variable and self-tuning controllers. In

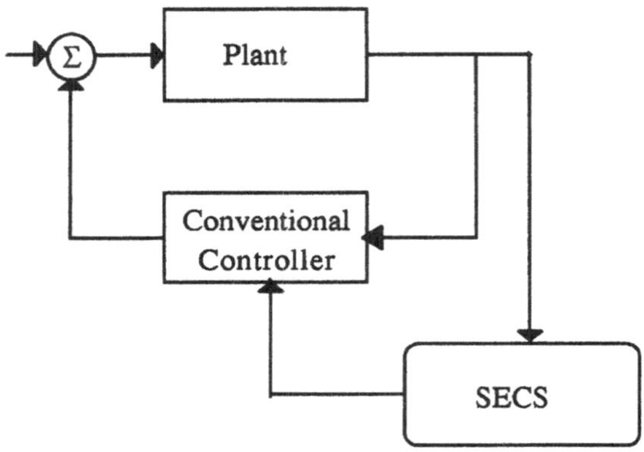

Fig. 1.3. The structure of a SECS.

these cases the fundamental control law is much more complicated than for example the control law of a PID-controller. To obtain a well-functioning adaptive control system it is also necessary to provide it with a considerable amount of heuristic logic. This goes under different names such as supervision safety nets or safety jackets [38, 103]. Experience has shown that it is quite time consuming to design and test this heuristic logic. On the other hand, the inherent ability of a KBS to make explicit the use of heuristic logic, and hence support incremental expansion of its knowledge representation and reasoning components, is extremely difficult to achieve with conventional programming languages. In this context the objective of a SECS is to use a KBS to encode knowledge representation and reasoning capabilities which will allow intelligent decisions and recommendations automatically rather than to pre-program logic which treats each problematic situation in an explicit way. The scope of such a KBS, judging by existing applications, is at the lower end of the scope characteristics and its nature is closed loop operation.

A knowledge based controller falls in the category of KBS for direct expert control systems. Thus it is a highly specialized KBS used for on-line, closed loop operation during the operation phase of a process control system. It either completely replaces the conventional control element or complements it in a supervisory mode. In particular, a fuzzy knowledge based controller (FKBC) replaces completely the conventional controller in the closed loop by employing a KBS built with a specific knowledge representation technique derived from fuzzy logic, and an inference engine in the form of fuzzy logic based production rules.

1.4 Knowledge Representation in KBCs

No design technique for closed-loop control can proceed without some implicit or explicit knowledge about the process to be controlled. Knowledge here means a model which provides a conceptual structure to capture those aspects of the process which accurately represent its behavior.

To any model there are two dimensions [122]: resolution and abstraction. The former represents the level of detail to which a model is employed and the latter determines the atomic primitives upon which the model is built. A shift between abstraction levels is usually accompanied by a change in the level of resolution. There is also a choice to be made as to the way in which knowledge is represented at a particular abstraction level, i.e., the knowledge (model) representation formalism.

In the analytic design of closed-loop control, the atomic primitive upon which a model is built is the continuous (or discrete) real valued function. The resolution level is given by the order of the process, and the knowledge representation formalisms include transfer functions, frequency response functions, state space representations, etc.

In this type of modelling, only one knowledge representation technique is used. Thus, the particular task of the controller is mapped into a specific control objective (performance criterion) which must be expressed at the level of abstraction of the process model. Once the performance criterion is decided upon and expressed at the appropriate level of abstraction, the design of the closed loop is relatively straightforward.

Although enormous progress has been achieved in linear and nonlinear analytic control theory, there are several instances [9] where the theoretical developments do not respond to the application needs. For example, to obtain a good PID-controller it is also necessary to consider factors outside its analytic formulation, such as operator interfaces, operational issues like smooth switching between manual and automatic control modes, maximum and minimum selectors, etc. An operational industrial PID-controller thus consists of two parts: (1) an implementation of the PID control law and (2) heuristic logic that provides for solutions in case a problematic situation is encountered. The latter part of the code may sometimes be much larger (in sequencing and adaptive controllers) than the code for the core PID control algorithm. In [9] Åström raises the following question: if one accepts that control algorithms contain heuristics then what can a more extensive use of heuristics contribute to the design of process control systems? To illustrate the issues raised by this question, in the next two sections we will consider the modelling and use of heuristics in the case of SECSs and DECSs.

1.4.1 Knowledge Representation in SECSs

The SECSs constitute a class of KBCs based on having the KBS monitor the closed loop. The control element of the closed loop is designed using standard techniques and involves an analytic process model. One use of a SECS is to extend the range of operation of the conventional control algorithm with respect to auto-tuning and adaptation [6, 106, 88, 183]. Another use of a SECS is to prevent the deterioration in the performance of adaptive controllers that results from degraded identification. The reason for this is that if the process excitation is not rich enough then the results of the identification may be very poor, which in turn is a cause for instabilities. In this context a SECS is used to decide on the trade-off between identification needs and control requirements [126]. A third use of a SECS is in the design of reconfigurable controllers that can behave optimally under a variety of circumstances. These may include changes in the process dynamics due to faults and failures, or changes in the control objectives. In this case the SECS is responsible for the determination of the best configuration of the monitoring, identification, and control algorithms [87, 99]. To illustrate the type of knowledge (model) used in the construction of a typical SECS we will describe a particular supervisory expert PI-tuner [168].

The task of the PI-tuner is to tune a PI-controller so that the closed loop transient step response lies within certain limits set by the commissioning engineer or process operator. The limits within which the PI-tuner works are maximum overshoot, maximum undershoot, and damping ratio. These limits are then used by the PI-tuner to compare the untuned closed loop transient step response shape and to update the controller parameters in accordance with existing disparities.

The operation of the PI-tuner proceeds as follows. Initially, an open-loop transient step response test is performed and an initial hypothesis concerning process characteristics is advanced. There are eight open-loop process characteristics (see Fig. 1.4):

1. no delay monotone
2. no delay oscillatory
3. short delay monotone
4. short delay oscillatory
5. medium delay monotone
6. medium delay oscillatory
7. long delay monotone
8. long delay oscillatory.

The description of the open-loop process as a "short", "medium", or "long" delay is done on the basis of the ratio of the pure time delay in the process and the dominant time constant.

Furthermore, each characteristic is identified in real time from closed loop data by employing an autoregressive moving average model using recursive least

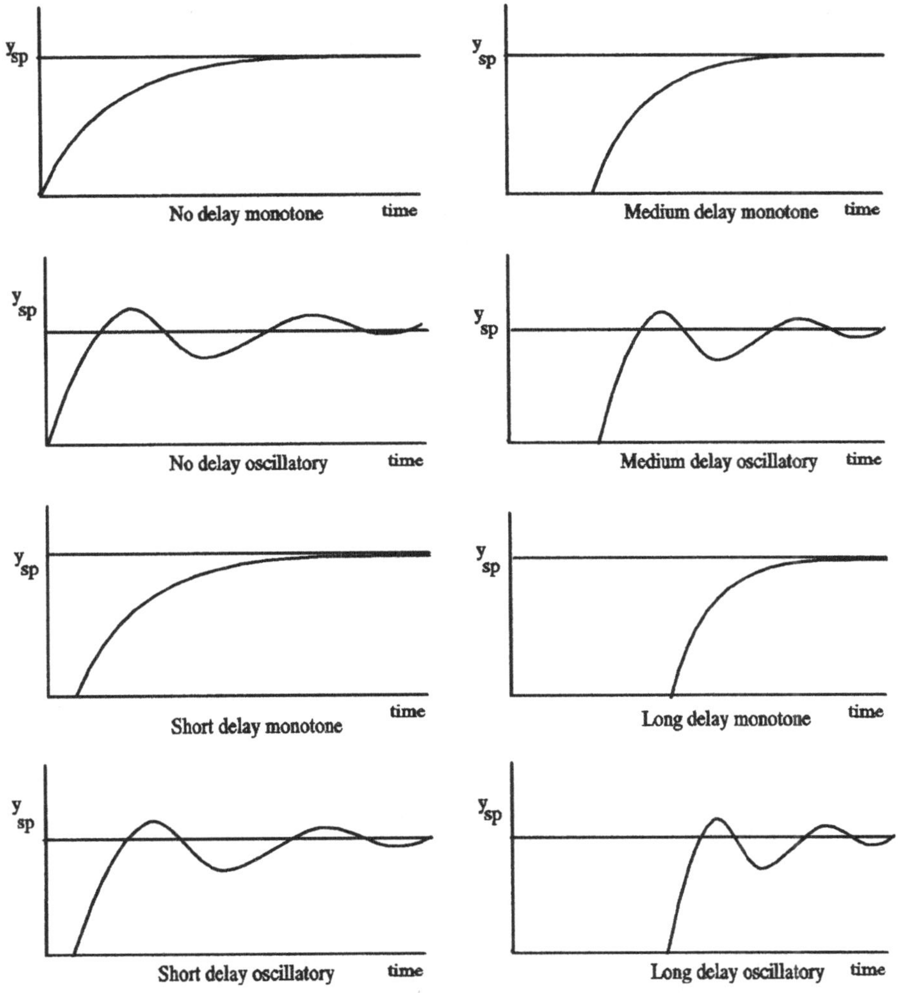

Fig. 1.4. Open-loop process characteristics.

squares [107]. The characteristics of the closed loop transient step response employ both of the following:

- The magnitude and time of the first overshoot and first undershoot for both the process output and control variable, in a response to a step change in the set-point.
- The integrals of both the process output and control variable evaluated at strategic instants of time along the transient.

These data are used to establish the following nine characteristics (properties) of the closed loop transient step response (see Fig. 1.5):

1. too low monotone
2. too low oscillatory
3. overshoot, undershoot
4. no overshoot, undershoot
5. no overshoot, no undershoot
6. overshoot, no undershoot
7. overshoot monotone
8. overshoot oscillatory
9. over safety limit.

These descriptions of the closed loop transient step response as "too low", "overshoot", or "undershoot" take into consideration the behavior outside the upper and lower limits as shown in Fig. 1.6.

After the hypothesis about the type (characteristic) of the open-loop process is advanced, this information is passed to the knowledge base of the PI-tuner. The knowledge base contains two types of production rules. The first type, called context rules, connects the closed loop transient step response characteristics to the open-loop process characteristics. For example, "*if* open-loop process is medium delay monotone, *then* closed loop transient step response is overshoot undershoot." This allows for differentiating between sets of closed loop transient step response characteristics which are superficially similar but which are in fact intrinsically different. The use of these rules is to establish a context for the consequent application of the second type of rules, namely the tuning rules. The context simply describes a fact, like "the closed loop transient step response is of overshoot-undershoot type and the open-loop process is of medium delay monotone type." Once this context is established it triggers a specific tuning rule.

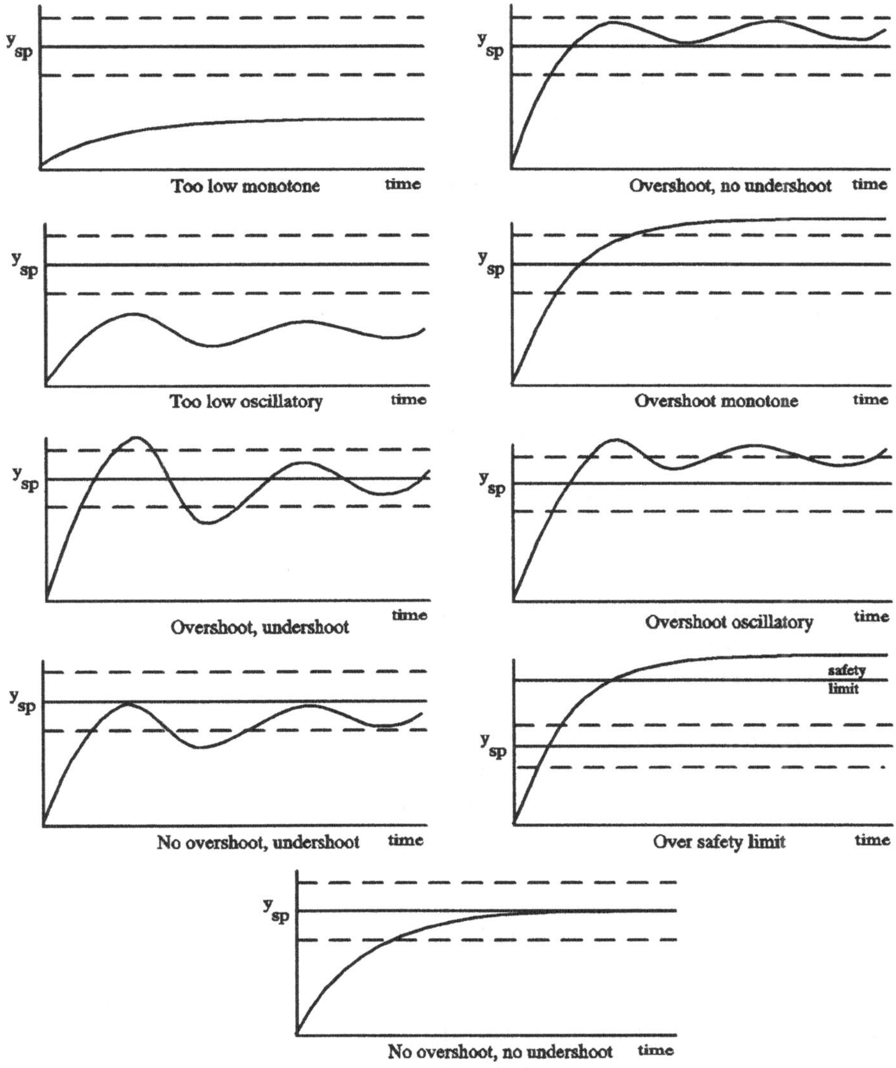

Fig. 1.5. Closed-loop process characteristics.

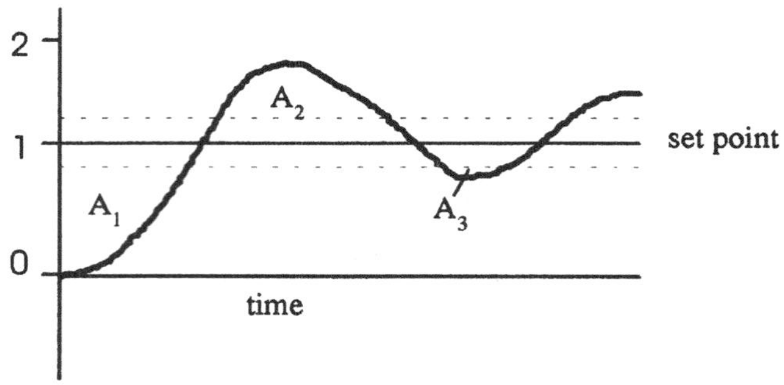

Fig. 1.6. Closed-loop performance limits. Zones A_1, A_2, and A_3 show undesired behavior outside the performance limits. (Redrawn from Porter et al. [168].)

The tuning rules are used to change the PI-controller gains. These rules are based on the experience of the control engineer and provide a quantitative judgement as to how the modification of the gains is to be done. Furthermore, these rules are constructed in relation to the context for the application of a rule and a matching of the present closed loop transient step responses to previous experiences. As already mentioned, the context is established by the context rules while the matching is effected by comparing the patterns of stored closed loop transient step responses (for which definite tuning rules are known) agains the present closed loop transient step response. For example,

context:
Closed-loop transient step response is overshoot undershoot.
Open-loop process is medium delay monotone.

tuning-rule:
if the control variable continues to increase after the transient first crosses the set-point
then reduce the integral-gain by $\mathrm{KI} = (A_1/(A_1 + A_2 + A_3)) \cdot \mathrm{KI}_{\mathrm{old}}$

Thus the execution of a tuning rule in the appropriate context changes the PI-controller gains. This information is kept in the so-called solution blackboard which contains the current hypothesis about the open-loop process characteristic, and the current tuning strategy together with its closed loop transient step response context.

After having changed the controller gains and when the process achieves a steady-state behavior, the KBS initiates a closed loop transient step response test. Using the results of this test the KBS either verifies or modifies the current open-loop process hypothesis. Once this is completed the whole procedure is repeated until the PI-controller is tuned according to specification.

From the modelling (knowledge representation) point of view, discussed in the previous section, we can say that this type of SECS uses a two-level knowledge representation hierarchy. For example, if we look at the closed loop transient step response, it is represented at the lower level as an analytic function describing the behavior of the closed loop over time with respect to certain performance limits (Fig. 1.6). At the higher metalevel this particular continuous behavior is stripped off its analytic form and is instead represented as an abstract (symbolic) object with certain properties. In this particular case the object is an abstract entity called "closed loop transient step response." Its properties are defined by considering only some qualitative aspects abstracted from its lower level analytic representation. These qualitative aspects include the following:

- The magnitude and time of the first overshoot and the first undershoot of the process output and control variable in response to a step change in the set-point.
- The integrals of both the weighted square of the control variable and the weighted square of the process output evaluated at certain strategic instants along the transient.

Both of the above are implicit in the analytic lower-level representation of the closed loop transient step response. At the second, higher level they are made explicit for the purpose of describing specific aspects of the overall transient step response behavior. Only the time of the first overshoot/undershoot, as well as integrals evaluated at certain strategic instants along the transient, are of interest from the point of view of the higher level. These qualitative aspects are then used to define the properties of objects at this higher level, e.g., the property "undershoot overshoot" for the object "closed loop transient step response".

In the case of a tuning rule, the higher-level description of the behavior of the control variable in its *if*-part is done again in qualitative terms representing an abstraction of its lower-level analytic description. For example, the control variable continues to increase after the transient first crosses the set-point. In other words, only the sign of its derivative is of interest and not its actual value. Furthermore, the sign of derivative is of interest only after the time of the first crossing of the set-point. This specific time point, which is implicit in the lower-level analytic description of the transient, is explicitly referred to at the higher level.

Thus, one can say that the objects at the higher level of knowledge representation, e.g., open-loop process, closed loop transient step-response, control variable, process output, gain coefficients, etc., are symbolic entities, whose actual meaning corresponds to certain analytic functions of real variables. These functions, all at the lower level, describe the quantitative behavior of the objects over time to the fullest extent possible. However, only certain aspects of this overall behavior are abstracted and made explicit at the higher level and then used to define properties of the objects at this level. Furthermore, logical rules are used to (causally) relate an object with certain properties to other objects and their corresponding properties. These rules provide an effective way to reason about whether certain relations between objects and their properties hold, and, if this is the case, appropriate actions are inferred. The actions change the properties of objects. These changes are transferred to the lower level by changing the parameters of the analytic representations of the objects which properties have to undergo a particular change. For example, such a rule may state the causal relationship between the integral term and the magnitute of the overshoot, namely that the integral term is responsible for the overshoot. Thus, if one decreases the integral gain at the time when the open-loop transient step response exceeds a certain value, then one can considerably reduce the overshoot. On the other hand, an increase in the integral gain during rise-time leads to an improvement in rise-time.

From the point of view of the knowledge representation at the lower level the above causal relationships cannot be made explicit. In its analytic representation the causal relationship is bidirectional. Furthermore, the analytic representation does not know what an overshoot or undershoot is. These are implicit in it. They can be reasoned about only if they are made explicit objects at the higher level with properties which reflect only certain qualitative aspects of their lower-level analytic description.

1.4.2 Knowledge Representation in FKBCs

The DECSs constitute a class of KBCs based on having the KBS in the closed loop, thus replacing completely the conventional control element. One particular subclass of DECSs are the FKBCs. To illustrate the basic knowledge representation issues involved in the construction of FKBCs, let us consider the problem of regulatory control and the corresponding FKBC. In the following we will present the approach of Sripada et al. presented in [197]. The authors consider the cases of servo control and regulatory control. The control objective in the case of servo control is to drive the process output from an initial steady state to a final steady state, i.e., open-loop, optimal state driving. Regulatory control compensates for disturbances that would appear to cause the control variable to deviate from the set-point once a steady state is achieved. The regulatory FKBC considered is built with the purpose of achieving the following three objectives;

- To remove any significant errors in process output $y(k)$ by appropriate adjustment of the control output $u(k)$.
- To prevent process output from exceeding some user-specified constraint y_c, i.e., for each k, $y(k)$ should be less or equal y_c.
- To produce smooth control action near the set-point, i.e., minor fluctuations in the process output are not passed further to the control output.

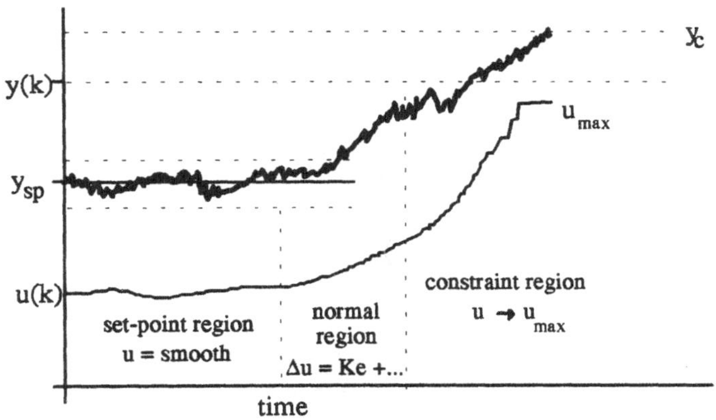

Fig. 1.7. Type of control action desired in different regions around the set-point. (Redrawn from Sripada et al. [197].)

Figure 1.7 illustrates the type of control action desired when the process output is in different regions around the set-point.

Although the above control objectives are typical of most industrial applications, a conventional PI(D)-controller cannot handle all three of them unless extended by some additional heuristic logic. This was the case described in the previous section. On the other hand, an experienced process operator can easily meet all three control objectives. Thus the idea behind a FKBC is to build a KBS which employs the process operator's experience in the form of *if-then* production rules.

A conventional PI-controller uses an analytical expression of the following form to compute the control action

$$u = K_P \cdot e + K_I \cdot \int_t edt. \tag{1.1}$$

When this expression is differentiated, one obtains

$$\dot{u} = K_P \cdot \dot{e} + K_I \cdot e. \tag{1.2}$$

The discrete-time version of this equation is written as

$$\Delta u(k) = K_P \cdot \Delta e(k) + K_I \cdot e(k), \tag{1.3}$$

where

- $\Delta u(k)$ is the change-of-control output, where $u(k)$ is the control output, and we have that

$$\Delta u(k) = u(k) - u(k-1). \tag{1.4}$$

- $e(k)$ is the error and we have that

$$e(k) = y_{sp} - y(k), \tag{1.5}$$

where $y(k)$ is the system output and y_{sp} the set-point or desired system output.

- $\Delta e(k)$ is change-of-error and we have that

$$\Delta e(k) = e(k) - e(k-1). \tag{1.6}$$

- k is the k-th sampling time.

The control objectives listed earlier would require variable gains near the set-point, i.e., small K_I near the y_{sp} and a very large K_I near the constraint. Thus a simple PI-controller is inherently incapable of achieving all of the above control objectives and has to be implemented with the help of additional heuristic logic which would allow the desired gain modification when required.

The FKBC employes a knowledge base consisting of production rules of the form

if ⟨process state⟩ *then* ⟨control output⟩

instead of the analytical expression defining the conventional PI-controller. The ⟨process state⟩ part of the rule is called the rule antecedent and contains a description of the process output at the k-th sampling instant. The description of this process output is done in terms of particular values of error, change-of-error, and the constraint.

The ⟨control output⟩ part of the rule is called the rule consequent and contains a description of the control output which should be produced given the particular process output in the rule antecedent. This description is in terms of the value of the change-in-control output.

Thus the FKBC is modeled at the same level of resolution as the conventional PI-controller. However, the abstraction levels of the two controllers are completely different.

In the analytic expression representing the conventional PI-controller, there is no explicit reference to the causal relation in the direction from the current process output to the corresponding control output because the analytic equation is simply bidirectional. In a production rule this causal relationship is made explicit via the rule antecedent and the rule consequent. From a modelling point of view this implies that we have moved up to a higher abstraction level. The analytic representation only states the relationship between certain variables, namely e, Δe, and Δu, without providing any information at all as to the purpose which this relationship serves. One may compute either Δu given e and Δe, or e given Δu and Δe, or Δe given e and Δu. In contrast, the production rule representation describes the relationship between the three variables by explicitly stating the directionality of this relationship, i.e., from e and Δe to

Δu. This directionality (causality) defines the purpose behind the relationship between the three variables, i.e., the computation of the value of Δu.

Another difference in the abstraction levels of the FKBC and the conventional PI-controller is observed when one takes into consideration the representation of the values which any of the three variables can take. The values which e takes in a production rule are verbally expressed as "negative big" (*NB*), "negative small" (*NS*), "zero" (*ZO*), "positive small" (*PS*), "positive big" (*PB*), and "constraint" (C). A value such as *PS* expresses the knowledge that the current value of the process output $y(k)$ is below the set-point, since we have that $e(k) = y_{sp} - y(k)$ and the difference between y_{sp} and $y(k)$ is rather small. A value of error such as *NB* expresses the fact that $y(k)$ is above the set-point and the difference between y_{sp} and $y(k)$ is rather large.

The values for change-of-error are "negative big" (*NB*), "small" (*S*), and "positive big" (*PB*). A value *PB* expresses the knowledge that the current process output $y(k)$ has significantly decreased its value compared to its previous value $y(k-1)$, since $\Delta e(k) = -(y(k) - y(k-1))$. Similarly, when $y(k)$ is *NB* this means that the current value of the process output is significantly bigger than its previous value. A "small" value of $\Delta e(k)$ means that the values of $y(k)$ and $y(k-1)$ are close enough to each other and $y(k)$ is either increasing or decreasing.

The values of the change-in-control output are expressed as "negative big" (*NB*), "negative small" (*NS*), "zero" (*ZO*), "positive small" (*PS*), "positive big" (*PB*), and "drastic change" (*DC*). These values express the magnitude of the current incremental change in control output which should be added, in the case of positive values, or subtracted, in the case of negative values, to/from the value of the previous control output $u(k-1)$.

Now one can see that in the case of a FKBC the values of error and change-of-error express explicitly a relation between two consecutive values of the process output, i.e., error is *NS* when the difference $y_{sp} - y(k)$ is small and $y(k) > y_{sp}$, and change-of-error is *PB* when the difference $y(k-1) - y(k)$ is big and $y(k) < y(k-1)$. In the case of the conventional PI-controller, an error value of -11.234 is just a real number and one cannot say much about its magnitude, i.e., "small", "big", etc., unless one puts it in the context of the whole range of the possible values of error. Neither can one say that -11.234 expresses the fact that $y_{sp} < y(k)$ since this relationship is not explicit in the analytic expression describing the controller.

The rule base of the FKBC is divided into two groups of rules:

- One group which is always active, i.e., the incremental change in control output determined by these rules is applied for every sampling time. These are the so-called active rules.

- One group which becomes active only when the process output is near or enters the constraint. The incremental change in control output determined from this group of rules is added/subtracted from the one already determined by the first group of rules. These are the so-called constraint rules.

A rule from the first group has the following form:

if value of $e(k)$ is ⟨verbal value⟩ *and* value of $\Delta e(k)$ is ⟨verbal value⟩
then value of $\Delta u(k)$ is ⟨verbal value⟩.

For example,

if value of $e(k)$ is *PB and* value of $\Delta e(k)$ is *PB*
then value of $\Delta u(k)$ is *PS*.

Let us see now what such a rule actually means. Consider first the rule antecedent. The possible combinations of positive/negative values of $e(k)$ and $\Delta e(k)$ are as follows: (positive e, positive Δe), (positive e, negative Δe), (negative e, positive Δe), and (negative e, negative Δe).

The combination (positive $e(k)$, negative $\Delta e(k)$) means that the current process output $y(k)$ is below the set-point, since $e(k) = y_{sp} - y(k) > 0$, and increasing, since $\Delta e(k) = -(y(k) - y(k-1)) < 0$. Thus, the current process output is approaching the set-point from below.

The combination (negative e(k), positive $\Delta e(k)$) means that the current process output is above the set-point and decreasing. Thus the process output is approaching the set-point from above.

The combination (negative $e(k)$, negative $\Delta e(k)$) means that the current process output is above the set-point and increasing. Thus the process output is moving further away from the set-point and approaching overshoot.

The combination (positive $e(k)$, positive $\Delta e(k)$) means that the current process output is below the set-point and decreasing. Thus the process output is moving further away from the set-point and approaching undershoot.

In this context, the rule antecedent from our example is of the type (positive $e(k)$, positive $\Delta e(k)$). The fact that both $e(k)$ and $\Delta e(k)$ are "large" means that the current process output is below the set-point at a large distance from it and it has settled at this particular position after having made a large step in the direction of the set-point. Thus, since the process output is moving in the direction of the set-point with a large step, the rule consequent prescribes a small $\Delta u(k)$ to be added to the previous control output so that $y(k)$ is moved further up in the direction of the set-point without going above it.

The rest of the active rules are given in Table 1.1.

In the table form of the rule base, the second row and the first column, for example, define the active rule

if value of $e(k)$ is *NS* and value of $\Delta e(k)$ is *NB*
then value of $\Delta u(k)$ is *ZO*

There are three rules in the second group of rules (constraint rules) which prescribe a control action when the error is in the constraint region, approaching it, or leaving it.

Table 1.1. The table form of the rule base

		Δe		
		small negative	small	large positive
	large negative	small negative	large negative	large negative
	small negative	zero	small negative	small negative
e	zero	—	zero	—
	small positive	small positive	small positive	zero
	large positive	large positive	large positive	small positive

- Constraint rule 1

  if value of $e(k)$ is in constraint region
  then value of $\Delta u(k)$ is *DC*.

This rule specifies the magnitude of the additional $\Delta U(k)$ to be added to the one already determined by the active rules when $e(k)$ is in the constraint region.

The second and the third rules are of the form

- Constraint rule 2

  if value of $e(k)$ enters constraint region
  then start summing up the values of $\Delta u(k)$ determined by constraint rule 1.

- Constraint rule 3

  if value of $e(k)$ leaves constraint region
  then subtract the total value of $\Delta u(k)$ determined by constraint rule 2.

While the value of $e(k)$ is in the constraint region, the latter rule is necessary to prevent the drastic change in the value of the control output from dominating the value of the control output computed for the region outside of the constraint. This is similar to reset wind-up protection.

One can, of course, ask the question of how it is that the purely experiential knowledge of the process operator is expressed in terms of such purely mathematical concepts such as error and its derivatives. Interestingly enough, the answer to this question is to be found in a publication by Bekey [15] that appeared a long time before the notion of a KBC was conceived. In the following we will reproduce the main findings of this publication.

Consider a tracking task in which the operator is required to follow a constant velocity input (a ramp signal) with zero error. It can be shown that the design of such a system requires at least two integrations in the forward loop

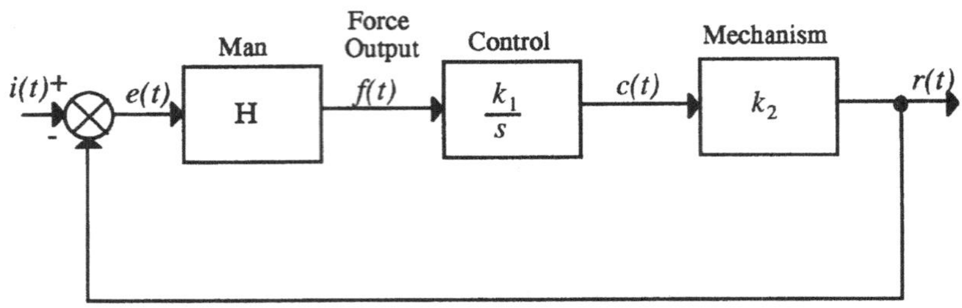

Fig. 1.8. Human-operator compensation: The basic system. (Redrawn from Bekey [15].)

[146]. Let us now assume that the controlled system has negligible dynamics and can be represented simply by a proportional factor, and that the operator is provided with a damped joystick that introduces a single integration into the system. This situation is reproduced in Fig. 1.8. The requirements on the operator's tracking strategy can be stated from a servo point of view as follows:

- The operator must introduce at least one integration to enable him to maintain zero error as desired.
- A system with two integrations and time delay (reaction time) can be shown to be unstable.
- The operator therefore also must introduce some anticipation, in the form of a derivative term that introduces lead into the control system. Hence the operator is required to introduce an integral and a derivative term to maintain the desired system performance (as shown in Fig. 1.9), in addition to whatever subjective other criteria he might use.

Evidently, a human operator does not literally perform the operations of differentiation and integration in a mathematical sense. Nevertheless, his tracking strategy, learned by experience and practice, results in control signals that can be closely approximated by devices having the required compensation characteristics. It is intuitively clear that the more complex the mathematical operations required of the operator, the more difficult the task will be, and the longer it will take to acquire the necessary skills.

Relieving the operator of the necessity of differentiating or integrating is generally known as "aiding". The removal of differentiations from the task is called "quickening". If there is a derivative term inserted in the feedback loop the operator would no longer see the actual system error but rather an error

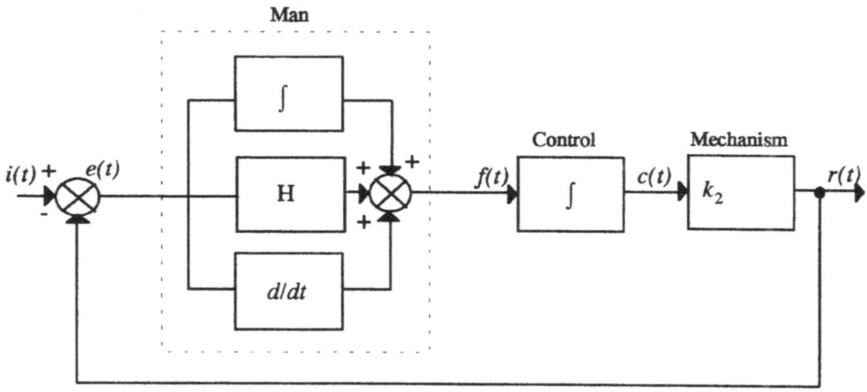

Fig. 1.9. Human-operator compensation. Compensation requirement. (Redrawn from Bekey [15].)

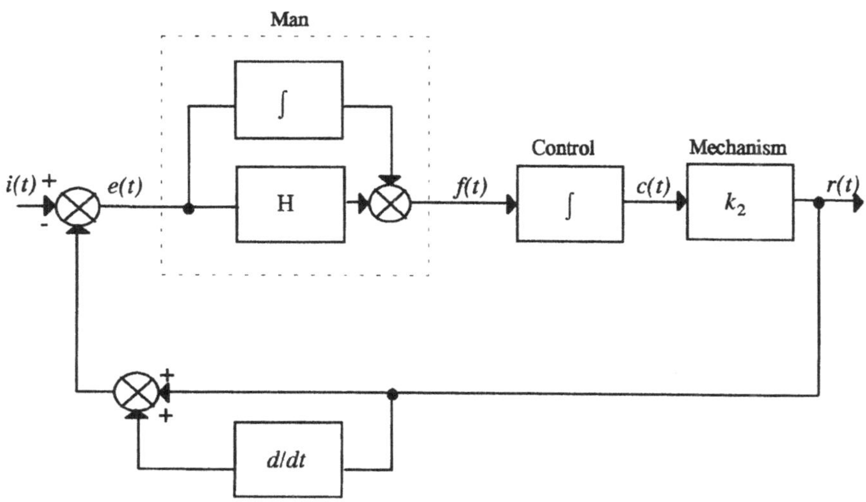

Fig. 1.10. Human-operator compensation. System with rate-aiding in feedback-loop. (Redrawn from Bekey [15].)

signal that includes some element proportional to the rate of change of the control variable. This situation is depicted in Fig. 1.10. In some cases, such derivative terms can be added directly to display devices, which are then known as rate-aided displays.

An aided display is anticipatory in the sense that it informs the operator of the results of his own control actions. This anticipation is not a true prediction, however, since it does not account for the dynamics of the system under control. The inclusion of error derivatives in the display simply indicates to the operator the trends resulting from his control actions and prevents excessive overshoots/undershoots.

If we return now to the table describing the FKBC for the servocontrol task we will see the close analogy between the role of an aided display and that of the FKBC.

2. The Mathematics of Fuzzy Control

2.1 Introduction: Fuzzy Sets

2.1.1 Vagueness

One of the major developments of fuzzy set theory, *fuzzy logic* was primarily designed to represent and reason with some particular form of knowledge. It was assumed that the knowledge would be expressed in a linguistic or verbal form, and also that the whole exercise should not be a mere intellectual undertaking, but must also be operationally powerful so that computers can be used. However, when using a language-oriented approach for representing knowledge about a certain system of interest, one is bound to encounter a number of non-trivial problems. Suppose, for example [119], that you are asked how strongly you agree that a given number $x \in [0, 20]$ is a *large* number. One way to answer this question is to say that if $x \geq d$ then you agree it is a *large* number and if $x < d$ then you disagree. Thus, if you place a mark on an *agree–disagree* scale, it might be distributed uniformly over the right half of the scale whenever $x \geq d$ and uniformly over the left half if $x < d$. Such a person is called a *threshold* person.

Another kind of person may try to place his mark close to the *agree* side of the scale according to how large he thinks x is – the larger x the closer to the *agree* end. This depends critically on the sample he is presented with, for if he has used up the scale by placing a mark very close to *agree* (his answer to the largest number just presented), then he will be squeezed if an even larger number is presented. However such a person will use the full continuum of the scale according to how large or how small he thinks the numbers are. Such a person is called an *estimator*.

A third kind of person might place his mark either on *agree* or *disagree*, or place no mark if he is uncertain as to whether x is *large*. The set of numbers presented to him will be then partitioned into three subsets: those he considers to be *large*, those he considers to be *not large* and those about which he is uncertain, i.e., neither *large* nor *not large*. Such people are called *conservatives*.

What we actually have here is three very different ways of representing one's knowledge about *large* numbers. The same fact that there are three possible ways of doing this already implies that the property *large* is inherently *vague*, meaning that the set of objects it applies to has no sharp boundaries, at least for the *estimators*. The *estimators* are the typical fuzzy-artificial-intelligence devotees: for them it is only to a certain degree that a property is characteristic of an object.

Thus, fuzzy logic was motivated in a large measure by the need for a conceptual framework which can come to grips with a certain central issue that was haunting the use of language for the purpose of knowledge representation, namely its inherent *vagueness* [23].

The fuzziness of a property lies in the lack of well defined boundaries of the set of objects to which this property applies. More specifically, let U be the field of reference, also-called *universe of discourse*, covering a definite range of objects. Now consider a subset F of U, where the transition between membership and non-membership is gradual rather than abrupt. This fuzzy subset F obviously has no well defined boundaries. For instance, F may be the set of *tall* men in a community U, i.e., the set of men having the property of being tall. Usually, there are members of U who are definitely *tall*, others who are *not tall* at all, but there exist also borderline cases. Then, a membership degree 1 is assigned to the objects that completely belong to F – here, the men who are definitely *tall*. Conversely, the objects that do not belong to F at all are assigned a membership degree of 0. Furthermore, the membership degrees of the borderline cases will naturally lie between 0 and 1. In other words, the more an element or object is characteristic of F, the closer to 1 is its degree of membership. Thus, the use of a numerical scale, as the interval $[0, 1]$, allows for a convenient representation of the gradation of membership.

However, three things should be made clear here. First, precise membership degrees do not exist by themselves, but are only tendency indices that are *subjectively* assigned by an individual or a group of individuals. Thus, from a psychological point of view, the membership degree is not a primitive object. It only reflects an ordering on the objects of the universe of discourse, induced by the property associated with F, e.g., *tall*. This ordering, when it exists, is more important than the membership degrees themselves. Second, the membership degrees are not absolutely defined, but are in most cases context dependent. For instance, Dimiter's height of only 175 cm may be classified as *tall* by an Italian, but a Swede would hardly judge him as being *tall* at all. Third, fuzziness differs from *imprecision* in that the latter refers to lack of knowledge about a value of a parameter, e.g., height, and is thus expressed as a crisp tolerance interval. This interval is the set of all possible values of a parameter. Fuzziness occurs when the interval has no sharp boundaries, i.e., it is a fuzzy set F.

2.1.2 Fuzzy Set Theory Versus Probability Theory

In classical probability theory an *event*, E, is defined as a crisp subset of a certain sample space, U. For example, when throwing a dice, the sample space, U, is the set of integer numbers $\{1, 2, 3, 4, 5, 6\}$, and an event such as "E = a number less than four," is a crisp subset of U given as, $\{1, 2, 3\}$. Furthermore, an event occurs if at least one of the elements of the subset that defines the event occurs. In our example, E will occur if the result of throwing a dice is any one of 1, 2, 3. Then, from the *frequentist* point of view, the probability of E, $P(E) \in [0,1]$, is the proportion of occurrences of E in a long series of experiments. Furthermore, a ***conditional probability***, denoted as $P(E \mid E')$, defines a probability of an event E given that we are told that the event E' has occurred. In our example, $P(E \mid 3) = 1$, since we know that 3 is an element of the subset U that defines the event E, and if 3 has occurred then, according to our definition of an event, this means that E has occurred as well. Then one can say that probability theory is concerned with setting up rules for manipulating these probabilities and for calculating the probabilities of more complex events within a certain system of axioms. However, let us consider the case when E is not a crisp subset of U but rather a fuzzy subset $\mathcal{E}$. For example, let "$\mathcal{E}$ = a *large* number" and its membership function (see Section 2.1.4) be defined as

$$\mathcal{E} = \mu_{\text{large}}(u) = 1/6 + 0.7/5 + 0.5/4 + 0.2/3 \tag{2.1}$$

and let the question be "What is the probability of throwing a *large* number?" According to Zadeh [234], the probability $P(\mathcal{E})$ is computed as

$$P(\mathcal{E}) = \sum_i P(u_i) \cdot \mu_{\text{large}}(u_i) = \frac{1}{6}(1 + 0.7 + 0.5 + 0.2) = 0.4, \tag{2.2}$$

where $\frac{1}{6}$ is the probability of each u_i. In the case when our event is a crisp set then we have that $\forall i : \mu_{\text{large}}(u_i) = 1$ and thus, the probability of the event would be computed to be 0.66.

Another type of probability is so-called *subjective* probability, which does not have a frequentist interpretation but is rather described as the amount of subjective belief that a certain event may occur. Here, a numerical probability only reflects an observer's incomplete (uncertain) knowledge about the true state of affairs. In knowledge representation and inference under uncertainty it is this concept of probability that is relevant. It has been shown by Cox [41] that if such a continuous measure of belief satisfies certain properties, then the standard axioms of probability theory follow logically. Among these properties is one, called the *clarity property*, which states that "every proposition that is a subject of a probability assertion must be decidable as either true or false."

An important issue which the proponents of the subjective school of thought in probability theory [34] claim to solve is that of vagueness or fuzziness. The proposal is that a membership function can be viewed as a probability function. Let $\mathcal{E}$ be a fuzzy set defined on the set of heights U, and let the meaning of $\mathcal{E}$ be *tall.* Then, provided that u is a precise value of height, $\mu_{\mathcal{E}}(u)$ is nothing but the conditional probability $P(\mathcal{E} \mid u, c)$ where c is the context within which the meaning of $\mathcal{E}$ is derived, e.g., the heights of persons who are male and in their twenties. Now the interpretation of $\mu_{\mathcal{E}}(u)$ in terms of probability faces the following problem [62]: $\mathcal{E}$ and u are defined over the same universe of discourse U, and $\{u\}$ is a singleton of that universe. Furthermore, since the clarity property requires a crisp representation of $\mathcal{E}$ (or *tall*), then $\mathcal{E}$ is a classical subset of heights (or u). But in this case either $u \in U$ and $P(\mathcal{E} \mid u) = 1$, since for $\mathcal{E}$

to be the case it is enough that at least one element of the classical set that defines $\mathcal{E}$ has to be the case, and we know that u is the case, or $u \notin U$ and $P(\mathcal{E} \mid u) = 0$. The fact that $P(\mathcal{E} \mid u)$ may lie outside the set $\{0, 1\}$ indicates that $\mathcal{E}$ is a fuzzy event, which violates the clarity property. In the situation where we know that the actual height of a person is u and we ask whether he/she is *tall*, the answer "The person is tall" may hold to a degree different from 0 or 1, because the fact we inquire about is fuzzy. We can model this degree by a number (up to a precise knowledge of the membership function) because the available information (the height u) is precise and certain.

In his paper "Fuzzy Sets as a Basis for a Theory of Possibility" [242] Zadeh gives an interesting survey of the differences between fuzzy set theory and probability theory. He states that a fuzzy set induces a *possibility distribution* on the universe of discourse. This means that the fuzzy set describing notion *large number* in the above dice example induces the following possibility distribution:

It is *impossible* that 1 or 2 are large numbers.
It is 0.2 *possible* that 3 is a large number.
It is 0.5 *possible* that 4 is a large number.
It is 0.7 *possible* that 5 is a large number.
It is *possible* that 6 is a large number.

Hence there is no need for a possibility distribution to satisfy the property that the sum of the possibilities is 1, as is the case in probability theory. A possibility distribution only describes the possible values induced by a fuzzy set.

An example of this is the so-called egg-eating example [242].[1]

Example 2.1 Consider the statement "Hans ate X eggs for breakfast," with X taking values in $U = \{1, 2, 3, 4, \ldots\}$. We may associate a possibility distribution with X by interpreting $\pi_X(u)$ as the degree of ease with which Hans can eat u eggs. We may also associate a probability distribution with X by interpreting $P_X(u)$ as the probability of Hans eating u eggs for breakfast. Assuming that we employ some explicit or implicit criterion for assessing the degree of ease with which Hans can eat u eggs for breakfast, the values of $\pi_X(u)$ and $P_X(u)$ might be as shown in Table 2.1.

We observe that, whereas the possibility that Hans may eat 3 eggs for breakfast is 1, the probability that he may do so might be quite small, e.g., 0.1. Thus a high degree of possibility does not imply a high degree of probability. However, if an event is probable, it must also be possible.

2.1.3 Classical Set Theory

A *classical set* is a collection of objects of any kind. The concept of a set has become one of the most fundamental notions of mathematics. So-called set

[1] According to Prof. Zadeh this example came into his mind when he had a breakfast together with Prof. Hans-J. Zimmermann.

Table 2.1. The possibility and probability distributions associated with X

u	1	2	3	4	5	6	7	8
$\pi_X(u)$	1	1	1	1	0.8	0.6	0.4	0.2
$P_X(u)$	0.1	0.8	0.1	0	0	0	0	0

theory was founded by the German mathematician Georg Cantor (1845–1918). In set theory the notions 'set' and 'element' are primitive. They are not defined in terms of other concepts. Letting A be a set, "$x \in A$" means that x is an element of the set A and "$x \notin A$" means that x does not belong to the set A. A set is fully specified by the elements it contains. The way in which these elements are specified is immaterial; for example, there is no difference between the set consisting of the elements 2, 3, 5, and 7, and the set of all prime numbers less than 11.

For any element in the universe of discourse it can unambiguously be determined whether it belongs to the set or not. A classical set may be finite, countable or uncountable. It can be described by either listing up the individual elements of the set, or by stating a property for membership, e.g., the property "$x \geq 0$" in Example 2.2.

Example 2.2 a. The set

$$R = \{\text{red, orange, yellow, green, blue}\} \tag{2.3}$$

is an example of a finite set that is described by its elements.
b. The set

$$T = \{x \in \mathbb{Z} \mid x \geq 0\}, \tag{2.4}$$

i.e., the set of positive integers, is an example of a countable set that is described by a property.
c. The real interval $[0, 1]$ is an example of an uncountable set.

Two sets are very important, namely, the universe U, containing all elements of the universe of discourse, and the empty set $\emptyset$, containing no elements at all. They complement each other.

When A and B are sets, then $A = B$ means that A and B contain the same elements, i.e., equality is defined extensionally. $A \subseteq B$ (A is a subset of B) means that all elements of A are also elements of B. The expression $A \subset B$ (A is a proper subset of B) means that all elements of A are also elements of B, but B contains other elements too. This is the same as $A \subseteq B$, but not $A = B$. $B \supseteq A$ is the same as $A \subseteq B$, and $B \supset A$ is the same as $A \subset B$.

If $P(x)$ is a predicate stating that x has the property P, then a set can also be denoted by $\{x \mid P(x)\}$. This notation was already used in Example 2.2 b.

Example 2.3 a. Let $P(x)$ be the predicate on the real numbers expressing that $x > 0$, then $\{x \mid P(x)\}$ is the set of positive real numbers.
b. Suppose $P(x)$ is true if and only if ($x \bmod 3 \equiv 0$), then $\{x \mid P(x)\}$ is the set of all multiples of three.
c. If length(x) denotes the length of a person x and $P(x)$ is the predicate stating that length(x) > 1.80 m, then $\{x \mid P(x)\}$ is the set of all persons larger than 1.80 m.

Classical set theory uses several operations like complement, intersection, difference, etc. Let A and B be two classical sets in a universe U, then the following set operations can be defined

- *Complement* of A, $A' = \{x \mid x \notin A\}$.
- *Intersection* of A and B, $A \cap B = \{x \mid x \in A \text{ and } x \in B\}$.
- *Union* of A and B, $A \cup B = \{x \mid x \in A \text{ or } x \in B\}$.
- *Difference* of A and B, $A - B = \{x \mid x \in A \text{ and } x \notin B\}$.
- *Symmetric difference* of A and B, $A + B = A - B \cup B - A$.
- *Power set* of A, $\mathcal{P}(A) = \{X \mid X \subseteq A\}$.
- *Cartesian product* of A and B, $A \times B = \{(x, y) \mid x \in A \text{ and } y \in B\}$.
- *Power n of A*, $A^n = \underbrace{A \times \ldots \times A}_{n \text{ times}}$.

The set operations 'intersection', 'union' and 'complement' are illustrated in Fig. 2.1.

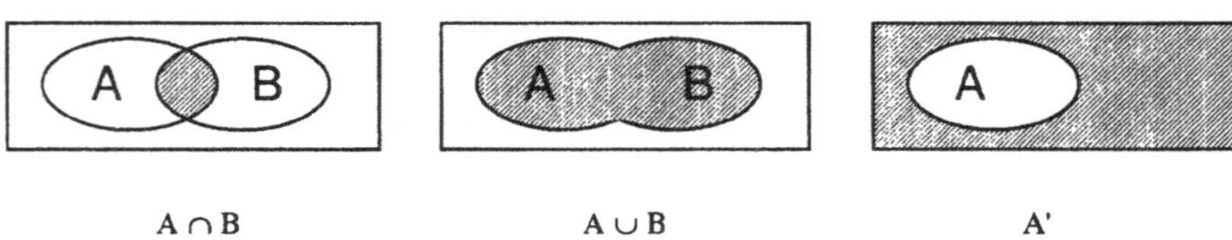

Fig. 2.1. The operations 'intersection', 'union' and 'complement'

The Cartesian product of n classical sets $A_1, \ldots, A_n$ is defined by

$$\mathop{\times}_{i=1}^{n} = A_1 \times \ldots \times A_n = \{(x_1, \ldots, x_n) \mid x_1 \in A_1, \ldots, x_n \in A_n\}. \tag{2.5}$$

The union and intersection of n sets $A_1, \ldots, A_n$ are

$$\bigcup_{i=1}^{n} A_i = A_1 \cup \ldots \cup A_n \qquad \text{and} \qquad \bigcap_{i=1}^{n} A_i = A_1 \cap \ldots \cap A_n. \tag{2.6}$$

The *cardinality* of a set A denotes the number of elements in A. Hence, the first set in Example 2.2 has cardinality 5, the second set has countably many elements, denoted by $\aleph_0$ (pronounced aleph sub zero), and the third set has uncountably many elements, denoted by $\aleph_1$.

The most important properties of the classical set-theoretic operations are summarized in Table 2.2 [118, 44].

Table 2.2. Properties of classical set operations

Involution	$(A')'$
Commutativity	$A \cup B = B \cup A$
	$A \cap B = B \cap A$
Associativity	$(A \cup B) \cup C = A \cup (B \cup C)$
	$(A \cap B) \cap C = A \cap (B \cap C)$
Distributivity	$A \cap (B \cup C) = (A \cap B) \cup (A \cap C)$
	$A \cup (B \cap C) = (A \cup B) \cap (A \cup C)$
Idempotence	$A \cup A = A$
	$A \cap A = A$
Absorption	$A \cup (A \cap B) = A$
	$A \cap (A \cup B) = A$
Absorption of complement	$A \cup (A' \cap B) = A \cup B$
	$A \cap (A' \cup B) = A \cap B$
Absorption by U and $\emptyset$	$A \cup U = U$
	$A \cap \emptyset = \emptyset$
Identity	$A \cup \emptyset = A$
	$A \cap U = A$
Law of contradiction	$A \cap A' = \emptyset$
Law of excluded middle	$A \cup A' = U$
De Morgan's laws	$(A \cap B)' = A' \cup B'$
	$(A \cup B)' = A' \cap B'$

We have seen that one way to define a set is to enumerate its elements and another way is to use a predicate $P(x)$, meaning that every element x of the set has a property P (Example 2.3). A third way that is interesting with respect to the extension from set theory to *fuzzy* set theory is the definition of a set A using its characteristic function μ_A. Let A be defined on the domain X.

Definition 2.4 $\mu_A : X \to [0, 1]$ *is a characteristic function of the set A iff for all x*

$$\mu_A(x) = \begin{cases} 1 & \textit{when } x \in A, \\ 0 & \textit{when } x \notin A. \end{cases} \tag{2.7}$$

Example 2.5 If A is the set of odd natural numbers, then μ_A assigns 1 to all odd, and 0 to all even numbers, e.g., $\mu_A(n) = n \bmod 2$.

Actually, there are only small differences between the predicate based definition and this characteristic function based definition. The predicate is truth-functional, i.e., assigns the values *true* and *false* to elements (if an element x has the property P, then $P(x)$ is true, otherwise false), while the characteristic function returns the values 1 and 0 to elements. To combine these two, one can define the extension or meaning of a predicate P in the following way: $P(x) \Leftrightarrow (\mu_P(x) = 1)$.

With this characteristic function, it is possible to redefine the operations 'complement', 'intersection' and 'union' in terms of functions, which will turn out to be useful when dealing with *fuzzy* set theory. Let A and B be two classical sets in a universe U, and let μ_A and μ_B be their membership functions. The set operations 'complement', 'intersection' and 'union' can alternatively be defined as

- *Complement* of A denoted as A', $\mu_{A'}(x) = 1 - \mu_A(x)$.
- *Intersection* of A and B denoted as $A \cap B$, $\mu_{A\cap B}(x) = \min(\mu_A(x), \mu_B(x))$.
- *Union* of A and B denoted as $A \cup B$, $\mu_{A\cup B}(x) = \max(\mu_A(x), \mu_B(x))$.

The choice of these three functions ('one minus' for complement, minimum for intersection, and maximum for union) is arbitrary. The only property, for example, the intersection operation has to satisfy is to return 1 if both arguments are 1, and to return 0 otherwise. The minimum operation does this, but alternatives for the intersection operation are, amongst many others:

$$\mu_{A\cap B}(x) = \begin{cases} \mu_A(x) \cdot \mu_B(x), & \text{or} \\ \max(0, \mu_A(x) + \mu_B(x) - 1), & \text{or} \\ \dfrac{\mu_A(x)\mu_B(x)}{\frac{3}{4} + \frac{1}{4}(\mu_A(x) + \mu_B(x) - \mu_A(x)\mu_B(x))}. \end{cases} \tag{2.8}$$

In set theory, all these complex operations are of little use because they all deliver the same results, but in fuzzy set theory we will assign to them a deeper meaning. That is the reason why we have introduced them here.

2.1.4 Fuzzy Sets

In fuzzy set theory, 'normal' sets are called *crisp sets*, in order to distinguish them from *fuzzy sets*. Let C be a crisp set defined on the universe U, then for any element u of U, either $u \in C$ or $u \notin C$. In fuzzy set theory this property is generalized, therefore in a *fuzzy set* F, it is not necessary that either $u \in F$ or $u \notin F$. In the last few years there have been many theories that presented a generalization of the membership property,[2] but fuzzy set theory seems to be the most intuitive among them.

The generalization is performed as follows. For any crisp set C it is possible to define a *characteristic function* $\mu_C : U \to \{0,1\}$ as in (2.7). In fuzzy set theory, the characteristic function is generalized to a *membership function* that assigns to every $u \in U$ a value from the *unit interval* $[0,1]$ instead from the two-element set $\{0,1\}$. The set that is defined on the basis of such an extended membership function is called a *fuzzy set.*

Definition 2.6 *The membership function* μ_F *of a fuzzy set* F *is a function*

$$\mu_F : U \to [0,1]. \tag{2.9}$$

So, every element u from U has a membership degree $\mu_F(u) \in [0,1]$. F is completely determined by the set of tuples

$$F = \{(u, \mu_F(u)) \mid u \in U\}. \tag{2.10}$$

Example 2.7 Suppose someone wants to describe the class of cars having the property of being expensive, by considering cars such as BMW, Buick, Ferrari, Fiat, Lada, Mercedes and Rolls Royce. Some cars, like Ferrari or Rolls Royce definitely belong to this class, while other cars, like Fiat or Lada do not belong to it. But there is a third group of cars, where it is difficult to state whether they are expensive or not. Using a *fuzzy set*, the fuzzy set of expensive cars is, e.g.,

$$\{(\text{Ferrari},1), (\text{Rolls Royce},1), (\text{Mercedes},0.8), (\text{BMW},0.7), (\text{Buick},0.4)\}, \tag{2.11}$$

i.e., Mercedes belongs to degree 0.8, BMW to 0.7 and Buick to 0.4 to the class of expensive cars.

Example 2.8 Suppose someone wants to define the set of *natural* numbers 'close to 6'. This can be expressed by

$$\tilde{6} = \{(3,0.1),(4,0.3),(5,0.6),(6,1.0),(7,0.6),(8,0.3),(9,0.1)\}. \tag{2.12}$$

The membership function of the fuzzy set of *real* numbers close to 6, is, e.g.

$$\mu_{\tilde{6}}(x) = \frac{1}{1+(x-6)^2}, \tag{2.13}$$

and the fuzzy set $\tilde{6}$ contains the elements $(6,1)$, $(5.5,0.8)$, $(100,0.000113161)$, etc.

---

[2]For example, rough set theory [158, 58], and also the Dempster-Shafer theory [189].

Zadeh proposed a more convenient notation for fuzzy sets. Suppose C is a finite crisp set $\{u_1, u_2, \ldots, u_n\}$, then an alternative notation is

$$C = u_1 + u_2 + \ldots + u_n, \tag{2.14}$$

where + denotes an enumeration. Furthermore, an alternative notation for $(u, \mu(u))$ is $\mu(u)/u$, where / denotes a tuple.

Example 2.9 The set of expensive cars (2.11) is described by

$$1/\text{Ferrari} + 1/\text{Rolls Royce} + 0.8/\text{Mercedes} + 0.7/\text{BMW} + 0.4/\text{Buick}. \tag{2.15}$$

The set $F = \{(u, \mu_F(u)) \mid u \in U\}$ is then described by

$$F = \mu_F(u_1)/u_1 + \cdots + \mu_F(u_n)/u_n = \sum_{i=1}^{n} \mu_F(u_i)/u_i, \tag{2.16}$$

where + satisfies $a/u + b/u = \max(a, b)/u$, i.e., if the same element has two different membership degrees 0.8 and 0.6, then its membership degree becomes 0.8. Any countable or discrete universe U allows a notation

$$F = \sum_{u \in U} \mu_F(u)/u, \tag{2.17}$$

but when U is uncountable or continuous, we will write

$$F = \int_U \mu_F(u)/u. \tag{2.18}$$

Hence, the $\int$-sign denotes an uncountable enumeration. In the classical notation it is also possible to write (2.16) and (2.18) as

$$\{\mu_F(u)/u \mid u \in U\} \tag{2.19}$$

To illustrate the theory given thus far, some examples follow:

Example 2.10 Let $U = \mathbb{N}$, then the alternative representation of $\tilde{6}$ is

$$\tilde{6} = \{0.1/3, 0.3/4, 0.6/5, 1.0/6, 0.6/7, 0.3/8, 0.1/9\}. \tag{2.20}$$

$\tilde{6}$ is a fuzzy set which can be interpreted as the set of integers approximately equal to 6.

Example 2.11 Let $U = \mathbb{R}$, and

$$\mu_F(u) = \begin{cases} 1 - \sqrt{\dfrac{|u-6|}{3}}, & \text{for } 3 \le u \le 9, \\ 0, & \text{otherwise;} \end{cases} \tag{2.21}$$

then

$$F = \int_{\mathbb{R}} \mu_F(u)/u \tag{2.22}$$

is the fuzzy set representing the real numbers approximately equal to 6. Note that this membership function is an alternative to (2.13).

Example 2.12 It is also possible to represent the set of old people as a fuzzy set. Let $U = [0, 120]$ represent the age of the people, and let

$$\mu_{\text{old}}(u) = \begin{cases} 0, & \text{for } 0 \leq u \leq 60, \\ \dfrac{u-60}{20}, & \text{for } 60 < u < 80, \\ 1, & \text{for } 80 \leq u \leq 120, \end{cases} \tag{2.23}$$

then old $= \int_0^{120} \mu_{\text{old}}(u)/u$. In this case we could have used the universe $\{0, \ldots, 120\}$ as well, and $\sum$ instead of an $\int$, but usually the more general case of a continuous domain is preferred.

In fuzzy control, there are usually four kinds of fuzzy sets that are dealt with, namely 'increasing', 'decreasing', and two kinds of 'approximating' notions. These kinds of fuzzy sets are called *convex* in fuzzy set theory. Non-convex fuzzy sets are fuzzy sets that are alternatively increasing and decreasing on the domain. Figure 2.12 (page 54) shows a convex and a non-convex set.

Increasing fuzzy sets are used to represent linguistic notions like 'tall', 'fat', 'hot' and 'old' on the domains of height (m), weight (kg), temperature (°C) and age ({0,1,...,120}) respectively. Examples of decreasing notions on these domains are 'small', 'thin', 'cold', and 'young'. Examples of approximating notions on these domains are often difficult to express in one word. On the domain of temperature it may be 'comfortable', on the domain of age it may be 'middle-aged'.

Suppose one wishes to build an air conditioner, then there may be the following linguistic notions which describe the temperature to be controlled: 'hot' (Fig. 2.2), 'cold' (Fig. 2.3), and 'comfortable' (Fig. 2.4). Furthermore, in an air-conditioning system, there can be the approximating notions 'cool' and 'warm'. These five fuzzy sets can be given in one figure, as in Fig. 2.5.

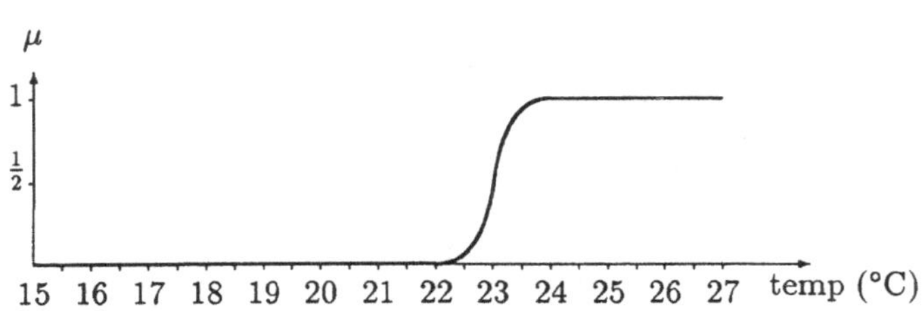

Fig. 2.2. The fuzzy notion *hot*.

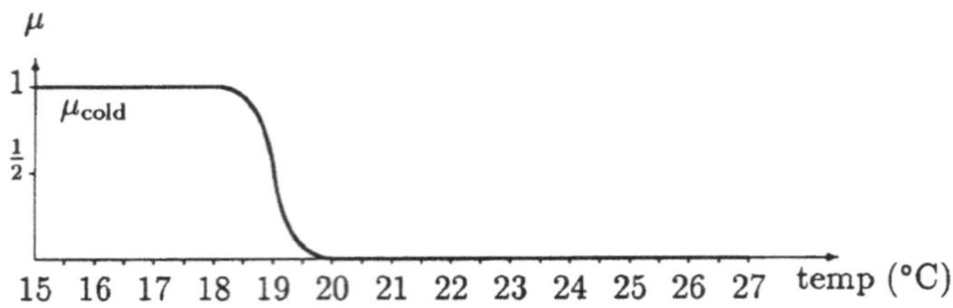

Fig. 2.3. The fuzzy notion *cold.*

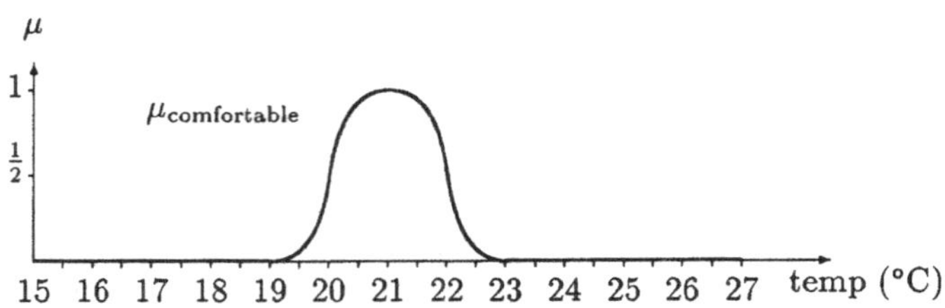

Fig. 2.4. The fuzzy notion *comfortable.*

These linguistic notions are meaning-dependent on the actual physical domain, for example, the domain of temperature. In fuzzy control it is usual to use linguistic notions that are meaning-independent of the particular physical domain. The domain described above consists of five elements. Consider now a domain consisting of seven elements which is usual in fuzzy control. The elements of the domain have names like 'Positive Big' (*PB*), 'Positive Medium' (*PM*), 'Positive Small' (*PS*), 'Zero' (*ZO*), 'Negative Small' (*NS*), 'Negative Medium' (*NM*), and 'Negative Big' (*NB*) (see Chapter 4). The domain used, is the domain of real numbers between -6 and 6. This is a frequently used *standard* domain. If the domain used is the real interval $[-1, 1]$, then we have the so-called *nor-*

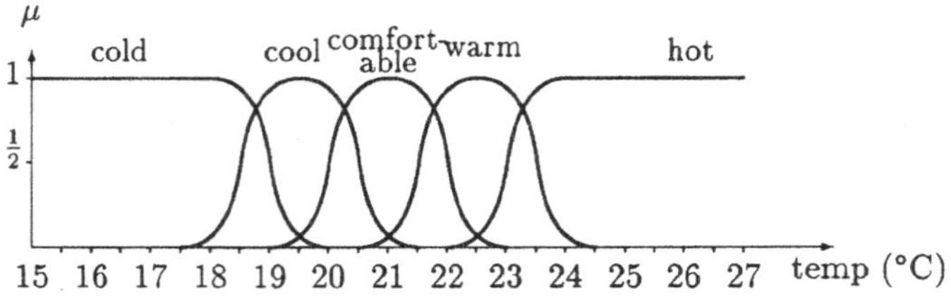

Fig. 2.5. The whole domain.

malized domain (Chapter 4). However, we will use the notion of *normalized* domain to refer to both normalized and standard domains. Furthermore, it is usual to have functions with straight lines, instead of functions with curves as were given in the temperature domain. The result of this is shown in Fig. 2.6.

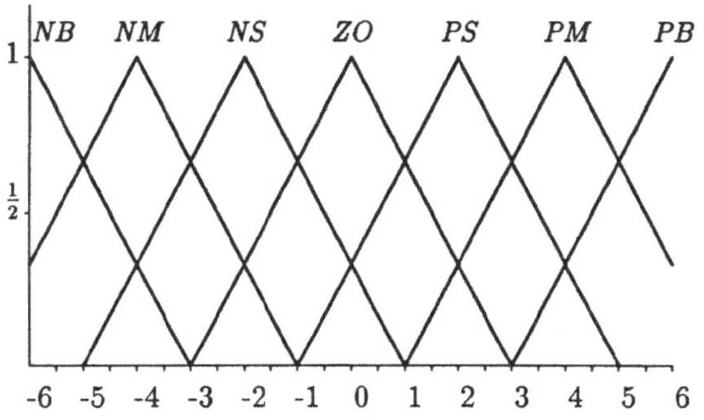

Fig. 2.6. The fuzzy sets *NB, NM, NS, ZO, PS, PM* and *PB* on the domain $[-6, 6]$.

Typical rules of a PI-like FKBC using these linguistic notions are:

if e is *NB and* $\dot{e}$ is *ZO then* $\dot{u}$ is *PB*
if e is *NM and* $\dot{e}$ is *ZO then* $\dot{u}$ is *PM*
if e is *ZO and* $\dot{e}$ is *PB then* $\dot{u}$ is *NB*

where e denotes error, $\dot{e}$ change-of-error, and $\dot{u}$ denotes the control output value.

We will use special names for these kinds of fuzzy sets with straight lines, because they return again and again in fuzzy control theory. They offer a simplified notation which can be compared to internal notations in fuzzy control systems or the LR-notation from Dubois and Prade [54]. The increasing membership functions with straight lines we will call Γ-functions, because of the similarity of these functions with this character. This function is a function of one variable and two parameters defined as follows (Fig. 2.7):

Definition 2.13 *The function* $\Gamma : U \to [0,1]$ *is a function with two parameters defined as*

$$\Gamma(u;\alpha,\beta) = \begin{cases} 0 & u < \alpha, \\ (u-\alpha)/(\beta-\alpha) & \alpha \le u \le \beta, \\ 1 & u > \beta. \end{cases} \tag{2.24}$$

Example 2.14 The notion *PB* in Fig. 2.6 can be described by the membership function $\Gamma(u;3,6)$, old can be described by the membership function $\Gamma(u;60,80)$, and tall by $\Gamma(u;170,190)$.

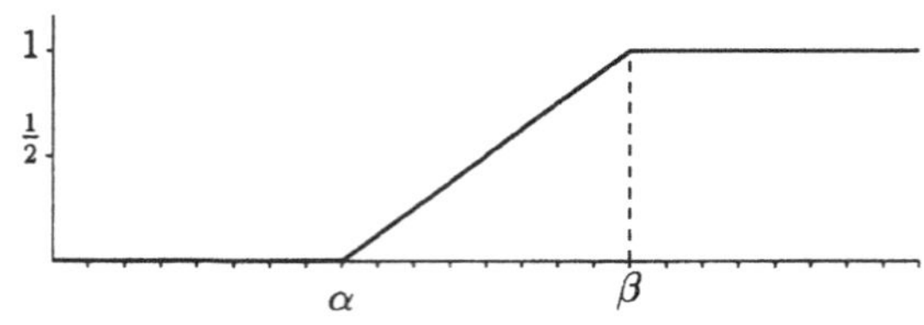

Fig. 2.7. An example of a Γ-function.

Zadeh's S-function, defined as

$$S(x;\alpha,\beta,\gamma) = \begin{cases} 0 & \text{for } x \leq \alpha, \\ 2\left(\dfrac{x-\alpha}{\gamma-\alpha}\right)^2 & \text{for } \alpha < x \leq \beta, \\ 1-2\left(\dfrac{x-\gamma}{\gamma-\alpha}\right)^2 & \text{for } \beta < x \leq \gamma, \\ 1 & \text{for } x > \gamma; \end{cases} \tag{2.25}$$

where $\beta = (\alpha + \gamma)/2$, can be considered as a more fluent variant of the Γ-function (Fig. 2.8). It is frequently used in fuzzy logic, but only seldomly in fuzzy control.

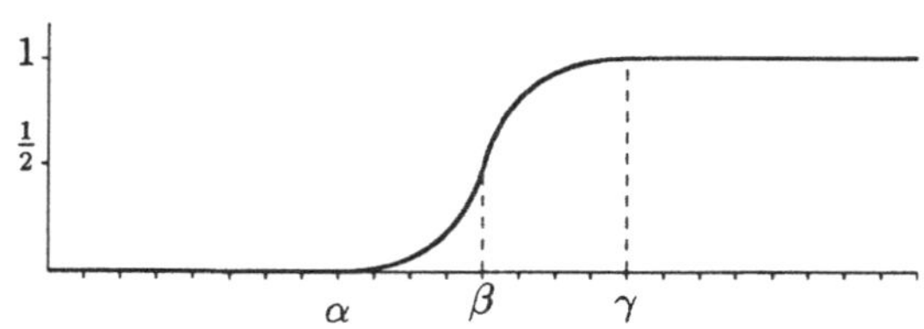

Fig. 2.8. An example of an S-function.

The decreasing membership functions with straight lines we will call L-functions; they are defined as follows (Fig. 2.9):

Definition 2.15 *The function* $L : U \to [0,1]$ *is a function with two parameters defined as*

$$L(u;\alpha,\beta) = \begin{cases} 1 & u < \alpha, \\ (\beta-u)/(\beta-\alpha) & \alpha \leq u \leq \beta, \\ 0 & u > \beta. \end{cases} \tag{2.26}$$

Example 2.16 The notion *NB* in Fig. 2.6 can be described by the membership function $L(u;-6,-3)$.

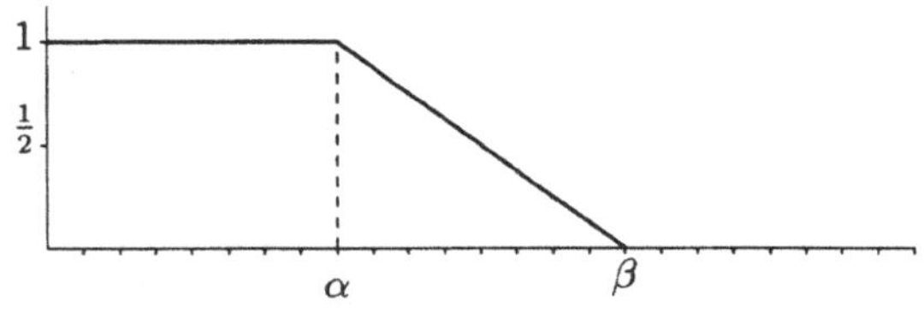

Fig. 2.9. An example of an L-function.

Bell-shaped membership functions with straight lines we will call Λ-functions or triangular functions; they are defined as follows (Fig. 2.10):

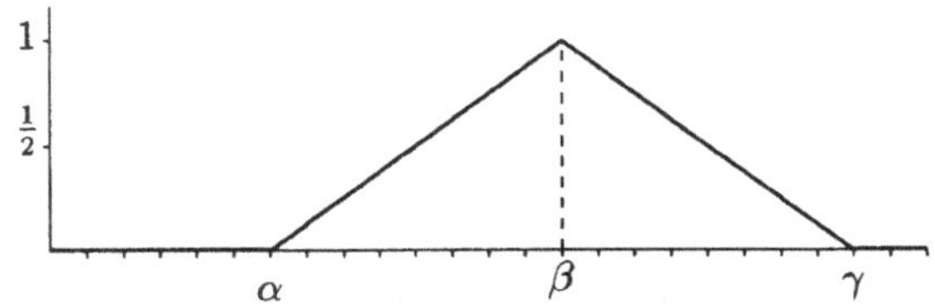

Fig. 2.10. An example of a Λ-function.

Definition 2.17 *The function $\Lambda : U \to [0,1]$ is a function with three parameters defined as*

$$\Lambda(u;\alpha,\beta,\gamma) = \begin{cases} 0 & u < \alpha, \\ (u-\alpha)/(\beta-\alpha) & \alpha \le u \le \beta, \\ (\beta-u)/(\beta-\alpha) & \beta \le u \le \gamma, \\ 0 & u > \gamma. \end{cases} \tag{2.27}$$

Example 2.18 The notion *ZO* in Fig. 2.6 can be described by the membership function $\Lambda(u;-3,0,3)$.

Zadeh's bell-shaped π-function, defined as

$$\pi(x;\beta,\gamma) = \begin{cases} \mathrm{S}(x;\gamma-\beta,\gamma-\beta/2,\gamma) & \text{for } x \le \gamma, \\ 1-\mathrm{S}(x;\gamma,\gamma+\beta/2,\gamma+\beta) & \text{for } x \ge \gamma. \end{cases} \tag{2.28}$$

can be considered as a more fluent variant of the Λ-function. Like the S-function given above, it is of low practical use for fuzzy control.

Approximating membership functions with straight lines, where the top is not one point but an interval, we will call Π-functions; these are defined as follows (Fig. 2.11):

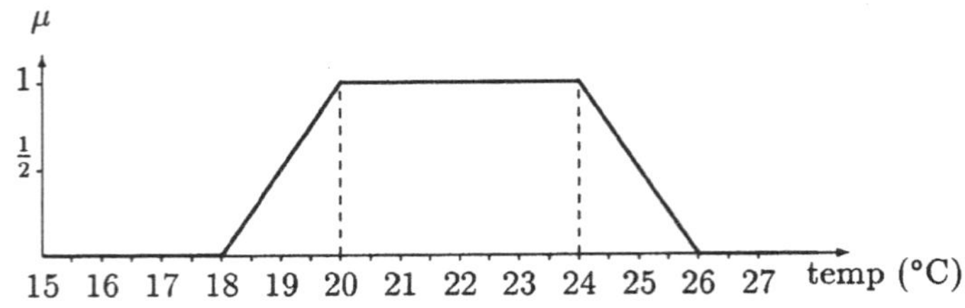

Fig. 2.11. An example of the Π-function $\Pi(u; 18, 20, 24, 26)$.

Definition 2.19 *The function $\Pi : U \to [0,1]$ is a function with four parameters defined as*

$$\Pi(u;\alpha,\beta,\gamma,\delta) = \begin{cases} 0 & u < \alpha, \\ (u-\alpha)/(\beta-\alpha) & \alpha \le u < \beta, \\ 1 & \beta \le u \le \gamma, \\ (\delta-u)/(\delta-\gamma) & \gamma < u \le \delta, \\ 0 & u > \delta. \end{cases} \tag{2.29}$$

This Π-function can be used to describe the three functions mentioned before (Γ, L, and Λ). Suppose the underlying domain is $[-6, 6]$, then the following equations hold

$$\begin{aligned} \Gamma(x;\alpha,\beta) &= \Pi(x;\alpha,\beta,6,6), \\ \mathrm{L}(x;\gamma,\delta) &= \Pi(x;-6,-6,\gamma,\delta), \\ \Lambda(x;\alpha,\beta,\delta) &= \Pi(x;\alpha,\beta,\beta,\delta). \end{aligned} \tag{2.30}$$

The TIL-Shell from Togai InfraLogic (Irvine, Cal.), a computer assisted software engineering tool for design of FKBC, uses a similar notation as the Π-function. $\Pi(u;\alpha,\beta,\gamma,\delta)$ in their system is described by

$$\text{POINTS}\ \ \alpha, 0\ \ \beta, 1\ \ \gamma, 1\ \ \delta, 0. \tag{2.31}$$

In the same way, $\Lambda(u;\alpha,\beta,\gamma)$ is described by

$$\text{POINTS}\ \ \alpha, 0\ \ \beta, 1\ \ \gamma, 0. \tag{2.32}$$

$\Gamma(u;\alpha,\beta)$ is described by the string "POINTS $\alpha, 0$ $\beta, 1$," and $\mathrm{L}(u;\alpha,\beta)$ is described by "POINTS $\alpha, 1$ $\beta, 0$."

2.1.5 Properties of Fuzzy Sets

In this subsection, let A and B be fuzzy sets, defined respectively on the universes X and Y, and let R be a fuzzy relation (see Section 2.2.2) defined on $X \times Y$. The *support* of a fuzzy set A is the crisp set that contains all elements of

A with non-zero membership degree. This is denoted by $S(A)$, formally defined as

Definition 2.20 *The support of a fuzzy set A is defined by*

$$S(A) = \{u \in X \mid \mu_A(u) > 0\}. \tag{2.33}$$

When one deals with *convex* fuzzy sets (Def. 2.26), as it is the case in fuzzy control theory, the support of a fuzzy set is an interval. Therefore, in fuzzy control theory the term *width* of a fuzzy set is used additionally to the term *support.*

Definition 2.21 *The width of a convex fuzzy set A with support set $S(A)$ is defined by*

$$\text{width}(A) = \sup(S(A)) - \inf(S(A)). \tag{2.34}$$

sup and inf denote the mathematical operations supremum and infimum. They are defined as follows:

$$\begin{aligned} \alpha = \sup(A) \quad &\text{iff} \quad \forall x \in A : x \leq \alpha \text{ and } \forall \epsilon > 0\, \exists x \in A : x > \alpha - \epsilon, \\ \beta = \inf(A) \quad &\text{iff} \quad \forall x \in A : x \geq \beta \text{ and } \forall \epsilon > 0\, \exists x \in A : x < \beta + \epsilon. \end{aligned} \tag{2.35}$$

If the support set $S(A)$ is bounded as is usual in fuzzy control, sup and inf can be replaced by max and min.

It is possible to talk about left and right width too, as in the following example.

Example 2.22 Consider a fuzzy set A with membership function $\Lambda(x; \alpha, \beta, \gamma)$ (see Fig. 2.10). Its support set is $S(A) = [\alpha, \gamma]$. Its width is $\text{width}(A) = \gamma - \alpha$. Its left width is $\text{left-width}(A) = \beta - \alpha$. Its right width is $\text{right-width}(A) = \gamma - \beta$.

Closely connected to the support of a fuzzy set is the *nucleus.* The nucleus of a fuzzy set A is the crisp set that contains all values with membership degree 1. Formally,

Definition 2.23 *The nucleus of a fuzzy set A is defined by*

$$\text{nucleus}(A) = \{u \in X \mid \mu_A(u) = 1\}. \tag{2.36}$$

If there is only one point with membership degree equal to 1, then this point is called the *peak value* of A.

Example 2.24 a. Consider the fuzzy set $F = \Pi(u; 18, 20, 24, 26)$ from Fig. 2.11. The support $S(F)$ of this fuzzy set is the interval $[18, 26]$, the nucleus is the

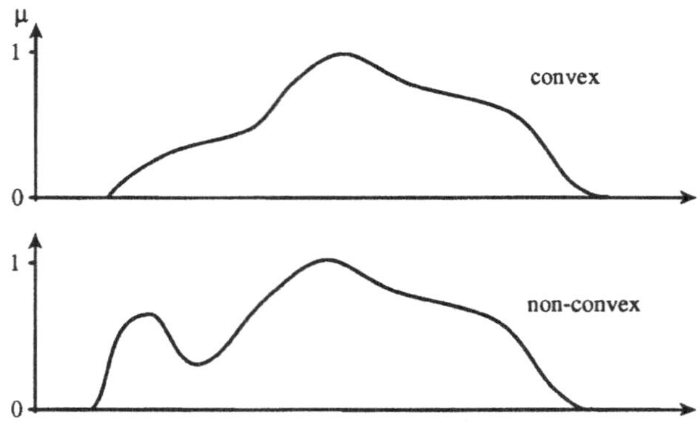

Fig. 2.12. An example of a convex and a non-convex fuzzy set.

interval [20, 24].
b. Consider the fuzzy set $F = \Lambda(u; 18, 20, 22)$. Its nucleus is the set $\{20\}$, its peak value is 20.

The *height* of a fuzzy set A is equal to the largest membership degree μ_A.

Definition 2.25 *The height of a fuzzy set A on X, denoted by* $\text{hgt}(A)$, *is defined as*

$$\text{hgt}(A) = \sup_{u \in X} \mu_A(u). \tag{2.37}$$

A fuzzy set A is called *normal*, if $\text{hgt}(A) = 1$, and *subnormal* if $\text{hgt}(A) < 1$.

In fuzzy control theory, it is usual to deal only with *convex* fuzzy sets. A fuzzy set is called convex if its membership function does not contain 'dips'. This means that the membership function is, for example, increasing, decreasing, or bell-shaped. Formally,

Definition 2.26 *A fuzzy set A is convex if and only if*

$$\forall x, y \in X \; \forall \lambda \in [0,1] : \mu_A(\lambda \cdot x + (1 - \lambda) \cdot y) \geq \min(\mu_A(x), \mu_A(y)). \tag{2.38}$$

It is easy to see that a fuzzy set A is convex if and only if its α-cuts are convex in the classical mathematical sense. Figure 2.12 shows a convex and a non-convex set. Convex sets are normally used to represent fuzzy numbers, and are also useful for the representation of linguistic notions.

2.1.6 Operations on Fuzzy Sets

Notions like *equality* and *inclusion* of two fuzzy sets are immediately derived from classical set theory. Two fuzzy sets are *equal* if every element of the universe has the same membership degree in each of them. A fuzzy set A is a *subset* of a fuzzy set B if every element of the universe has a lower membership degree in A than in B.

Definition 2.27 *Two fuzzy sets are equal* ($A = B$) *if and only if*

$$\forall x \in S : \mu_A(x) = \mu_B(x). \tag{2.39}$$

Definition 2.28 A *is a subset of* B ($A \subseteq B$) *if and only if*

$$\forall x \in X : \mu_A(x) \leq \mu_B(x). \tag{2.40}$$

Correspondingly, there are also the notions of *strict subset*, *superset*, and *strict superset*.

In classical set theory the *union*, *intersection* and *complement* of sets are simple operations that are unambiguously defined (see page 44). The unambiguity follows from the fact that the logical operators *and*, *or*, and *not* have a well defined semantics based on propositional logic. For example, "ϕ *and* ψ" in propositional logic is true if and only if both expressions ϕ and ψ are true. In fuzzy set theory their interpretation is not so simple, because graded characteristic functions are used. Zadeh proposed [233]:

$$\forall x \in X : \quad \mu_{A \cap B}(x) = \min(\mu_A(x), \mu_B(x)), \tag{2.41}$$

$$\forall x \in X : \quad \mu_{A \cup B}(x) = \max(\mu_A(x), \mu_B(x)), \tag{2.42}$$

$$\forall x \in X : \quad \mu_{A'}(x) = 1 - \mu_A(x). \tag{2.43}$$

This is a very simple extension of the classical operations. Other extensions are possible as well, e.g., $\mu_{A \cap B}(x) = \mu_A(x) \cdot \mu_B(x)$ or $\mu_{A \cup B}(x) = \min(1, \mu_A(x) + \mu_B(x))$ (compare the alternative definitions of the crisp intersection operation in equation (2.8)).

Example 2.29 To illustrate this we use an example from Van Nauta Lemke that he presented in a Dutch course on fuzzy control. First consider Fig. 2.13a. There are two questions: (1) how easy is it for the person to escape from his prison cell, and (2) how easy is it for the sunlight to come into the prison cell. Suppose window w_1 is very difficult to escape through (0.1 easy) where window w_2 is simpler (0.3 easy). Suppose also that window w_1 is dirty and thick such that it is not so easy for the sunlight to penetrate it (0.6 easy) while w_2 is clean (0.9 easy).

In case (a) the question on the first answer will be $\min(0.1, 0.3) = 0.1$, where

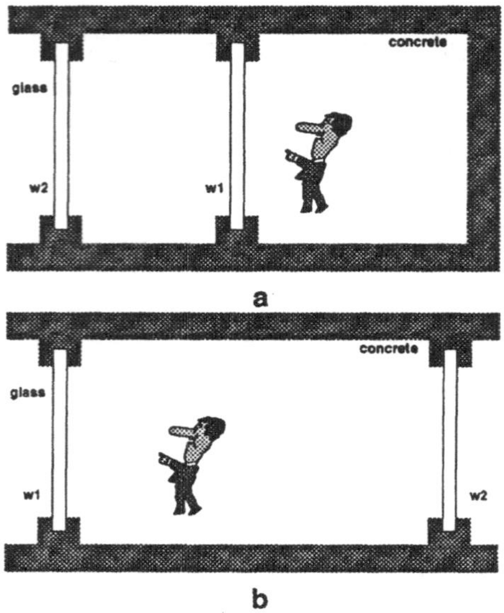

Fig. 2.13. Two different prison cells. Case (a) represents the 'and' operation, case (b) represents the 'or' operation.

we do not take into account the fact that the person has to be quick or gets tired, which makes the result probably less than 0.1. The ease with which the sunlight penetrates both windows will be $0.6 \cdot 0.9 = 0.54$, a product operation. So the 'and' operation can be described by minimum and product, depending on the application.
Case (b) represents the 'or' operator. Question (1) can be answered by using a maximum operation, question (2) demands an operation like $0.6+0.9-0.6\cdot0.9 = 0.96$.

More generally, *triangular norms* (T-norms and S-norms) are used to represent intersection, union and complement; see Schweitzer and Sklar [187, 188],

Table 2.3. The two windows w_1 and w_2 and their degrees of easiness for a person to escape through them and for the sunlight to penetrate them

	easiness to	
window	escape	penetrate
w_1	0.1	0.6
w_2	0.3	0.9

and Weber [225]. Let $a, b, c, d \in [0,1]$, then $T(a,b)$ is the T-norm of a and b. A T-norm can be considered to be the most general intersection operator. Other representations are $I(a,b)$, from Kandel [110], and $k(a,b)$, from Dombi [49]. We will use $a \mathbin{\overset{\wedge}{\star}} b$, following Dubois and Prade [56], to make it clear that a T-norm is a binary operator.

Definition 2.30 *A triangular norm or T-norm* $\overset{\wedge}{\star}$ *denotes a class of binary functions that can represent the intersection operation. It satisfies the following criteria:*

T-1: $a \mathbin{\overset{\wedge}{\star}} b = b \mathbin{\overset{\wedge}{\star}} a$;

T-2: $(a \mathbin{\overset{\wedge}{\star}} b) \mathbin{\overset{\wedge}{\star}} c = a \mathbin{\overset{\wedge}{\star}} (b \mathbin{\overset{\wedge}{\star}} c)$;

T-3: $a \le c$ *and* $b \le d$ *implies* $a \mathbin{\overset{\wedge}{\star}} b \le c \mathbin{\overset{\wedge}{\star}} d$;

T-4: $a \mathbin{\overset{\wedge}{\star}} 1 = a$.

From T-3 it follows that for any a, $0 \mathbin{\overset{\wedge}{\star}} a \le 0 \mathbin{\overset{\wedge}{\star}} 1$, hence, with T-4, $0 \mathbin{\overset{\wedge}{\star}} a = 0$. Examples of T-norms are $\min(a,b)$, ab, $\max(0, a+b-1)$, and many others based on these three.

Another property that is sometimes used is the *Archimedean property*, defined as:

Definition 2.31 *A T-norm is said to be Archimedean if it satisfies T-1 to T-4, together with*

T-5: $\forall a \in (0,1) : a \mathbin{\overset{\wedge}{\star}} a < a$.

T-norms satisfying the Archimedean property form a mathematically very interesting class. In the case of T-norms that are strictly increasing in each of their arguments, Schweitzer and Sklar [188] have shown that there exists a continuous and decreasing function $f : [0,1] \to [0,\infty)$ such that $f(0) = \infty$, $f(1) = 0$ and $a \mathbin{\overset{\wedge}{\star}} b = f^{-1}(f(a) + f(b))$. When the T-norms are non-strict, the inverse function f^{-1} does not exist, and a similar but more difficult additive property can be proved. While this book is not the right place to deal with these functions, we refer the interested reader to [188, 49, 55] and other papers.

$S(a,b)$ is an S-norm, also notated by $U(a,b)$, $d(a,b)$ or $a \mathbin{\overset{\star}{\vee}} b$. An S-norm can be considered to be the most general union operator.

Definition 2.32 *A triangular co-norm or S-norm* $\overset{\star}{\vee}$ *denotes a class of binary functions that can represent the union operation. It satisfies the first three criteria of the T-norms, i.e., S-1 to S-3 are equal to T-1 to T-3, but*

S-4: $a \mathbin{\overset{\star}{\vee}} 0 = a$.

It easily follows from this criterion, together with criterion *T-3*, that for any a, $1 \underset{\vee}{\star} a = 1$. There is the following general relation between T- and S-norms

$$a \overset{\wedge}{\star} b = 1 - ((1 - a) \underset{\vee}{\star} (1 - b)). \tag{2.44}$$

When a T- and S-norm satisfy this property, then each is the other's conjugate. One example is minimum and maximum. Another one is product and the operation $a \underset{\vee}{\star} b = a + b - a \cdot b$.

A more extensive discussion of criteria like *T-1* to *T-4* and *S-1* to *S-4*, can be found in Bellman and Giertz [17].

The complement is denoted by $c(a)$, where c is the most general complement operation.

Definition 2.33 *The complement operation c should at least satisfy*

c-1: $c(0) = 1$;

c-2: $a < b$ *implies* $c(a) > c(b)$;

c-3: $c(c(a)) = a$.

Hence, $c(1) = c(c(0)) = 0$.

This means that instead of the 'minimum', 'maximum' and 'one minus' operations for intersection, union and complement, respectively, the following operations are used:

$$\forall x \in X: \quad \mu_{A \cap B}(x) = \mu_A(x) \overset{\wedge}{\star} \mu_B(x) \tag{2.45}$$

$$\forall x \in X: \quad \mu_{A \cup B}(x) = \mu_A(x) \underset{\vee}{\star} \mu_B(x) \tag{2.46}$$

$$\forall x \in X: \quad \mu_{A'}(x) = c(\mu_A(x)). \tag{2.47}$$

These T-norms and S-norms incorporate many operators, even classes of operators, e.g., the family T_p from Schweitzer and Sklar [187].

Definition 2.34 *The T_p family of T-norms is defined as*

$$T_p(a,b) = 1 - [(1-a)^p + (1-b)^p - (1-a)^p \cdot (1-b)^p]^{1/p} \quad p \in \mathbb{R}. \tag{2.48}$$

For $p = 1$, $T_1(a,b) = a \cdot b$, and

$$\lim_{p \downarrow 0} T_p(a,b) = T_W(a,b), \tag{2.49}$$

$$\lim_{p \to \infty} T_p(a,b) = \min(a,b), \tag{2.50}$$

where T_W is the function on the unit square given by

$$T_W(a,b) = \begin{cases} a & \text{for } b = 1, \\ b & \text{for } a = 1, \\ 0 & \text{otherwise.} \end{cases} \tag{2.51}$$

This T_W operation can be considered as the most 'pessimistic' T-norm operation.

Definition 2.35 *The H_λ family of T-norms (the Hamacher product* [86]*) is defined as*

$$H_\lambda(a,b) = \frac{a \cdot b}{\gamma + (1-\gamma) \cdot (a + b - a \cdot b)} \quad \gamma \geq 0. \tag{2.52}$$

So, H_1 is the product, and $\lim_{\lambda \to \infty} H_\lambda(a,b) = T_W(a,b)$.

This T-norm is used in the fuzzy classification system FUCS [24] to combine properties in a multidimensional environment. They use H_0, i.e.,

$$H_0(a,b) = \frac{1}{\frac{1}{a} + \frac{1}{b} - 1} = \frac{a \cdot b}{a + b - a \cdot b}, \tag{2.53}$$

where the purpose is to combine several properties less optimistically than the minimum operation.

Other families are the Sugeno S_λ family of S-norms, the Frank F_s family and Yager Y_q family of T-norms and the *non-Archimedean* σ_α family (Def. 2.31) of T-norms from Dubois and Prade:

$$\begin{aligned} S_\lambda(a,b) &= \min(1, a + b + \lambda \cdot a \cdot b), \quad \lambda \geq -1, & (2.54)\\ F_s(a,b) &= F_s(a,b) = \log_s\left(1 + \frac{(s^a - 1)\cdot(s^b - 1)}{s-1}\right), \quad s > 0, & (2.55)\\ Y_q(a,b) &= 1 - \min\left(1, ((1-a)^q + (1-b)^q)^{1/q}\right), \quad q \geq 0, & (2.56)\\ \sigma_\alpha(a,b) &= \frac{a \cdot b}{\max(a,b,\alpha)}. & (2.57) \end{aligned}$$

This last T-norm is interesting, because it uses a kind of boundary α. For $\alpha = 0$, $\sigma_\alpha = \min$, and for $\alpha = 1$, $\sigma_\alpha(a,b) = a \cdot b$. For values of α between 0 and 1, it is a function between minimum and product. Note that

$$\sigma_\alpha(a,b) = \begin{cases} a \cdot b/\alpha & \text{for } a, b < \alpha, \\ \min(a,b) & \text{otherwise.} \end{cases} \tag{2.58}$$

Example 2.36 Suppose $\alpha = 0.4$. Then $\sigma_{0.4}(0.7, 0.6) = 0.6$, $\sigma_{0.4}(0.7, 0.2) = 0.2$ and $\sigma_{0.4}(0.3, 0.2) = 0.06/0.4 = 0.15$. This clearly shows that above the boundary 0.4, the minimum operation is chosen, and below this boundary, the product operation is chosen.

This T-norm is used in several fuzzy classification systems.

An overview of classes like these can be found in Dombi, and Dubois and Prade [49, 55, 59]. It can be proved [187, 188] that for any T-norm $\hat{\star}$

$$T_W(a,b) \leq a \mathbin{\hat{\star}} b \leq \min(a,b), \tag{2.59}$$

and for any S-norm $\underset{\vee}{\star}$

$$\max(a,b) \leq a \overset{\star}{\vee} b \leq S_W(a,b), \tag{2.60}$$

where $S_W(a,b)$ is defined

$$S_W(a,b) = \begin{cases} a, & \text{for } b = 0, \\ b, & \text{for } a = 0, \\ 1, & \text{otherwise.} \end{cases} \tag{2.61}$$

There are several operations described in literature that do not satisfy properties (2.59) and (2.60), i.e., they do not satisfy criteria *T-1* to *T-4* or *S-1* to *S-4* respectively. Usually the problem is criterion *T-2* or *S-2*. First, there is the class of so-called *averaging operators* that form a convex combination of usual fuzzy operators, e.g., minimum or maximum, and the arithmetic mean.

Definition 2.37 *The averaging operator for 'and' is defined by*

$$\forall x : \mu_{A \,\mathrm{and}\, B}(x) = \gamma \cdot \min(\mu_A(x), \mu_B(x)) + (1-\gamma) \cdot \frac{\mu_A(x) + \mu_B(x)}{2}, \tag{2.62}$$

where $\gamma \in [0,1]$. *In the same way, the averaging operator for 'or' is defined by*

$$\forall x : \mu_{A \,\mathrm{or}\, B}(x) = \gamma \cdot \max(\mu_A(x), \mu_B(x)) + (1-\gamma) \cdot \frac{\mu_A(x) + \mu_B(x)}{2}, \tag{2.63}$$

where $\gamma \in [0,1]$.

This means that for $\gamma < 1$ the averaging operation is more optimistic than the minimum operation, which was the most optimistic one in the T-norm case. These two operations are not associative, as can easily be checked with a small numerical example.

Example 2.38 Suppose $\gamma = 0.6$ and $a = 0.2$, $b = 0.4$ and $c = 0.6$. Then (a and b) and c is unequal to a and (b and c), because

$$\begin{aligned} a \text{ and } b &= 0.6 \cdot \min(a,b) + 0.4 \cdot (a+b)/2 = 0.24 && (2.64) \\ b \text{ and } c &= 0.6 \cdot \min(b,c) + 0.4 \cdot (b+c)/2 = 0.44 && (2.65) \\ (a \text{ and } b) \text{ and } c &= 0.6 \cdot \min(.24,c) + 0.4 \cdot (.24+c)/2 = 0.312 && (2.66) \\ a \text{ and } (b \text{ and } c) &= 0.6 \cdot \min(a,.44) + 0.4 \cdot (a+.44)/2 = 0.248 && (2.67) \end{aligned}$$

Secondly, there is the class of so-called *compensatory operators*, which are located somewhere between intersection and union operators. They are defined thus:

Definition 2.39 *The compensatory operator or γ-operator* [247] *of n membership functions μ_1, ..., μ_n is defined by*

$$\forall x : \mu(x) = \left(\prod_{i+1}^{n} \mu_i(x) \right)^{(1-\gamma)} \cdot \left(1 - \prod_{i+1}^{n} (1 - \mu_i(x)) \right)^{\gamma}, \quad \gamma \in [0,1]. \tag{2.68}$$

The first part is the product, i.e., a T-norm, the second part is its corresponding S-norm, by equation (2.44). Another example of an averaging operator, for sake of simplicity given for the binary case only, is defined by:

$$\forall x : \mu_{A \,\mathrm{and}\, B} = \gamma \cdot \min(\mu_A(x), \mu_B(x)) + (1-\gamma) \cdot \max(\mu_A(x), \mu_B(x)). \tag{2.69}$$

These operators also do not satisfy associativity.

Operations that are infrequently used in fuzzy control theory are the following four. First, the *bounded sum* or *bold union* of A and B.

Definition 2.40 *The bounded sum of A and B ($A \oplus B$) is defined by*

$$A \oplus B = \int_X \min(1, \mu_A(x) + \mu_B(x))/x, \tag{2.70}$$

where $+$ is the arithmetic sum.

A similar operation is the *bounded difference.*

Definition 2.41 *The bounded difference of A and B ($A \ominus B$) is defined by*

$$A \ominus B = \int_X \max(0, \mu_A(x) - \mu_B(x))/x, \tag{2.71}$$

where $-$ is the arithmetic difference.

The third and the fourth definition are the *product* operation and its immediate extension *involution.*

Definition 2.42 *The product $A \cdot B$ of two fuzzy sets A and B is defined by*

$$A \cdot B = \int_X \mu_A(x) \cdot \mu_B(x)/x. \tag{2.72}$$

Definition 2.43 *The involution or power α of a fuzzy set A, where α is a real number, is defined*

$$A^\alpha = \int_X (\mu_A(x))^\alpha /x. \tag{2.73}$$

2.2 Fuzzy Relations

2.2.1 Classical Relations

A relation can be considered as a set of *tuples,* where a tuple is an ordered pair. A binary tuple is denoted as (x, y), an example of a ternary tuple is (x, y, z), and an example of an n-ary tuple is $(x_1, \ldots, x_n)$.

Example 2.44 a. Let X be the domain of men {Adam, Bernhard, Charles} and Y the domain of women {Diana, Eva, Francis}, then the relation 'married to' on $X \times Y$ is, for example,

$$\{(\text{Adam, Eva}), (\text{Bernhard, Francis}), (\text{Charles, Diana})\}. \tag{2.74}$$

b. Consider the domain of natural numbers, then $\leq$ is the relation on $\mathbb{N} \times \mathbb{N}$ defined as

$$\{(m, n) \mid m \leq n\}. \tag{2.75}$$

Hence, this relations contains pairs like $(5, 7)$, $(1110, 1111)$ and $(30, 30)$. The pair $(4, 1)$ is not in this relation.
c. A ternary relation on $\mathbb{N}^3$ is

$$\{(p, q, r) \mid p^2 + q^2 = r^2\} \tag{2.76}$$

which contains those triples where p, q and r are the edges of a perpendicular triangle. This relation contains elements $(3, 4, 5)$ and $(5, 12, 13)$, etc.

It is clear now that the Cartesian product mentioned in Section 2.1.3 is a binary relation, and that the power of a set, mentioned there, is an n-ary relation.

Just like classical sets, classical relations can be described by a characteristic function. Suppose R is an n-ary relation defined on $X_1 \times \ldots \times X_n$, then

Definition 2.45 *$\mu_R : X_1 \times \ldots \times X_n \to [0, 1]$ is a characteristic function of the set A iff for all $x_1, \ldots, x_n$,*

$$\mu_R(x_1, \ldots, x_n) = \begin{cases} 1 & \textit{when } (x_1, \ldots, x_n) \in R, \\ 0 & \textit{when } (x_1, \ldots, x_n) \notin R. \end{cases} \tag{2.77}$$

In a fuzzy relation this characteristic function is extended to the interval $[0, 1]$.

Relations can have several properties, among which the following are the most frequently used.

Definition 2.46 *Let R be a relation defined on $U \times U$, then this relation is called*

reflexive, if for all $u \in U$, it holds that $(u, u) \in R$,

anti-reflexive, if for all $u \in U$, it holds that $(u, u) \notin R$,

symmetric, if for all $u, v \in U$, if $(u, v) \in R$, then $(v, u) \in R$ too,

anti-symmetric, if for all $u, v \in U$, if (u, v) and (v, u) are in R, then $u = v$,

transitive, if for all $u, v, w \in U$, if (u, v) and (v, w) are in R, then (u, w) is in R too.

Example 2.47 The relation $\leq$ is an example of a reflexive, anti-symmetric and transitive relation. The relation $<$ is anti-reflexive. The relation 'married to' is an example of an anti-reflexive, symmetric and *in*transitive relation.

Other important properties concerning binary relations are:

R is an *equivalence relation*, if R is reflexive, symmetric and transitive,

R is an *partial order relation*, if R is reflexive, anti-symmetric and transitive,

R is an *total order relation*, if R is a partial order relation, and for every u, v it holds that either (u, v) or (v, u) are in R,

Example 2.48 a. The equality relation on natural numbers $(=)$ is an example of an equivalence relation
b. Consider the relation R on $\mathbb{N} \times \mathbb{N}$ defined as:

$$\{(m, n) \mid (n - m) \bmod 3 \equiv 0\}, \tag{2.78}$$

i.e., the set of pairs (n, m) such that $|n - m|$ is divisible by 3. This relation is obviously reflexive ($n - n$ is divisible by 3), symmetric (if $n - m$ is divisible by 3, then also $m - n = -(n - m)$), and transitive (if both $n - m$ and $m - \ell$ are divisible by 3, then $n - \ell$ is divisible by 3 too).
c. The subset relation on sets $(\subseteq)$ is a partial order relation.
d. The relation $\leq$ on natural numbers is a total order relation.

2.2.2 Fuzzy Relations

The former subsection stated that a relation can be considered as a set of tuples. In the same way, a fuzzy relation is a *fuzzy set of tuples*, i.e., each tuple has a membership degree between 0 and 1. Its definition is

Definition 2.49 *Let U and V be uncountable (continuous) universes, and $\mu_R : U \times V \to [0, 1]$, then*

$$R = \int_{U \times V} \mu_R(u, v)/(u, v) \tag{2.79}$$

is a binary fuzzy relation on $U \times V$. If U and V are countable (discrete) universes, then

$$R = \sum_{U \times V} \mu_R(u, v)/(u, v) \tag{2.80}$$

The integral symbol denotes the set of all tuples $\mu_R(u, v)/(u, v)$ on $U \times V$. It is also possible to express (2.79) with $\int_U \int_V \mu_R(u, v)/(u, v)$, i.e., with a double integral.

Example 2.50 A simple example: when $U = \{1, 2, 3\}$, then 'approximately equal' is the binary fuzzy relation

$$\begin{gathered} 1/(1,1) + 1/(2,2) + 1/(3,3) \\ + 0.8/(1,2) + 0.8/(2,3) + 0.8/(2,1) + 0.8/(3,2) \\ + 0.3/(1,3) + 0.3/(3,1). \end{gathered} \tag{2.81}$$

The membership function μ_R of this relation can be described by

$$\mu_R(x, y) = \begin{cases} 1 & \text{when } x = y, \\ 0.8 & \text{when } |x - y| = 1, \\ 0.3 & \text{when } |x - y| = 2. \end{cases} \tag{2.82}$$

In matrix notation this can be represented as

(2.83)

		Y: 1	Y: 2	Y: 3
X	1	1	0.8	0.3
X	2	0.8	1	0.8
X	3	0.3	0.8	1

Example 2.51 Let $U = [0, 250]$ be the interval of height of persons, and suppose that

$$\mu_R(x, y) = \begin{cases} 0, & \text{for } x - y \le 0, \\ \dfrac{x - y}{20}, & \text{for } 0 < x - y < 20, \\ 1, & \text{for } x - y \ge 20, \end{cases} \tag{2.84}$$

that is, $\mu_R(x, y) = \Gamma(x - y; 0, 20)$, then

$$R = \int_{U \times U} \mu_R(x, y)/(x, y) \tag{2.85}$$

represents the notion 'much taller than'.

In the same way as in classical relations, it is possible to define n-ary fuzzy relations as fuzzy sets of n-tuples. In general, it is a relation of pairs

$$\mu_R(x_1, \ldots, x_n)/(x_1, \ldots, x_n), \tag{2.86}$$

or in the binary case, of pairs $\mu_R(x, y)/(x, y)$.

Example 2.52 Consider the FKBC rule:

if e is *PB and* $\dot{e}$ is *PS*, *then* $\dot{u}$ is *NM*, where *PB*, *PS* and *NM* are fuzzy sets, defined on universes of discourse (domains) E, ΔE and ΔU respectively. This rule is often represented by the *ternary* fuzzy relation R defined as:

$$R = \int_{E\times\Delta E\times\Delta U} \min\left(\mu_{PB}(e), \mu_{PS}(\dot{e}), \mu_{NM}(\dot{u})\right) / (e, \dot{e}, \dot{u}). \tag{2.87}$$

i.e., each triple $(e, \dot{e}, \dot{u})$ has a membership degree equal to the minimum of $\mu_{PB}(e)$, $\mu_{PS}(\dot{e})$ and $\mu_{NM}(\dot{u})$.

2.2.3 Operations on Fuzzy Relations

In this section we will deal with several important operations on fuzzy relations. Fuzzy relations are very important in fuzzy control because they can describe interactions between variables. This is particularly interesting in if-then rules. We will describe how such a relation can be described by a fuzzy relation, and how this is usually done in fuzzy control. But before, we will have to give several other definitions. Let us first define two binary relations R and S that are used in the examples in this section (see [247]).

R = "x considerably larger than y": (2.88)

	y_1	y_2	y_3	y_4
x_1	.8	1	.1	.7
x_2	0	.8	0	0
x_3	.9	1	.7	.8

S = "y very close to x": (2.89)

	y_1	y_2	y_3	y_4
x_1	.4	0	.9	.6
x_2	.9	.4	.5	.7
x_3	.3	0	.8	.5

Two operations on fuzzy relations are the intersection and the union operations. They are defined as follows:

Definition 2.53 *Let R and S be binary relations defined on $X \times Y$. The intersection of R and S is defined by*

$$\forall (x, y) \in X \times Y : \mu_{R\cap S}(x, y) = \min(\mu_R(x, y), \mu_S(x, y)). \tag{2.90}$$

Instead of the minimum, any T-norm can be used.

Definition 2.54 *The union of R and S is defined by*

$$\forall (x, y) \in X \times Y : \mu_{R\cup S}(x, y) = \max(\mu_R(x, y), \mu_S(x, y)). \tag{2.91}$$

Instead of the maximum, any S-norm can be used.

These definitions can be extended for n-ary relations.

Example 2.55 The intersection of the relations R and S denotes "x considerably larger than y **and** y very close to x." For the relations from (2.88) and (2.89), it is given by the relation

$R \cap S =$ (2.92)

	y_1	y_2	y_3	y_4
x_1	.4	0	.1	.6
x_2	0	.4	0	0
x_3	.3	0	.7	.5

Suppose the H_0 T-norm from Definition 2.35 and (2.53) is used instead of the minimum operation to represent the intersection. In this case the result is

$R \cap S =$ (2.93)

	y_1	y_2	y_3	y_4
x_1	.3636	0	.0989	.4773
x_2	0	.3636	0	0
x_3	.2903	0	.5957	.4444

Example 2.56 The union of the relations R and S denotes "x considerably larger than y **or** y very close to x." It is given by the relation

$R \cup S =$ (2.94)

	y_1	y_2	y_3	y_4
x_1	.8	1	.9	.7
x_2	.9	.8	.5	.7
x_3	.9	1	.8	.8

Suppose the simple Sugeno S_{-1} (from equation (2.54): $S_{-1}(a,b) = a + b - a \cdot b$) is used, then the resulting matrix is

$R \cup S =$ (2.95)

	y_1	y_2	y_3	y_4
x_1	.88	1	.91	.88
x_2	.9	.88	.5	.7
x_3	.93	1	.94	.9

so this operation is obviously more optimistic than the maximum operation: all the membership degrees are at least as high as in the maximum operation, while it is assumed that "a or b" does not mean "either a or b exclusively" but "a or b, or a and b." For other Sugeno S-norms S_λ, with λ greater than -1, it is even more optimistic.

Two very important operations on fuzzy sets and fuzzy relation are *projection* and *cylindrical extension*. The projection operation brings a ternary relation back to a binary relation, or a binary relation to a fuzzy set, or a fuzzy set to a single crisp value.

The whole definition of projection is quite complicated, but we present it for the sake of completeness. Let R be a relation on $U = \times_{i=1}^{n} U_i$. Let $(i_1, \ldots, i_k)$ be a subsequence of $(1, \ldots, n)$, and let $(j_1, \ldots, j_l)$ be the complementary subsequence of $(1, \ldots, n)$. Let $V = \times_{m=1}^{k} U_{i_m}$.

Definition 2.57 *The projection of R on V is defined by* [240]*:*

$$\text{proj } R \text{ on } V = \int_V \sup_{x_{j_1}, \ldots, x_{j_l}} \mu_R(x_1, \ldots, x_n)/(x_{i_1}, \ldots, x_{i_k}). \tag{2.96}$$

In the binary case it is far simpler (let R be defined on $X \times Y$):

$$\text{proj } R \text{ on } Y = \int_Y \sup_x \mu_R(x, y)/y. \tag{2.97}$$

Instead of the supremum which is necessary when X and Y are continuous, it is usual to deal with discrete domains using the maximum operation.

Example 2.58 Consider the relation R as given in (2.88),

	y_1	y_2	y_3	y_4
x_1	.8	1	.1	.7
x_2	0	.8	0	0
x_3	.9	1	.7	.8

(2.98)

then the projection on X means that

x_1 is assigned the highest membership degree from the tuples (x_1, y_1), (x_1, y_2), (x_1, y_3) and (x_1, y_4), i.e., 1 (the maximum of the first row)

x_2 is assigned the highest membership degree from the tuples (x_2, y_1), (x_2, y_2), (x_2, y_3) and (x_2, y_4), i.e., .8 and (the maximum of the second row)

x_3 is assigned the highest membership degree from the tuples (x_3, y_1), (x_3, y_2), (x_3, y_3) and (x_3, y_4), i.e., 1 (the maximum of the third row).

So one obtains the fuzzy set

$$\text{proj } R \text{ on } X = 1/x_1 + .8/x_2 + 1/x_3. \tag{2.99}$$

In the same way, the projection on Y can be taken by searching for the maxima of the four columns. This gives the fuzzy set

$$\text{proj } R \text{ on } Y = .9/y_1 + .8/y_2 + .7/y_3 + .8/y_4. \tag{2.100}$$

The total projection of this relation can also be taken. It is equal to 1, the maximal membership degree in this relation. In table form this can be illustrated by

	y_1	y_2	y_3	y_4	Proj. on X
x_1	.8	1	.1	.7	1
x_2	0	.8	0	0	.8
x_3	.9	1	.7	.8	1
Proj. on Y	.9	1	.7	.8	Total Projection
					1

(2.101)

The projection operation is almost always used in combination with the cylindrical extension. The *cylindrical extension* is more or less the opposite of the projection. It extends fuzzy sets to fuzzy binary relations, fuzzy binary relations to fuzzy ternary relations, etc. It mainly serves the following goal: let A be a fuzzy set defined on X, and let R be a fuzzy relation defined on $X \times Y$, then it is, of course, not possible to take the intersection of A and R, but when A is extended to $X \times Y$, this is possible. Let S be a relation on $V = \times_{m=1}^{k} U_{i_m}$ mentioned above, and use again $U = \times_{i=1}^{n} U_i$, compare Definition 2.57.

Definition 2.59 *The cylindrical extension of S into U*

$$ce(S) = \int_U \mu_S(x_{i_1}, \ldots, x_{i_k})/(x_1, \ldots, x_n). \tag{2.102}$$

In the binary case (let F be a fuzzy set defined on Y), the cylindrical extension of F on $X \times Y$ is the set of all tuples $(x, y) \in X \times Y$ with membership degree equal to $\mu_F(y)$, i.e.,

$$ce(F) = \int_{X \times Y} \mu_F(y)/(x, y). \tag{2.103}$$

Hence, proj $ce(S)$ on $V = S$, but, in general, ce(proj R on V) $\neq R$.

Example 2.60 Consider the fuzzy set

$$A = \text{Proj. of } R \text{ on } X = 1/x_1 + .8/x_2 + 1/x_3, \tag{2.104}$$

that was derived in (2.101). Its cylindrical extension on the domain $X \times Y$ is given by

$ce(A) =$

	y_1	y_2	y_3	y_4
x_1	1	1	1	1
x_2	.8	.8	.8	.8
x_3	1	1	1	1

(2.105)

Consider the fuzzy set

$$B = \text{Proj. of } R \text{ on } Y = .9/y_1 + .8/y_2 + .7/y_3 + .8/y_4, \tag{2.106}$$

that was derived in (2.101). Its cylindrical extension on $X \times Y$ is given by

$ce(B) =$ (2.107)

	y_1	y_2	y_3	y_4
x_1	.9	1	.7	.8
x_2	.9	1	.7	.8
x_3	.9	1	.7	.8

The following example shows the use of the cylindrical extension operation for the intersection of fuzzy sets and relations.

Example 2.61 Consider again the relation R expressing "x is considerably larger than y" as given in matrix notation in (2.88), and suppose it is known that x is small, expressed by the fuzzy set

$$A = 0.3/x_1 + 1/x_2 + 0.8/x_3. \tag{2.108}$$

The combination of the fuzzy relation R and the fuzzy set A, expressed by "x is considerably larger than y **and** x is small" can be given by the intersection of the relation and the extension of A. The extension of A into $X \times Y$ is

$ce(A) =$ (2.109)

	y_1	y_2	y_3	y_4
x_1	.3	.3	.3	.3
x_2	1	1	1	1
x_3	.8	.8	.8	.8

The intersection of R and $ce(A)$ is

$R \cap ce(A) =$ (2.110)

	y_1	y_2	y_3	y_4
x_1	.3	.3	.1	.3
x_2	0	.8	0	0
x_3	.8	.8	.7	.8

The combination of fuzzy sets and fuzzy relation with the aid of cylindrical extension and projection is called *composition*. It is denoted by $\circ$.

Definition 2.62 *Let A be a fuzzy set defined on X and R be a fuzzy relation defined on $X \times Y$. Then the composition of A and R resulting in a fuzzy set B defined on Y is given by*

$$B = A \circ R = \text{proj}\,(ce(A) \cap R) \ \textit{on } Y \tag{2.111}$$

or, if intersection is performed with the minimum operation and projection with maximum,

$$\mu_B(y) = \max_x \min(\mu_A(x), \mu_R(x,y)). \tag{2.112}$$

This is called max-min composition. If intersection is performed with product and projection with maximum, we have

$$\mu_B(y) = \max_x \mu_A(x) \cdot \mu_R(x,y). \tag{2.113}$$

This is called max-dot or max-product composition.

The following example shows the use of the composition operation.

Example 2.63 Let us now consider a more complex example. Consider the relation on the domain of height (in centimeters)

$$\{170.0, 172.5, 175.0, 177.5, 180.0, 182.5, 185.0\} \tag{2.114}$$

described by the linguistically expressed relation

R: Agnes is *somewhat taller* than Olga.

in which *somewhat taller* is a binary fuzzy relation with membership function $\mu_R(x,y)$ denoting the degree to which x is somewhat taller than y. This membership function is described by the following matrix

Y \ X	170.0	172.5	175.0	177.5	180.0	182.5	185.0
185.0	0	.1	.4	.7	1	.7	.4
182.5	.1	.4	.7	1	.7	.4	.1
180.0	.4	.7	1	.7	.4	.1	0
177.5	.7	1	.7	.4	.1	0	0
175.0	1	.7	.4	.1	0	0	0
172.5	.7	.4	.1	0	0	0	0
170.0	.4	.1	0	0	0	0	0

(2.115)

So, *somewhat taller* is well described by "x is approximately 5 centimeter taller than y." If the difference is 15 centimeter or more, a low membership degree is assigned; in that case the notion *much taller* is more appropriate. Consider also the fact that "Agnes is rather tall," described by

$$A = 0/170.0 + .1/172.5 + .4/175.0 + .7/177.5 + .9/180.0 + 1/182.5 + 1/185. \tag{2.116}$$

We have to take the composition of A and R, $A \circ R$. A can be extended to

	185.0	0	.1	.4	.7	.9	1	1
	182.5	0	.1	.4	.7	.9	1	1
	180.0	0	.1	.4	.7	.9	1	1
$ce(A) = Y$	177.5	0	.1	.4	.7	.9	1	1
	175.0	0	.1	.4	.7	.9	1	1
	172.5	0	.1	.4	.7	.9	1	1
	170.0	0	.1	.4	.7	.9	1	1
		170.0	172.5	175.0	177.5	180.0	182.5	185.0
					X			

(2.117)

To combine these two matrices ((2.115) and (2.117)), one has to take their intersection, i.e., to apply the minimum operation to the relation R and the extension of the fact A. This delivers a new binary relation, but we are only interested in the height of Olga, i.e., the values of y in this relation, and not in this new relation with membership degrees for *all* (x, y). Therefore, we take the projection of $R \cap ce(A)$ into Y, the domain of the height of Olga. This results in

									Proj. on Y
	185.0	0	.1	.4	.7	.9	.7	.4	.9
	182.5	0	.1	.4	.7	.7	.4	.1	.7
	180.0	0	.1	.4	.7	.4	.1	0	.7
Y	177.5	0	.1	.4	.4	.1	0	0	.4
	175.0	0	.1	.4	.1	0	0	0	.4
	172.5	0	.1	.1	0	0	0	0	.1
	170.0	0	.1	0	0	0	0	0	.1
		170.0	172.5	175.0	177.5	180.0	182.5	185.0	
					X				

(2.118)

i.e., the fuzzy set O denoting Olga's height is:

$$O = .1/170.0 + .1/172.5 + .4/175.0 + .4/177.5 + .7/180.0 + .7/182.5 + .9/185.0 \tag{2.119}$$

which shows that "Olga is shorter than Agnes."

Suppose there are two relations R and S, where R is defined on $X \times Y$ and S is defined on $Y \times Z$. It is of course not possible to take the intersection of R and S, because they are defined on different domains. In this case one has to extend both relations to $X \times Y \times Z$. When this has happened, one can take the intersection. This intersection has to be projected onto $X \times Z$. Formally, the intersection of R and S is

$$\text{Proj of } (ce(R) \cap ce(S)) \text{ on } X \times Z. \tag{2.120}$$

Example 2.64 Consider the relations R and S of (2.88) and (2.89), where R is defined on $X \times Y$, but suppose that S is defined on $Z \times Y$. Thus, S states that "y is very close to x." In table form (note that S is transposed)

	y_1	y_2	y_3	y_4
x_1	.8	1	.1	.7
x_2	0	.8	0	0
x_3	.9	1	.7	.8

$\cap$

	z_1	z_2	z_3
y_1	.4	.9	.3
y_2	0	.4	0
y_3	.9	.5	.8
y_4	.6	.7	.5

(2.121)

Now one has to extend these two relations to $X \times Y \times Z$, and find their intersection. The resulting relation has to be projected onto $X \times Z$, which results in

	z_1	z_2	z_3
x_1	.6	.8	.5
x_2	0	.4	0
x_3	.7	.9	.7

(2.122)

Theoretically, these operations proceed as follows:

$$R = \int_{X \times Y} \mu_R(x_i, y_j)/(x_i, y_j) \qquad S = \int_{Y \times Z} \mu_S(y_j, z_k)/(y_j, z_k). \tag{2.123}$$

The extensions of these relations onto $X \times Y \times Z$ is mathematically given by

$$ce(R) = \int_{X \times Y \times Z} \mu_R(x_i, y_j)/(x_i, y_j, z_k) \qquad ce(S) = \int_{X \times Y \times Z} \mu_S(y_j, z_k)/(x_i, y_j, z_k). \tag{2.124}$$

The intersection of $ce(R)$ and $ce(S)$ is equal to

$$\int_{X \times Y \times Z} \min\left(\mu_R(x_i, y_j), \mu_S(y_j, z_k)\right)/(x_i, y_j, z_k). \tag{2.125}$$

This new ternary relation is projected onto $X \times Z$ by

$$\int_{X \times Z} \sup_j \min\left(\mu_R(x_i, y_j), \mu_S(y_j, z_k)\right)/(x_i, z_k). \tag{2.126}$$

This last operation with a combination of minimum and maximum operations is extremely important in fuzzy set theory. It is denoted by the sup-min composition $R \circ S$,

$$R \circ S = \int_{X \times Z} \sup_j \min\left(\mu_R(x_i, y_j), \mu_S(y_j, z_k)\right)/(x_i, z_k). \tag{2.127}$$

When A is a fuzzy set on X and R is a binary relation on $X \times Y$, then the sup-min composition $A \circ R$ is expressed by

$$A \circ R = \int_Y \sup_i \min\left(\mu_A(x_i), \mu_R(x_i, y_j)\right)/y_j, \tag{2.128}$$

i.e., first the intersection of R and the extension of A is taken, then this intersection is projected onto the Y-axis.

2.2.4 The Extension Principle

One of the most important notions in fuzzy set theory is the *extension principle.* The extension principle provides a general method for combining non-fuzzy and fuzzy concepts of all kinds, e.g., for combining fuzzy sets and relations, but also for the operation of a mathematical function on fuzzy sets. Fuzzy sets can also be interpreted as fuzzy numbers. In this case one can use the extension principle to add or multiply these fuzzy numbers.

Let $A_1, \ldots, A_n$ be fuzzy sets, defined respectively on $U_1, \ldots, U_n$, and let f be a non-fuzzy function $f : U_1 \times \ldots \times U_n \to V$. The aim is to extend f such that it operates on $A_1, \ldots, A_n$, and returns a fuzzy set F on V. This is done by using the sup-min composition as follows.

Definition 2.65 *The extension of f, operating on $A_1, \ldots, A_n$ results in the following membership function for F*

$$\mu_F(v) = \sup_{\substack{u_1,\ldots,u_n \\ f(u_1,\ldots,u_n)=v}} \min(\mu_{A_1}(u_1), \ldots, \mu_{A_n}(u_n)), \tag{2.129}$$

when $f^{-1}(v)$ exists. Otherwise $\mu_A(v) = 0$.

Another way to obtain this result is

$$F = \int_{U_1 \times \ldots \times U_n} \min(\mu_{A_1}(u_1), \ldots, \mu_{A_n}(u_n))/f(u_1, \ldots, u_n). \tag{2.130}$$

In the binary case and on a discrete or compact domain, equation (2.129) is given by

$$\mu_{f(A_1,A_2)}(y) = \max_{\substack{x_1,x_2 \\ y=f(x_1,x_2)}} \min(\mu_{A_1}(x_1), \mu_{A_2}(x_2)). \tag{2.131}$$

So the function f is extended from the domain of real numbers to the domain of fuzzy numbers.

Example 2.66 The extension principle is often used in fuzzy arithmetic. Let $\widetilde{m}$ and $\tilde{n}$ be two fuzzy numbers (fuzzy sets defined on $\mathbb{R}$), then the membership function of the sum of $\widetilde{m}$ and $\tilde{n}$ (denoted $\widetilde{m} \oplus \tilde{n}$) is

$$\mu_{\widetilde{m}\oplus\tilde{n}}(z) = \sup_{\substack{x,y \\ z=x+y}} \min(\mu_{\widetilde{m}}(x), \mu_{\tilde{n}}(y)). \tag{2.132}$$

So, the original function, in this case +, is given in the subscript of the supremum operation. It is worthwhile to work this out. Let $\tilde{6}$ be the fuzzy number with membership degree $\Lambda(x;4,6,8)$ and $\tilde{4}$ the fuzzy number with membership degree $\Lambda(y;3,4,5)$, then $\tilde{6}\oplus\tilde{4}$ has membership degree $\Lambda(z;7,10,13)$, i.e., approximately 10, but quite fuzzy. As a formula,

$$\tilde{6} \oplus \tilde{4} = \int_{\mathbb{R}} \max_{z=x+y} \min(\mu_{\tilde{6}}(x), \mu_{\tilde{4}}(y))/z. = \int_{\mathbb{R}} \Lambda(z;7,10,13)/z \tag{2.133}$$

In the same way, subtraction is defined by

$$\mu_{\tilde{m}\ominus\tilde{n}}(z) = \sup_{\substack{x,y \\ z=x-y}} \min(\mu_{\tilde{m}}(x), \mu_{\tilde{n}}(y)) \tag{2.134}$$

and multiplication by

$$\mu_{\tilde{m}\otimes\tilde{n}}(z) = \sup_{\substack{x,y \\ z=x\times y}} \min(\mu_{\tilde{m}}(x), \mu_{\tilde{n}}(y)). \tag{2.135}$$

While the domain in fuzzy control usually is either discrete or compact, one can almost always use the max-min composition instead of the sup-min composition. An 'extended' extension operator can be found by using the $\overset{\star}{\vee}$-$\overset{\wedge}{\star}$ (S-norm–T-norm) composition, i.e.,

$$\mu_F(v) = \overset{\star}{\bigvee_{\substack{u_1,\ldots,u_n \\ f(u_1,\ldots,u_n)=v}}} \mu_{A_1}(u_1) \overset{\wedge}{\star} \cdots \overset{\wedge}{\star} \mu_{A_n}(u_n). \tag{2.136}$$

On continuous domains one has to be careful with the choice of $\overset{\star}{\vee}$. Normally, only the supremum operation and the S_W operation can be used. The system FUCS (FUzzy Classification System [24]) uses a max-H_0 composition (see equation (2.54)). Another well known composition operation is the max-product or max-dot composition, i.e.,

$$\mu_F(v) = \max_{\substack{u_1,\ldots,u_n \\ f(u_1,\ldots,u_n)=v}} \mu_{A_1}(u_1) \cdot \mu_{A_2}(u_2) \cdot \ldots \cdot \mu_{A_n}(u_n). \tag{2.137}$$

2.3 Approximate Reasoning

2.3.1 Introduction

Approximate reasoning is the best-known form of fuzzy logic and covers a variety of inference rules whose premises contain fuzzy propositions. Inference in approximate reasoning is in sharp contrast to inference in classical logic – in the former the consequence of a given set of fuzzy propositions depends in an essential way on the meaning attached to these fuzzy propositions. Thus, inference in approximate reasoning is computation with fuzzy sets that represent the meaning of a certain set of fuzzy propositions. For example, given the membership functions of μ_A and μ_R, representing the meaning of a fuzzy proposition "X is A" and the meaning of a fuzzy conditional "*if* X is A *then* Y is B," we can compute the membership function representing the meaning of the conclusion "Y is B."

2.3.2 Linguistic Variables

Before we start a survey on the various inference rules in approximate reasoning, we have to make some remarks about knowledge representation. The fundamental knowledge representation unit in approximate reasoning is the notion of a *linguistic variable*. In [240], Zadeh states:

> "By a *linguistic variable* we mean a variable whose values are words or sentences in a natural or artificial language. For example, *Age* is a linguistic variable if its values are linguistic rather than numerical, i.e., *young, not young, very young, quite young, old, not very old and not very young*, etc., rather than 20, 21 22, 23, ...

It is usual in approximate reasoning to have the following framework associated with the notion of a linguistic variable

$$\langle X, \mathcal{L}X, \mathcal{X}, M_X \rangle . \tag{2.138}$$

Here X denotes the symbolic name of a linguistic variable, e.g., *age, height, speed, temperature, error, change-of-error*, etc. $\mathcal{L}X$ is the set of *linguistic values* that X can take. A linguistic value denotes a symbol for a particular property of X. In the case of the linguistic variable *temperature* T we have

$$\mathcal{L}T = \{cold, cool, comfortable, warm, hot\}, \tag{2.139}$$

in the case of *error* or *change-of-error* in FKBC applications it usually is the set {*NB, NM, NS, ZO, PS, PM, PB*} or a subset from this set. $\mathcal{L}X$ is also-called the term-set of X or the reference-set of X. We denote an arbitrary element of this set by LX. $\mathcal{X}$ is the actual physical domain over which the linguistic variable X takes its quantitative (crisp) values. In the case of the linguistic variable *temperature* it can be the interval $[-10°\text{C}, 35°\text{C}]$. In the case of *speed* it can be $[0 \text{ km/h}, 200 \text{ km/h}]$. In the case of *error* and *change-of-error*, though there is an actual physical domain, one often uses a 'normalized domain' $[-6, 6]$ (see Fig. 2.6). It is usual to have also U (for universe of discourse) instead of $\mathcal{X}$. $\mathcal{X}$ can be discrete or continuous. M_X is a semantic function which gives a 'meaning' (interpretation) of a linguistic value in terms of the quantitative elements of $\mathcal{X}$, i.e.,

$$M_X : LX \to \widetilde{LX}, \tag{2.140}$$

where $\widetilde{LX}$ is a denotation for a fuzzy set defined over $\mathcal{X}$, i.e.,

$$\widetilde{LX} = \sum_{\mathcal{X}} \mu_{LX}(x)/x \quad \text{in the case of discrete } \mathcal{X} \tag{2.141}$$

$$\widetilde{LX} = \int_{\mathcal{X}} \mu_{LX}(x)/x \quad \text{in the case of continuous } \mathcal{X} \tag{2.142}$$

In other words, M_X is a function which takes a symbol as its argument, e.g., *cold*, and returns the 'meaning' or the *cold-symbol* in terms of a fuzzy set. Instead of $\widetilde{LX}$ we will also use μ_{LX}, i.e., the membership function without an argument.

Example 2.67 In the case of a FKBC, consider the linguistic variable E, denoting error. The frame of E is

$$\langle E, \mathcal{L}E, \mathcal{E}, M_E \rangle \tag{2.143}$$

where $\mathcal{L}E$ is $\{NB, NM, NS, ZO, PS, PM, PB\}$, $\mathcal{E} = [-6, 6]$ and $M_E : LE \rightarrow \widetilde{LE}$. This frame is shown in Fig. 2.6. So,

$$\widetilde{NB} = \widetilde{LE}_1 = \int_{-6}^{6} \mathrm{L}(x; -6, -3)/x \tag{2.144}$$

$$\widetilde{NM} = \widetilde{LE}_2 = \int_{-6}^{6} \Lambda(x; -7, -4, -1)/x \tag{2.145}$$

$$\widetilde{NS} = \widetilde{LE}_3 = \int_{-6}^{6} \Lambda(x; -5, -2, 1)/x \tag{2.146}$$

$$\vdots$$

$$PB = \widetilde{LE}_7 = \int_{-6}^{6} \Gamma(x; 3, 6)/x \tag{2.147}$$

The value -7 in *NM* only plays a formal role, because the $\int$-symbol is only defined on $[-6, 6]$.

From now on we will distinguish between symbols denoting linguistic values and their *meaning* described by a fuzzy set. *Fuzzy sets are always written with a '~'.*

2.3.3 Fuzzy Propositions

Approximate reasoning is used to represent and reason with knowledge expressed in *atomic primitives*, which are expressed in a natural language form, e.g.,

"Error has the value negative-big."

The formal, symbolic translation of this natural language expression in terms of linguistic variables proceeds as follows

1. a symbol E is chosen to denote the physical variable 'error',
2. a symbol *NB* is chosen to denote the particular value 'negative-big' of 'error',
3. the above natural language expression is rewritten as

"Error has the property of being negative-big,"

4. the symbolic representation of the latter is then given as

   E is NB,

   where 'is' stands for 'has the property of being'.

Such an expression is called an atomic fuzzy proposition. The 'meaning' (interpretation) of an atomic expression is then defined by a fuzzy set $\widetilde{NB}$ or a membership function μ_{NB}, defined on the normalized physical domain $\mathcal{E} = [-6, 6]$ of the physical variable 'error',

$$\forall e \in \mathcal{E} : \widetilde{NB} = \mu_{NB} = \text{"membership function."} \tag{2.148}$$

$\mu_{NB}(e)$ specifies the degree to which a particular quantitative crisp value of the physical variable error, e, belongs to the set wise defined physical value of error negative-big.

The 'meaning' of the symbolic expression "E is NB" helps us decide the degree to which this symbolic expression is satisfied given a specific physical value of error.

1. A crisp value is assigned to the symbol E. This is called a variable assignment (VA), e.g.,
$$\text{VA}(E) = -3.2 \in [-6, 6]. \tag{2.149}$$
   Thus the variable assignment function takes the symbol E, denoting the physical variable error, and returns a physical value from $\mathcal{E}$.

2. The degree of membership of -3.2 in μ_{NB} is found, e.g., $\mu_{NB}(-3.2) = 0.7$. This degree of membership represents the degree to which the symbolic expression "E is NB" is satisfied given some particular circumstances, i.e.,

   - NB is interpreted as μ_{NB},
   - E takes the value -3.2.

In control applications a symbolic expression such as "E is NB" is usually written as "e is NB," where e also denotes an arbitrary element of the domain $\mathcal{E}$. This may introduce some confusion between the purely symbolic level in terms of E and NB and the 'meaning' level in terms of $\mathcal{E}$ and $\widetilde{NB}$ or μ_{NB}. However, throughout this book we will use the notation "e is NB," so that it is easier for the reader to relate our presentation to other sources in the fuzzy control literature.

Based on the notion of atomic fuzzy propositions and linguistic connectives such as 'and', 'or', 'not' and 'if-then' one can form more complex fuzzy propositions called compound fuzzy propositions, e.g.,

X is A *and* X is B,
X is A *or* X is B,
X is *not* A,
(X is A *and* X is *not* B) *or* X is C,
if X is A *then* X is B, etc.

The meaning of these compound fuzzy propositions is given by interpreting the connectives 'and', 'or' and 'not' as 'conjunction', 'disjunction' and 'negation' respectively.

Conjunction

Let p and q be the following two atomic fuzzy propositions: p: "X is A" and q: "X is B" where A and B are fuzzy sets defined on the same universe of discourse U. Then the *conjunction* ($\cap$) is defined as follows

Symbolic	Meaning
X is A,	μ_A or $\widetilde{A}$,
X is B,	μ_B or $\widetilde{B}$,
$\therefore$ X is $A \cap B$,	$\therefore$ $\mu_{A \cap B}$ or $\widetilde{A} \cap \widetilde{B}$;

the meaning of the conclusion "X is $A \cap B$" is given by $\mu_{A\cap B}$ which is defined by a T-norm or perhaps a compensatory operator. For example, if p = "Pressure is *not very high*" and q = "Pressure is *not low*," then the conclusion produced by the rule will be, "Pressure is *not very high* and *not low*," and its meaning will be given as

$$\mu_{(\text{not very high}) \cap (\text{not low})} \tag{2.150}$$

Disjunction

The *disjunction* ($\cup$) is given by

Symbolic	Meaning
X is A,	μ_A or $\widetilde{A}$,
X is B,	μ_B or $\widetilde{B}$,
$\therefore$ X is $A \cup B$,	$\therefore$ $\mu_{A \cup B}$ or $\widetilde{A} \cup \widetilde{B}$;

the meaning of the conclusion X is $A \cup B$ is given by $\mu_{A\cup B}$ which is defined by an S-norm or a compensatory operator.

Addition to Conjunction and Disjunction

Let us consider the two fuzzy propositions

p: E is NB; $\qquad$ p: $\dot{E}$ is PS

where their meaning is represented as follows

$$\widetilde{NB} = \int_{\mathcal{E}} \mu_{NB}(e)/e \qquad \widetilde{PS} = \int_{\dot{\mathcal{E}}} \mu_{PS}(\dot{e})/\dot{e}. \tag{2.151}$$

In this case, in contrast to the previously described case of conjunction, the linguistic variables e and $\dot{e}$ are defined on *different* domains $\mathcal{E}$ and $\dot{\mathcal{E}}$.

The meaning of 'and' in

r: E is *NB and* $\dot{E}$ is *PS*

is represented as a fuzzy relation defined on $\mathcal{E} \times \dot{\mathcal{E}}$ as follows

$$\mu_r(e, \dot{e}) = \int_{\mathcal{E}\times\dot{\mathcal{E}}} \min(\mu_{NB}(e), \mu_{PS}(\dot{e}))/(e, \dot{e}). \tag{2.152}$$

In a similar way, the meaning of 'or' in

s: E is *NB or* $\dot{E}$ is *PS*

is represented as a fuzzy relation

$$\mu_s(e, \dot{e}) = \int_{\mathcal{E}\times\dot{\mathcal{E}}} \max(\mu_{NB}(e), \mu_{PS}(\dot{e}))/(e, \dot{e}). \tag{2.153}$$

Negation

The *negation* "X is *not* A of a fuzzy proposition "X is A" is given by

Symbolic	Meaning
X is A,	μ_A or $\widetilde{A}$,
$\therefore$ X is A',	$\therefore$ $\mu_{A'}$ or $\widetilde{A}'$;

where A' is the complement of A, which can be defined by any c-norm.

Linguistic Approximation

In connection with conjunction and disjunction the following problem arises: Suppose we have a linguistic variable X defined over the universe of discourse U and a collection of linguistic values of X. For example, *very low*, *low*, *above low*, *medium*, *above medium*, *high* and *very high*. Furthermore, by using the conjunction and negation operations on the fuzzy sets defining the meaning of *very low* and *above low*, we can compute a fuzzy set representing the meaning of the composite fuzzy set, *not (very low) and not (above low)*. Then the linguistic approximation problem is to find a single linguistic value whose meaning, i.e., membership function, is the same as, or the closest possible, to the meaning of the fuzzy set generated using the above operations. One idea which comes immediately to the mind is to use an appropriate measure of *distance* between fuzzy sets. However, when the set of linguistic values is very large, a simple enumerative matching procedure requires too much computation time: determination of the distance between two fuzzy sets involves all the elements of their discrete supports. To cope with this difficulty, Bonissone [26], has proposed a fast pattern recognition approach. The method proceeds in two steps. Four features are precalculated for all the elements of the initial set of membership functions: the power set, a Shannon-like entropy, the first moment, and the third moment. The first step consists in evaluating the four features of the unlabeled fuzzy set and to prescreen the initial set of membership functions in order to keep the closest terms, in the sense of a quadratic weighted distance in the four-feature space. In the second step, an appropriate measure of distance between the unlabeled fuzzy set and the meaning of each selected membership function is determined. The linguistic value of the closest labeled fuzzy set is then assigned to the unlabeled one. In the case of our example, the single linguistic value closest to *not (very low) and not (above low)* will turn out to be that of *low.*

2.3.4 Fuzzy If-then Statements

A fuzzy conditional or a fuzzy if-then production rule is symbolically expressed as

if ⟨fuzzy proposition⟩ *then* ⟨fuzzy proposition⟩

where ⟨fuzzy proposition⟩ is either an atomic or a compound fuzzy proposition. As already discussed in Chapter 1 an if-then production rule describes the causal relationship between process state and control output variables. For example, if e and $\dot{e}$ are process state variables and $\dot{u}$ is the control output variable then

if e is *NB and* $\dot{e}$ is *PB then* $\dot{u}$ is *NS*

is a symbolic expression of the following causal relationship stated in a natural language form:

> *if* it is the case that the current value of e is negative big *and* the current value of $\dot{e}$ is positive big *then* this is a cause for a small decrease in the previous value of the control output,

or

> *if* it is the case that the current error has the property negative big *and* the current last change-of-error has the property positive big *then* this is a cause for the incremental change in control output to have the property negative small.

Now suppose that X is a linguistic variable $\langle X, \mathcal{L}X, \mathcal{X}, M_X\rangle$, where the term set $\mathcal{L}X = \{Z, S, M, B\}$ (Zero, Small, Medium and Big) and Y is another linguistic variable with term set $\mathcal{L}Y = \{Z, S, M, B\}$. Furthermore, each linguistic value is interpreted by a membership function defined on $\mathcal{X}$ and $\mathcal{Y}$, e.g.,

$$\forall x \in \mathcal{X} : \mu_B(x) : \mathcal{X} \to [0, 1], \qquad \forall y \in \mathcal{Y} : \mu_S(y) : \mathcal{Y} \to [0, 1]. \tag{2.154}$$

Suppose that there is the following functional relationship between crisp values of X and crisp values of Y given graphically in Fig. 2.14 by the function $f(x)$. This functional relationship between X and Y is not a causal one, it is bidirectional, i.e.,

if a value of x is given, *then* $y = f(x)$,
if a value of y is given, *then* $x = f^{-1}(y)$.

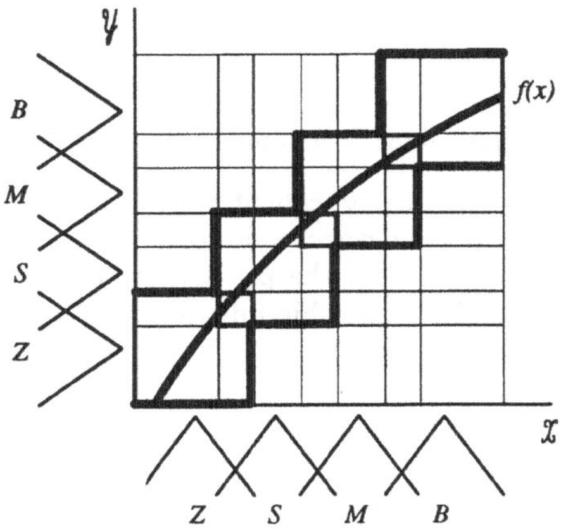

Fig. 2.14. A functional relationship between X and Y.

With the use of if-then rules one approximates the analytic function f as follows

if x is Z, *then* y is Z,
if x is S, *then* y is S,
if x is M, *then* y is M,
if x is B, *then* y is B.

The result of this approximation is shown in Fig. 2.14. Furthermore, each of the above production rules (if-then rules) states explicitly that one can compute a value of Y only when a value of X is already present.

If a value of Y is given then there is no explicit rule telling how a value of X is to be determined. Thus, the causality is in the direction of X to Y.

The meaning of the symbolic expression

if X is A, *then* Y is B,

is represented as a fuzzy relation defined on $\mathcal{X} \times \mathcal{Y}$ where $\mathcal{X}$ and $\mathcal{Y}$ are the domains of the linguistic variables X and Y. The construction of this fuzzy relation proceeds as follows:

1. the meaning of "X is A," called the *rule antecedent*, is represented by a fuzzy set $\widetilde{A} = \int_{\mathcal{X}} \mu_A(x)/x$,

2. the meaning of "Y is B," called the *rule consequent*, is represented by a fuzzy set $\widetilde{B} = \int_{\mathcal{Y}} \mu_A(y)/y$,

3. the meaning of the fuzzy conditional is then a fuzzy relation μ_R such that

$$\forall x \in \mathcal{X} \forall y \in \mathcal{Y} : \mu_R(x, y) = \mu_A(x) * \mu_B(y), \tag{2.155}$$

where '$*$' can be either Cartesian product or any fuzzy implication operator. The latter two choices for '$*$' will be discussed in detail in Section 2.3.7.

When the rule antecedent or rule consequent are compound fuzzy propositions then first the membership function corresponding to each such compound proposition is determined as described in the previous two sections. Finally, '$*$' is applied to the so-determined membership functions.

2.3.5 Inference Rules

In approximate reasoning, two inference rules are of major importance, viz. the *compositional rule of inference* and the *generalized modus ponens*. The first rule uses a fuzzy relation to represent explicitly the connection between two fuzzy propositions, the second uses an if-then rule that implicitly represents a fuzzy relation. The generalized modus ponens has the symbolic inference scheme

(I) S_1 is Q_1,
if S_1 is P_1 then S_2 is P_2,
∴ S_2 is Q_2;

where S_1 and S_2 are symbolic names for objects, and P_1, P_2, Q_1 and Q_2 are object properties. One example of the generalized modus ponens is the following

(II) The tomato is very red,
if the tomato is red *then* the tomato is ripe,
∴ the tomato is very ripe.

The symbolic name 'tomato' stands for the real-world object tomato. In this example S_1 and S_2 are the same. Red and ripe are symbolic names for properties, thus corresponding to P_1 and P_2. The meaning or the symbol 'red' is described by a fuzzy set with membership function $\Pi(x; 640\text{Nm}, 680\text{Nm}, 720\text{Nm}, 760\text{Nm})$. Due to the representation of the meaning of the properties in terms of fuzzy sets, a conclusion can be derived even when the input is 'very red' instead of 'red'. In fuzzy set theory the membership functions representing the meaning of 'red' and 'very red' will overlap each other, i.e., there are lots of values in the domain that have membership degrees greater than zero in both fuzzy sets.

The *compositional rule of inference* can be considered to be a special case of the generalized modus ponens. Its general symbolic form is

(III) S_1 is Q_1,
S_1 R S_2,
∴ S_2 is Q_2;

where $S_1\ R\ S_2$ reads as "S_1 is in relation R to S_2" and its meaning is represented as a fuzzy relation μ_R. Hence, instead of the if-then rule, there is a fuzzy relation R. An example of the compositional rule of inference is

(IV) $\widetilde{m}$ is a *small* fuzzy number,
$\widetilde{m}$ is *somewhat smaller* than $\tilde{n}$,
∴ $\tilde{n}$ is a *rather small* fuzzy number;

where *somewhat smaller* is a fuzzy relation R.

2.3.6 The Compositional Rule of Inference

In Zadeh's 1971 paper on fuzzy semantics [235] and in his 1972 paper on hedges [236], there is still no mention of the compositional rule of inference. Perhaps the first time this inference rule was used was in 1973 [237]. But in this paper there is some confusion about the difference between the compositional rule of inference and the generalized modus ponens. Here Zadeh starts with the treatment of the if-then rule, called the fuzzy conditional statement: *if* X is A *then* Y is B, and the *if ... then ... else* rule, *if* X is A *then* Y is B *else* Z is C. In this paper he tries to combine the following two symbolic statements

1. x is *very small*
2. *if* x is *small then* y is *large else* y is *not very large*

Here Zadeh introduces the idea that the meaning of the second statement should be defined as a fuzzy relation $\widetilde{R}$. With this relation in hand he introduces the compositional rule of inference:

> "If $\widetilde{R}$ is a fuzzy relation from U to V, and $\widetilde{A}$ is a fuzzy subset of U, then the fuzzy subset $\widetilde{B}$ of V which is induced by $\widetilde{A}$ is given by the composition (...) of R and $\widetilde{A}$; that is,
> $$\widetilde{B} = \widetilde{A} \circ \widetilde{R},$$
> in which $\widetilde{A}$ plays the role of a unary relation." [237, page 37]

The $\circ$ operation (composition) was described in the former section. In this case one should take the cylindrical extension of $\widetilde{A}$, take the intersection with $\widetilde{R}$, and project the result onto the Y-axis.

When R is built up from $\widetilde{A}$ and $\widetilde{B}$ with the rule

$$\widetilde{R} = ce(\widetilde{A}) \cup \widetilde{B}, \quad \text{i.e.,} \quad \mu_R(x,y) = \max(1-\mu_A(x), \mu_B(y)), \qquad (2.156)$$

then it should be noted that $\widetilde{A} \circ \widetilde{R} = \widetilde{B}$ does not hold in general.

In later publications, e.g., [239], Zadeh makes an explicit discrimination between the compositional rule of inference and what is there called *compositional modus ponens* (generalized modus ponens here). When the compositional rule of inference is used, an explicit fuzzy relation R must be given a priori, which is not necessary in the generalized modus ponens.

The compositional rule of inference always requires an *explicit relation*, for example in

(V) Agnes is *tall*,
Olga is *a bit shorter than* Agnes,
∴ Olga is *more or less tall*;

where the relation R is *a bit shorter than*, and its meaning $\tilde{R}$ can be described by

$$\tilde{R} = \int_{[0,250]^2} \Lambda(y - x; -20, -10, 0)/(x, y), \tag{2.157}$$

i.e., Olga is a bit shorter than Agnes if they differ by approximately 10 cm in height.

In [239] and many other papers Zadeh proposes the following way to handle the compositional rule of inference. When the inference scheme is:

(VI) S_1 is P,
$S_1 \; R \; S_2$,
∴ S_2 is Q;

where $\tilde{P}$ and $\tilde{Q}$ are fuzzy sets representing the meaning of P and Q and $\tilde{R}$ is a fuzzy relation defining the meaning of R, and $\tilde{P}$ is defined on X, and $\tilde{R}$ on $X \times Y$, then

$$\tilde{Q} = \text{proj}(\tilde{P} \circ \tilde{R}) \text{ on } Y, \qquad \text{i.e.,} \qquad \mu_Q(y) = \max_x \min(\mu_P(x), \mu_R(x, y)). \tag{2.158}$$

Example 2.68 A small example on the discrete domain $X = Y = \{1, 2, 3, 4\}$. Suppose $\mu_{\text{small}} = 1/1 + 0.6/2 + 0.2/3$ and the binary relation $\mu_{\text{approximately equal}}$ is given by

$$\begin{aligned} &1/\left((1,1) + (2,2) + (3,3) + (4,4)\right) + \\ &0.5/\left((1,2) + (2,1) + (2,3) + (3,2) + (3,4) + (4,3)\right). \end{aligned} \tag{2.159}$$

In this case $\mu_{\text{small}} \circ \mu_{\text{approximately equal}}$ may be expressed as the max-min product of their relation matrices. Thus

$$\begin{aligned} \widetilde{\text{small}} \circ \widetilde{\text{approximately equal}} &= (1 \quad 0.6 \quad 0.2 \quad 0) \circ \begin{pmatrix} 1 & 0.5 & 0 & 0 \\ 0.5 & 1 & 0.5 & 0 \\ 0 & 0.5 & 1 & 0.5 \\ 0 & 0 & 0.5 & 1 \end{pmatrix} \\ &= (1 \quad 0.6 \quad 0.5 \quad 0.2). \end{aligned} \tag{2.160}$$

Hence, the result is the fuzzy set $\mu_Q = 1/1 + 0.6/2 + 0.5/3 + 0.2/4$, which can be denoted by the linguistic label 'more or less small'.

Example 2.69 Compare Zadeh [237, pages 37,38]. Suppose $X = \{1, 2, 3, 4, 5\}$, and *small* is interpreted as $\mu_{\text{small}} = 1/1 + 0.8/2 + 0.6/3 + 0.4/4 + 0.2/5$ and *large* as $\mu_{\text{large}} = 0.2/1 + 0.4/2 + 0.6/3 + 0.8/4 + 1/5$, and *not very large* as

$\mu_{\text{not very large}}(x) = 1 - \mu^2_{\text{large}}(x)$. Then the meaning of "*if* x is small *then* y is large *else* y is not very large" is represented by the relation $\tilde{R}$ below, where

$$\mu_R(x,y) = \max\left(\min\left(\mu_{\text{small}}(x), \mu_{\text{large}}(y)\right), \min\left(1 - \mu_{\text{small}}(x), \mu_{\text{not very large}}(y)\right)\right), \tag{2.161}$$

which states symbolically "(small and large) or (not small and not very large)." This if-then rule induces the following operation

$$\overset{\tilde{A}}{\begin{pmatrix}1 & 0.64 & 0.36 & 0.16 & 0.04\end{pmatrix}} \circ \overset{\tilde{R}}{\begin{pmatrix}0.2 & 0.4 & 0.6 & 0.8 & 1\\ 0.2 & 0.4 & 0.6 & 0.8 & 0.8\\ 0.4 & 0.4 & 0.6 & 0.6 & 0.6\\ 0.6 & 0.6 & 0.6 & 0.4 & 0.4\\ 0.8 & 0.8 & 0.64 & 0.36 & 0.2\end{pmatrix}}$$

$$= \overset{\tilde{B}}{\begin{pmatrix}0.36 & 0.4 & 0.6 & 0.8 & 1\end{pmatrix}}. \tag{2.162}$$

2.3.7 Representing the Meaning of If-then Rules

In this subsection we will consider a number of relations that can be used to represent the meaning of *if* X is A *then* Y is B. These relations are usually derived from *multi-valued logic* (see, e.g., Rescher [175] or Bolc [25]). Historically, many-valued logic was initiated by Łukasiewicz. Infinite-valued logic is often called $Ł_{\aleph_0}$ or $Ł_{\aleph_1}$, depending on the countability of the domain. Another name is $Ł_\infty$. Concerning the relation between multi-valued logic and fuzzy logic, Giles states "Now it turns out that (...) Łukasiewicz logic is exactly appropriate for the formulation of 'fuzzy set theory' (...); indeed, it is not too much to claim that $Ł_\infty$ is related to fuzzy set theory exactly as classical logic to ordinary set theory." [74, page 117]

In $Ł_\infty$, there are infinitely many truth values, usually numbered from 0 to 1, i.e., the interval $[0,1]$. Let p and q be two propositions in a multi-valued logic, where $v(p) = 0.5$ (the truth value of p is one half) and $v(q) = 0.7$. The truth value of p *and* q ($v(p \wedge q)$) can be calculated with a certain T-norm, also $v(p \vee q)$ can be calculated with an S-norm. The truth value of $p \to q$ is usually determined by the fact that $p \to q$ in two-valued logic is the same as $\neg p \vee q$ and as $(p \wedge q) \vee \neg p$. In the first case bounded sum is normally used to represent $\vee$, in the second case maximum and minimum are used. This results in

$$v(p \to q) = \min(1, 1 - v(p) + v(q)) \tag{2.163}$$

or

$$v(p \to q) = \max(\min(v(p), v(q)), v(q)). \tag{2.164}$$

In fuzzy logic this works out as follows.

Example 2.70 Suppose there is the rule "*if X is A then Y is B*," where the meanings of X is A and Y is B are given as

$$\widetilde{A} = .1/x_1 + .4/x_2 + .7/x_3 + 1/x_4 \quad \text{and} \quad \widetilde{B} = .2/y_1 + .5/y_2 + .9/y_3. \tag{2.165}$$

The meaning of the rule "*if A then B*" is considered to be the same as the compound fuzzy proposition "*not A or B*." The 'or' operation is performed by the union operation, but to take the union one needs the extensions of '*not A*' and of 'B', the negation of A is performed with the $1 - \cdot$ operation and the union with the maximum operation. As a formula,

$$\begin{aligned} \mu_{\textit{not A or B}} &= \max\left(\mu_{ce(A')}(x,y), \mu_{ce(B)}(x,y)\right) \\ &= \max\left(1 - \mu_{ce(A)}(x,y), \mu_{ce(B)}(x,y)\right). \end{aligned} \tag{2.166}$$

Numerically, in terms of the extensions $\mu_{ce(A')}$ and $\mu_{ce(B)}$,

$$ce(\widetilde{A}') \;\; \begin{array}{c|ccc} & y_1 & y_2 & y_3 \\ \hline x_1 & .9 & .9 & .9 \\ x_2 & .6 & .6 & .6 \\ x_3 & .3 & .3 & .3 \\ x_4 & 0 & 0 & 0 \end{array} \;\cup\; ce(\widetilde{B}) \;\; \begin{array}{c|ccc} & y_1 & y_2 & y_3 \\ \hline x_1 & .2 & .5 & .9 \\ x_2 & .2 & .5 & .9 \\ x_3 & .2 & .5 & .9 \\ x_4 & .2 & .5 & .9 \end{array} \;=\; ce(\widetilde{A}') \cup ce(\widetilde{B}) \;\; \begin{array}{c|ccc} & y_1 & y_2 & y_3 \\ \hline x_1 & .9 & .9 & .9 \\ x_2 & .6 & .6 & .9 \\ x_3 & .3 & .5 & .9 \\ x_4 & .2 & .5 & .9 \end{array} \tag{2.167}$$

One would expect that the composition of $\widetilde{A}$ and $\widetilde{R} = \widetilde{A}' \cup \widetilde{B}$ delivers $\widetilde{B}$. That this is not the case follows from

$$\overset{\widetilde{A}}{\begin{pmatrix} .1 & .4 & .7 & 1 \end{pmatrix}} \circ \overset{\widetilde{R}}{\begin{pmatrix} .9 & .9 & .9 \\ .6 & .6 & .9 \\ .3 & .5 & .9 \\ .2 & .5 & .9 \end{pmatrix}} = \overset{\widetilde{B}^*}{\begin{pmatrix} .46 & .5 & .9 \end{pmatrix}}. \tag{2.168}$$

This is one of the well known side effects of fuzzy inference rules. If one uses the Gödel implication rule (see below), then this inequality does not take place, but then the following happens: if $\widetilde{A}$ is changed into $\widetilde{A}^* \subset \widetilde{A}$, $\widetilde{B}$ remains unchanged, which does not fit into, for example, the tomato example in Scheme (II). In Hellendoorn [92, 93] it was proved that every relation has such disadvantages. This effect is often called *interaction*.

In the following we will give a number of relations that represent fuzzy implications like the one in the example above.

Kleene-Dienes implication This relation is the one that was already introduced above, and that is based on Kleene's three-valued logic operator. This one is denoted R_b defined as

$$\widetilde{R}_b = ce(\widetilde{A}') \cup ce(\widetilde{B}) \qquad \mu_{R_b}(x,y) = \max(1 - \mu_A(x), \mu_B(y)). \tag{2.169}$$

This implication is also-called Dienes-Rescher implication.

Lukasiewicz implication Both this implication and the former are based on the equivalence $p \rightarrow q \equiv \neg p \vee q$. To represent 'or' it is also possible to use the bounded sum $\min(1, 1 - p + q)$ instead of the maximum $\max(p, q)$. This gives the relation here called R_a, defined as

$$\widetilde{R}_a = ce(\widetilde{A}') \oplus ce(\widetilde{B}) \qquad \mu_{R_a}(x, y) = \min(1, 1 - \mu_A(x) + \mu_B(y)). \tag{2.170}$$

This one was also proposed by Zadeh. It is clear that always, for all x and y, $\mu_{R_b}(x, y) \leq \mu_{R_a}(x, y)$, i.e., it is always the case that $\widetilde{R}_b \subseteq \widetilde{R}_a$: R_b is a stronger implication than R_a.

Example 2.71

$$\begin{array}{c|c|c|c|}
\multicolumn{4}{c}{ce(\widetilde{A}')} \\
 & y_1 & y_2 & y_3 \\ \hline
x_1 & .9 & .9 & .9 \\ \hline
x_2 & .6 & .6 & .6 \\ \hline
x_3 & .3 & .3 & .3 \\ \hline
x_4 & 0 & 0 & 0 \\ \hline
\end{array}
\oplus
\begin{array}{c|c|c|c|}
\multicolumn{4}{c}{ce(\widetilde{B})} \\
 & y_1 & y_2 & y_3 \\ \hline
x_1 & .2 & .5 & .9 \\ \hline
x_2 & .2 & .5 & .9 \\ \hline
x_3 & .2 & .5 & .9 \\ \hline
x_4 & .2 & .5 & .9 \\ \hline
\end{array}
=
\begin{array}{c|c|c|c|}
\multicolumn{4}{c}{ce(\widetilde{A}') \cup ce(\widetilde{B})} \\
 & y_1 & y_2 & y_3 \\ \hline
x_1 & 1 & 1 & 1 \\ \hline
x_2 & .8 & 1 & 1 \\ \hline
x_3 & .5 & .8 & 1 \\ \hline
x_4 & .2 & .5 & .9 \\ \hline
\end{array} \tag{2.171}$$

Zadeh implication It was already stated above that in two-valued logic $p \rightarrow q$ has the same truth values as $(p \wedge q) \vee \neg p$. This equivalence was used by Zadeh as follows

$$\begin{aligned}
\widetilde{R}_m &= (ce(\widetilde{A}) \cap ce(\widetilde{B})) \cup ce(\widetilde{A}') \\
\mu_{R_m}(x, y) &= \max(\min(\mu_A(x), \mu_B(y)), 1 - \mu_A(x)).
\end{aligned} \tag{2.172}$$

Example 2.72

$$\begin{array}{c|c|c|c|}
\multicolumn{4}{c}{ce(\widetilde{A})} \\
 & y_1 & y_2 & y_3 \\ \hline
x_1 & .1 & .1 & .1 \\ \hline
x_2 & .4 & .4 & .4 \\ \hline
x_3 & .7 & .7 & .7 \\ \hline
x_4 & 1 & 1 & 1 \\ \hline
\end{array}
\cap
\begin{array}{c|c|c|c|}
\multicolumn{4}{c}{ce(\widetilde{B})} \\
 & y_1 & y_2 & y_3 \\ \hline
x_1 & .2 & .5 & .9 \\ \hline
x_2 & .2 & .5 & .9 \\ \hline
x_3 & .2 & .5 & .9 \\ \hline
x_4 & .2 & .5 & .9 \\ \hline
\end{array}
\cup
\begin{array}{c|c|c|c|}
\multicolumn{4}{c}{ce(\widetilde{A}')} \\
 & y_1 & y_2 & y_3 \\ \hline
x_1 & .9 & .9 & .9 \\ \hline
x_2 & .6 & .6 & .6 \\ \hline
x_3 & .3 & .3 & .3 \\ \hline
x_4 & 0 & 0 & 0 \\ \hline
\end{array}
=
\begin{array}{c|c|c|c|}
\multicolumn{4}{c}{ce(\widetilde{A}') \cup ce(\widetilde{B})} \\
 & y_1 & y_2 & y_3 \\ \hline
x_1 & .9 & .9 & .9 \\ \hline
x_2 & .6 & .6 & .6 \\ \hline
x_3 & .2 & .5 & .7 \\ \hline
x_4 & .2 & .5 & .9 \\ \hline
\end{array} \tag{2.173}$$

Stochastic implication The so-called stochastic implication is based on the equality $P(B \mid A) = 1 - P(A) + P(A)P(B)$. Note that a product operation is used to represent the intersection, i.e.,

$$\begin{aligned}
\widetilde{R}_* &= ce(\widetilde{A}') \cup (ce(\widetilde{A}) \cap ce(\widetilde{B})) \\
\mu_{R_*}(x, y) &= \min(1, 1 - \mu_A(x) + \mu_A(x)\mu_B(y)).
\end{aligned} \tag{2.174}$$

Example 2.73

$ce(\tilde{A}')$

	y_1	y_2	y_3
x_1	.9	.9	.9
x_2	.6	.6	.6
x_3	.3	.3	.3
x_4	0	0	0

$\cup$ $ce(\tilde{A})$

	y_1	y_2	y_3
x_1	.1	.1	.1
x_2	.4	.4	.4
x_3	.7	.7	.7
x_4	1	1	1

$\cap$ $ce(\tilde{B})$

	y_1	y_2	y_3
x_1	.2	.5	.9
x_2	.2	.5	.9
x_3	.2	.5	.9
x_4	.2	.5	.9

$=$ $ce(\tilde{A}') \cup ce(\tilde{B})$

	y_1	y_2	y_3
x_1	.9	.9	.9
x_2	.6	.6	.6
x_3	.3	.35	.63
x_4	.2	.5	.9

(2.175)

Goguen implication One requirement in multi-valued logic is that $p \rightarrow q$ should satisfy

$$v(p) \times v(p \rightarrow q) \leq v(q). \tag{2.176}$$

This goal is satisfied when one uses the definition $v(p \rightarrow q) = \min(1, v(q)/v(p))$ that was proposed by Goguen. This gives the following fuzzy relation:

$$\mu_{R_\Delta}(x,y) = \min(1, \frac{\mu_A(x)}{\mu_B(y)}). \tag{2.177}$$

Example 2.74 Using the same fuzzy sets $\tilde{A}$ and $\tilde{B}$ as in the examples before, one obtains

$ce(\tilde{A}') \cup ce(\tilde{B})$

	y_1	y_2	y_3
x_1	.5	.2	.11
x_2	1	.8	.44
x_3	1	1	.77
x_4	1	1	1

(2.178)

Gödel implication The Gödel implication is one of the best-known implication formulas in multi-valued logic. It is defined as

$$v(p \underset{g}{\rightarrow} q) = \begin{cases} 1, & \text{if } v(p) \leq v(q), \\ v(q), & \text{otherwise;} \end{cases} \tag{2.179}$$

e.g., $v(0.6 \underset{g}{\rightarrow} 0.8) = 1$ and $v(0.6 \underset{g}{\rightarrow} 0.4) = 0.4$, which results in the following fuzzy relation that is frequently used in fuzzy logic:

$$\mu_{R_g}(x,y) = (\mu_A(x) \underset{g}{\rightarrow} \mu_B(y)). \tag{2.180}$$

Sharp implication The so-called Sharp implication looks like the Gödel implication but is more restrictive. It is defined as

$$v(p \underset{s}{\rightarrow} q) = \begin{cases} 1, & \text{if } v(p) \leq v(q); \\ 0, & \text{otherwise;} \end{cases} \tag{2.181}$$

e.g., $v(0.6 \underset{g}{\rightarrow} 0.8) = 1$ and $v(0.6 \underset{g}{\rightarrow} 0.4) = 0$, which results in the following fuzzy relation:

$$\mu_{R_s}(x,y) = (\mu_A(x) \underset{s}{\rightarrow} \mu_B(y)). \tag{2.182}$$

Example 2.75 The Gödel and the Sharp implication for the case *if* X is A *then* Y is B are given by

Gödel

	y_1	y_2	y_3
x_1	1	1	1
x_2	.2	1	1
x_3	.2	.5	1
x_4	.2	.5	.9

Sharp

	y_1	y_2	y_3
x_1	1	1	1
x_2	0	1	1
x_3	0	0	1
x_4	0	0	0

(2.183)

General implication A very general implication which does not have an explicit name in the literature may be considered as a combination of the Gödel implication and the Sharp implication. It is based on the following formula:

$$p \leftrightarrow q \equiv (p \underset{\alpha}{\rightarrow} q) \wedge (\neg p \underset{\beta}{\rightarrow} \neg q), \tag{2.184}$$

where α and β may be s or g. In fuzzy terms this gives the relation $R_{\alpha\beta}$ defined as

$$\mu_{R_{\alpha\beta}}(x,y) = \min((\mu_A(x) \underset{\alpha}{\rightarrow} \mu_B(y)), ((1-\mu_A(x)) \underset{\beta}{\rightarrow} (1-\mu_B(y))). \tag{2.185}$$

This relation is hardly ever used in the literature.

Mamdani implication With respect to fuzzy control, this is the *most important* implication known in the literature. Its definition is based on the intersection operation, i.e., $p \rightarrow q \equiv p \wedge q$. The relation R_c (c from conjunction) is defined as

$$\tilde{R}_c = ce(\tilde{A}) \cap ce(\tilde{B}) \qquad \mu_{R_c}(x,y) = \min(\mu_A(x), \mu_B(y)). \tag{2.186}$$

Example 2.76 Consider again the rule "*if* X is A *then* Y is B" from Example 2.70, where

$$\tilde{A} = .1/x_1 + .4/x_2 + .7/x_3 + 1/x_4 \quad \text{and} \quad \tilde{B} = .2/y_1 + .5/y_2 + .9/y_3. \tag{2.187}$$

The Mamdani relation R_c is now given by

Mamdani

	y_1	y_2	y_3
x_1	.1	.1	.1
x_2	.2	.4	.4
x_3	.2	.5	.7
x_4	.2	.5	.9

(2.188)

All these relations can be combined with the projection principle. Dubois and Prade [54] use the combination of these relations and the max-min composition to give an ordering.

$$\tilde{R}_a \supseteq \tilde{R}_* \supseteq \tilde{R}_b \supseteq \tilde{R}_m \supseteq \tilde{R}_c;$$

$$\tilde{R}_a \supseteq \tilde{R}_\Delta \supseteq R_g \supseteq \tilde{R}_c;$$

$$\tilde{R}_g \supseteq \tilde{R}_s.$$

In the following large example, we will illustrate the theory given thus far with a fuzzy control application.

Example 2.77 Consider the seven fuzzy sets as given in Fig. 2.6, together with the following simple if-then control rule

if e is PM *then* u is NS,

where e is defined on the domain $E = [-6, 6]$ and u is defined on the domain $U = [-6, 6]$. Both domains contain the same term set $\{NB, NM, \ldots, PB\}$. In practical applications of fuzzy control theory there are basically two approaches to handle such a rule. The first approach is to use a functional representation, i.e., if an input value comes in, the fuzzy set $\widetilde{PM}$ is considered as a function, and the membership degree is returned. The second approach is to use a look-up table. As these approaches are not principally different, we will use the second one, because it is easier for illustration. Our domain is chosen very coarse, because otherwise the example becomes too big. So $\widetilde{PM}$ and $\widetilde{NS}$ are defined as

$$\widetilde{PM} = \tfrac{1}{3}/2 + \tfrac{2}{3}/3 + 1/4 + \tfrac{2}{3}/5 + \tfrac{1}{3}/6 \tag{2.189}$$
$$\widetilde{NS} = \tfrac{1}{3}/\text{-}4 + \tfrac{2}{3}/\text{-}3 + 1/\text{-}2 + \tfrac{2}{3}/\text{-}1 + \tfrac{1}{3}/0. \tag{2.190}$$

We have seen above that there are several ways to represent the meaning of the if-then rule, and that in fuzzy control theory usually the Mamdani implication is used. When we use this implication, we get the following relation on $E \times U$:

$$U \quad \begin{array}{c|ccccccccc}
\vdots & \ddots & \vdots & \vdots & \vdots & \vdots & \vdots & \vdots & \vdots & \vdots \\
3 & \ldots & 0 & 0 & 0 & 0 & 0 & 0 & 0 & 0 \\
2 & \ldots & 0 & 0 & 0 & 0 & 0 & 0 & 0 & 0 \\
1 & \ldots & 0 & 0 & 0 & 0 & 0 & 0 & 0 & 0 \\
0 & \ldots & 0 & 0 & 0 & \frac{1}{3} & \frac{1}{3} & \frac{1}{3} & \frac{1}{3} & \frac{1}{3} \\
-1 & \ldots & 0 & 0 & 0 & \frac{1}{3} & \frac{2}{3} & \frac{2}{3} & \frac{2}{3} & \frac{1}{3} \\
-2 & \ldots & 0 & 0 & 0 & \frac{1}{3} & \frac{2}{3} & 1 & \frac{2}{3} & \frac{1}{3} \\
-3 & \ldots & 0 & 0 & 0 & \frac{1}{3} & \frac{2}{3} & \frac{2}{3} & \frac{2}{3} & \frac{1}{3} \\
-4 & \ldots & 0 & 0 & 0 & \frac{1}{3} & \frac{1}{3} & \frac{1}{3} & \frac{1}{3} & \frac{1}{3} \\
-5 & \ldots & 0 & 0 & 0 & 0 & 0 & 0 & 0 & 0 \\
-6 & \ldots & 0 & 0 & 0 & 0 & 0 & 0 & 0 & 0 \\
\hline
 & \ldots & -1 & 0 & 1 & 2 & 3 & 4 & 5 & 6
\end{array} \tag{2.191}$$
$$E$$

Suppose the input to the controller or the actual value of E corresponds to 3 on the error domain $[-6, 6]$. This value alone cannot be combined with the relation given above, therefore it has to be *fuzzified*. Fuzzification here means that the *crisp* value -3 is transformed into the fuzzy set $1/3$, which on the domain $[-6, 6]$ is the same as

$$0/-6 + \ldots + 0/-1 + 0/0 + 0/1 + 0/2 + 1/3 + 0/4 + 0/5 + 0/6. \qquad (2.192)$$

If we now use the generalized modus ponens, i.e., perform *composition* between the above fuzzy set and the relation representing the meaning or the rule, we obtain the relation

$$\begin{array}{cc|ccccccccc}
 & \vdots & \ddots & \vdots & \vdots & \vdots & \vdots & \vdots & \vdots & \vdots & \vdots \\
 & 3 & \cdots & 0 & 0 & 0 & 0 & 0 & 0 & 0 & 0 \\
 & 2 & \cdots & 0 & 0 & 0 & 0 & 0 & 0 & 0 & 0 \\
 & 1 & \cdots & 0 & 0 & 0 & 0 & 0 & 0 & 0 & 0 \\
 & 0 & \cdots & 0 & 0 & 0 & 0 & \frac{1}{3} & 0 & 0 & 0 \\
 & -1 & \cdots & 0 & 0 & 0 & 0 & \frac{2}{3} & 0 & 0 & 0 \\
U & -2 & \cdots & 0 & 0 & 0 & 0 & \frac{2}{3} & 0 & 0 & 0 \\
 & -3 & \cdots & 0 & 0 & 0 & 0 & \frac{2}{3} & 0 & 0 & 0 \\
 & -4 & \cdots & 0 & 0 & 0 & 0 & \frac{1}{3} & 0 & 0 & 0 \\
 & -5 & \cdots & 0 & 0 & 0 & 0 & 0 & 0 & 0 & 0 \\
 & -6 & \cdots & 0 & 0 & 0 & 0 & 0 & 0 & 0 & 0 \\
\hline
 & & \cdots & -1 & 0 & 1 & 2 & 3 & 4 & 5 & 6 \\
 & & & & & & E & & & &
\end{array} \qquad (2.193)$$

Finally, the projection of this relation on the U-axis is

$$0/-6 + 0/-5 + \tfrac{1}{3}/-4 + \tfrac{2}{3}/-3 + \tfrac{2}{3}/-2 + \tfrac{1}{3}/-1 + 0/0 + \ldots + 0/6, \qquad (2.194)$$

i.e., the column for $e = 3$ in the first relation. For other input values of e, another column is taken. This means that each column in the Mamdani relation given above corresponds to a 'clipped' fuzzy set in the graphical representation of FKBC.

Let us see now whether the above fuzzy set can be obtained in another way. This can be done as follows:

1. For the crisp input $e = 3$, determine its degree of membership in μ_{PM}. It is $\mu_{PM}(3) = \frac{2}{3}$.

2. Form a modified version of μ_{NS} as follows:

$$\forall u \in \mathcal{U} : \mu_{CNS}(u) = \begin{cases} \mu_{NS}(u) & \text{if } \mu_{NS}(u) \leq \mu_{PM}(3) = \frac{2}{3}, \\ \mu_{PM}(3) = \frac{2}{3} & \text{otherwise;} \end{cases} \qquad (2.195)$$

   this is the same as

$$\forall u \in \mathcal{U} : \mu_{CNS}(u) = \min\left(\mu_{PM}(3), \mu_{NS}(u)\right) = \min\left(\tfrac{2}{3}, \mu_{NS}(u)\right). \qquad (2.196)$$

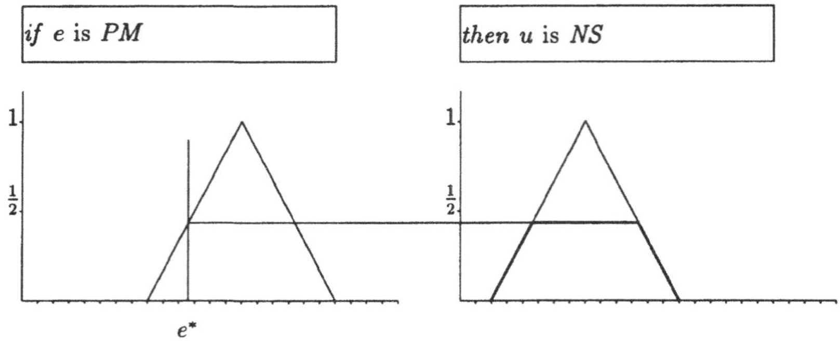

Fig. 2.15. The graphical representation of the firing of the rule if e is PM, then u is NS. The thick lines in NS denote the clipped fuzzy set CNS.

The modified version of μ_{NS} representing the fuzzy value of the control output U is called *clipped* μ_{NS}, defined as μ_{CNS}.

Graphically, (1) and (2), which constitute the *firing* of "*if* e is PM *then* u is NS" given $\text{VA}(E) = 3$, can be illustrated as in Fig. 2.15.

Thus, in the case of a Mamdani implication used for interpreting the meaning of a rule we have that for a crisp input value e^*,

$$\begin{aligned} \forall u : \mu_{CNS}(u) &= \int_{\mathcal{E}} \mu^*(e)/e \circ \int_{\mathcal{E}\times\mathcal{U}} \mu_{R_m}(e,u)/(e,u) \\ &= \min\left(\mu_{PM}(e^*), \mu_{NS}(u)\right) \\ &= \mu_{CNS}(u), \end{aligned} \tag{2.197}$$

where $\mu^*(e)$ is the membership function obtained after fuzzification of the crisp input e^*,

$$\forall e : \mu^*(e) = \begin{cases} 1 & \text{if } e = e^*, \\ 0 & \text{otherwise.} \end{cases} \tag{2.198}$$

The process described by (2.197) is called *firing*.

In the fuzzy control literature, there exists another way of obtaining a modified version of μ_{NS} from our example. This is done as follows

1. For the crisp input $e = 3$, determine its degree of membership in μ_{PM}. It is $\mu_{PM}(3) = \frac{2}{3}$.

2. Form a modified version of μ_{NS} denoted as μ_{SNS} (scaled-NS) as follows

$$\forall u \in \mathcal{U} : \mu_{SNS}(u) = \mu_{PM}(3) \cdot \mu_{NS}(u) = \tfrac{2}{3} \cdot \mu_{NS}(u). \tag{2.199}$$

This type of firing of a single rule or individual-rule based inference, which results in a scaled fuzzy set representing the control output, will be called by us *scaled inference.* However, scaled inference is called max-dot or max-product inference in the fuzzy control literature. It should be stressed here, that this so-called max-dot or max-product inference has *nothing* in common with max-dot or max-product composition as defined in Definition 2.62.

2.4 Representing a Set of Rules

2.4.1 Mamdani Versus Gödel

We have considered a system with a single rule "*if* e is *PM* *then* u is *NS*" and crisp input, in combination with the Mamdani type of inference. In this section we will consider the case of multiple rules, with crisp input, where we will mainly focus on Mamdani type of inference, because this is normally used in fuzzy control. The Mamdani type of inference uses the Mamdani implication to represent the meaning of the if-then rule.

In the example from the previous section we have seen that the result of *firing* the one rule was a clipped fuzzy set, which we denoted $\widetilde{CNS}$. In a system of n rules each rule is symbolically represented as

$$\textit{if } e \text{ is } LE^{(k)} \textit{ then } u \text{ is } LU^{(k)}, \quad k = 1, \ldots, n,$$

where

- $LE^{(k)}$ is the linguistic value of E in the kth rule. $LE^{(k)} \in \mathcal{L}E$ and "e is $LE^{(k)}$" is interpreted as $\widetilde{LE}^{(k)} = \int_{\mathcal{E}} \mu_{LE^{(k)}}(e)/e$
- $LU^{(k)}$ is the linguistic value of U in the kth rule. $LU^{(k)} \in \mathcal{L}U$ and "u is $LU^{(k)}$" is interpreted as $\widetilde{LU}^{(k)} = \int_{\mathcal{U}} \mu_{LU^{(k)}}(u)/u$

The Mamdani interpretation of the rule is defined as

$$\forall k : \tilde{R}_m^{(k)} = \int_{\mathcal{E}\times\mathcal{U}} \min\left(\mu_{LE^{(k)}}(e), \mu_{LU^{(k)}}(u)\right)/(e,u). \tag{2.200}$$

The membership function of the fuzzified crisp input e^* is

$$\forall e : \mu^*(e) = \begin{cases} 1 & \text{for } e = e^*, \\ 0 & \text{otherwise.} \end{cases} \tag{2.201}$$

Now the meaning of the *whole set of rules* is defined as

$$\tilde{R}_m = \bigcup_{k=1}^{n} \tilde{R}_m^{(k)} \tag{2.202}$$

which means that

$$\begin{aligned} \forall e,u : \mu_{R_m}(e,u) &= \max_k \mu_{R_m^{(k)}}(e,u) \\ &= \max_k \min\left(\mu_{LE^{(k)}}(e), \mu_{LU^{(k)}}(u)\right). \end{aligned} \tag{2.203}$$

Then the firing of a set of rules can be expressed according to (2.197) as

$$\widetilde{U} = \mu^* \circ \widetilde{R}_m, \quad \text{i.e.,} \quad \forall u : \mu_U(u) = \max_k \min\left(\mu_{LE^{(k)}}(e^*), \mu_{LU^{(k)}}(u)\right). \quad (2.204)$$

The firing of a set of rules via the operation composition will from now on be called *composition based inference.*

If we fire each rule separately, then the result of firing will be n clipped fuzzy sets $\widetilde{CLU}^{(1)}, \ldots, \widetilde{CLU}^{(n)}$, one for each rule, where

$$\forall u \forall k : \mu_{CLU^{(k)}}(u) = \min(\mu_{LE^{(k)}}(e^*), \mu_{LU^{(k)}}(u)). \quad (2.205)$$

Firing each rule separately is from now on called *individual-rule based inference.* Thus, if we form the union via maximum of all outputs then we obtain

$$\forall u : \max_k \mu_{CLU^{(k)}}(u) = \max_k \min(\mu_{LE^{(k)}}(e^*), \mu_{LU^{(k)}}(u)), \quad (2.206)$$

which is the same as (2.204). The output fuzzy set is the result of the operation $e \circ R_m$, i.e., the max-min composition rule, which simplifies to (2.206), because the input is non-fuzzy.

So we may derive the following result:

1. in *composition based inference* one first combines all rules into $\widetilde{R}_m$ and then 'fires' with fuzzy input μ^* via operation composition;
2. in *individual-rule based inference* one first 'fires' each rules individually with crisp input e^* and thus obtains n clipped fuzzy sets that are combined into one overall fuzzy set.
3. The above two are equivalent for the Mamdani type of inference. This even holds in the case of fuzzy inputs instead of crisp inputs e^*.

If we use Gödel implication (or the Gödel type of inference) instead of Mamdani implication, then we get different results. The Gödel implication representing the meaning of the k-th rule is given by

$$\forall k : \widetilde{R}_g^{(k)} = \int_{\mathcal{E} \times \mathcal{U}} \mu_{R_g^{(k)}}(e, u)/(e, u), \quad (2.207)$$

where

$$\forall e \forall u : \mu_{R_g^{(k)}}(e, u) = \begin{cases} 1 & \text{if } \mu_{LE^{(k)}}(e) \leq \mu_{LU^{(k)}}(u), \\ \mu_{LU^{(k)}}(u) & \text{otherwise.} \end{cases} \quad (2.208)$$

In this case, the firing of a single rule k with a crisp input e^* (individual-rule based inference) is equal to

$$\mu_{R_g^{(k)}}(\mu_{LE^{(k)}}(e^*), \mu_{LU^{(k)}}(u)) = \begin{cases} 1 & \text{if } \mu_{LE^{(k)}}(e^*) \leq \mu_{LU^{(k)}}(u), \\ \mu_{LU^{(k)}}(u) & \text{otherwise.} \end{cases} \quad (2.209)$$

Figure 2.16 compares the output membership functions for Mamdani and Gödel types of implication in the case of firing a single rule.

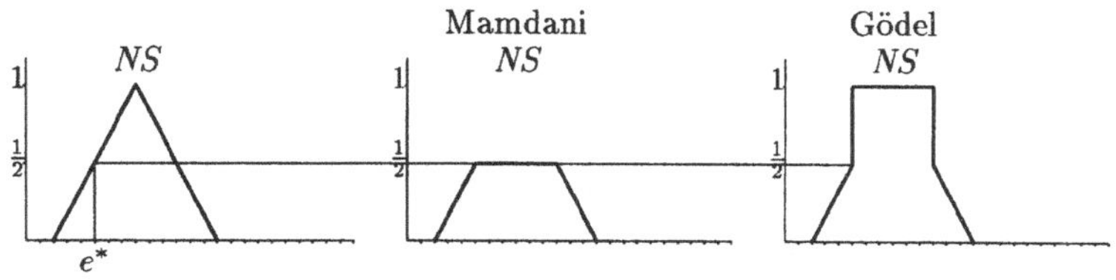

Fig. 2.16. The fuzzy set *NS* (left), its clipped version after the Mamdani implication (middle), and the version after Gödel implication (right).

The meaning of the whole set of rules in the case of Gödel implication is given by

$$\tilde{R}_g = \bigcap_{k=1}^{n} \tilde{R}_g^{(k)} \tag{2.210}$$

i.e., an intersection operation instead of the union as in the Mamdani type of inference. This means that

$$\begin{aligned} \forall e, u : \mu_{R_g}(e,u) &= \min_k \mu_{R_g^{(k)}}(e,u) \\ &= \min_k \left(\begin{cases} 1 & \text{if } \mu_{LE^{(k)}}(e) \leq \mu_{LU^{(k)}}(u), \\ \mu_{LU^{(k)}}(u) & \text{otherwise.} \end{cases} \right) \end{aligned} \tag{2.211}$$

Then the firing of a set of rules (composition based inference) can be expressed by

$$\tilde{U} = \mu^* \circ \tilde{R}_g, \tag{2.212}$$

i.e.,

$$\begin{aligned} \forall u : \mu_U(u) &= \min_k \left(\begin{cases} 1 & \text{if } \mu_{LE^{(k)}}(e^*) \leq \mu_{LU^{(k)}}(u), \\ \mu_{LU^{(k)}}(u) & \text{otherwise.} \end{cases} \right) \\ &= \begin{cases} 1 & \text{if } \forall k : \mu_{LE^{(k)}}(e^*) \leq \mu_{LU^{(k)}}(u), \\ \min_k \mu_{LU^{(k)}}(u) & \text{otherwise.} \end{cases} \end{aligned} \tag{2.213}$$

In the case of the Gödel type of inference with fuzzy inputs, there is a difference between individual-rule based inference and composition based inference.

Considering a set of n rules and the Mamdani type of individual-rule based inference, one obtains the clipped fuzzy sets $\widetilde{CLU}^{(1)}, \ldots \widetilde{CLU}^{(n)}$ that have to be combined to obtain one output fuzzy set $\tilde{U}$,

$$\tilde{U} = \bigcup_{k=1}^{n} \widetilde{CLU}^{(k)}. \tag{2.214}$$

The easiest way to perform the union operation is with maximum, as given in (2.206), but in principle it is possible to use any S-norm to perform the union operation $\bigcup_{k=1}^{n}$. However, there are some practical constraints due to the fact that sometimes many rules may fire in parallel, i.e., n can be large.

If we use the maximum (or supremum) operation for the union, there is no problem, but for example the Sugeno S_λ family of S-norms, with $\lambda = 1$, i.e.,

$$S_1(a, b) = \min(1, a + b - ab) \tag{2.215}$$

(compare (2.54)) can become fairly complicated in the case of many arguments, e.g.,

$$\begin{aligned} S(a, b, c, d) &= \min(1, a + b + c + d - ab - ac - ad - bc - bd - cd \\ &\quad + abc + abd + acd + bcd - abcd). \end{aligned} \tag{2.216}$$

It is a well known fact in fuzzy logic that in the case of *infinitely many rules*, only the maximum (or supremum) and S_W operation, i.e., the two S-norms extremes (equation (2.60)), can be used.

Dubois and Prade [63] have carefully studied the differences between these Mamdani and Gödel approaches. They state that a control law can be described by (1) a *functional description* $y = f(x)$ and (2) a *graph description* $(x, y) \in f$, where f also denotes the relation which holds between x and u if and only if $y = f(x)$. These views seem equivalent, but are not in the case where one deals with fuzzy values for x and y. For a rule

$$\textit{if } e \text{ is } LE^{(k)} \textit{ then } \dot{e} \text{ is } LU^{(k)}, \quad k = 1, \ldots, n$$

in a FKBC, this means that the graph view corresponds to Mamdani implication and the functional view corresponds to Gödel implication. The graph view leads one to consider the pairs $(LE^{(k)}, LU^{(k)})$ as elements of a relation

$$\forall k : \widetilde{LE}^{(k)} \times \widetilde{LU}^{(k)} \subseteq \tilde{R}^{(k)}, \quad \text{i.e.,} \tilde{R}_c^{(k)} = \bigcup_{i=1}^{n} \widetilde{LE}^{(k)} \times \widetilde{LU}^{(k)}. \tag{2.217}$$

The functional view leads one to consider $LU^{(k)}$ as the image of $LE^{(k)}$ by a relation R

$$\exists R \forall k : \widetilde{LE}^{(k)} \circ R = \widetilde{LU}^{(k)}, \quad \text{i.e.,} \tilde{R}_g^{(k)} \subseteq \widetilde{LE}^{(k)} \to \widetilde{LU}^{(k)}. \tag{2.218}$$

Suppose the domains of E and U are divided into four linguistic values LE_1, ..., LE_4, and LU_1, ..., LU_4, and there are the following four rules in the rule base

if e is LE_1 *then* u is LU_1,
if e is LE_2 *then* u is LU_2,
if e is LE_3 *then* u is LU_3,
if e is LE_4 *then* u is LU_4.

In Fig. 2.17 these four rules are shown and the difference between these two approaches is made clear. It follows immediately from these figures that for an input value $e^* \in \widetilde{LE}_2 \cap \widetilde{LE}_3$, the functional view returns the correct conclusion $u \in \widetilde{LU}_2 \cap \widetilde{LU}_3$, whereas the graph view (Mamdani) yields the more imprecise conclusion $u \in \widetilde{LU}_2 \cup \widetilde{LU}_3$ [63].

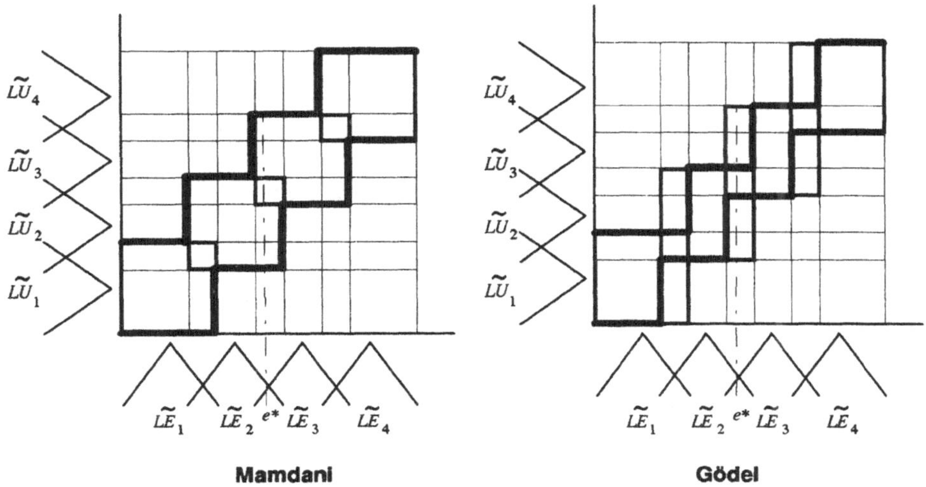

Fig. 2.17. For an input value e^* in the intersection of $\widetilde{LE}_2$ and $\widetilde{LE}_3$ the conclusions for Mamdani and Gödel implication are different. In the Mamdani case, the conclusion is a clipped version of the union of $\widetilde{LE}_2$ and $\widetilde{LE}_3$. In the Gödel case, the conclusion lies in the intersection interval of $\widetilde{LE}_2$ and $\widetilde{LE}_3$.

2.4.2 Properties of a Set of Rules

We have considered several fuzzy control rules and their meaning in the Mamdani and Gödel case. An example of a FKBC is a system with two input values e and $\dot{e}$ (error and change of error) and one output value u, with rules like

if e is NB *and* $\dot{e}$ is NB *then* u is PB,
if e is ZO *and* $\dot{e}$ is NS *then* u is NS.

Figure 2.18 shows the domains of E, $\dot{E}$ and all the rules of this controller; the domain of U is equal to that of E and $\dot{E}$. In the case that e is NB and $\dot{e}$ is PB for example, the output field for u is empty: this means that there is no rule for this case. Figure 2.19 shows the behavior of the controller for this rule base, by showing for each crisp value of e and $\dot{e}$, the exact (non-fuzzy) value of u. Figures like this one are often used to study the behavior of the controller.

Important properties for a set of rules are

- completeness,
- consistency,
- continuity, and
- interaction.

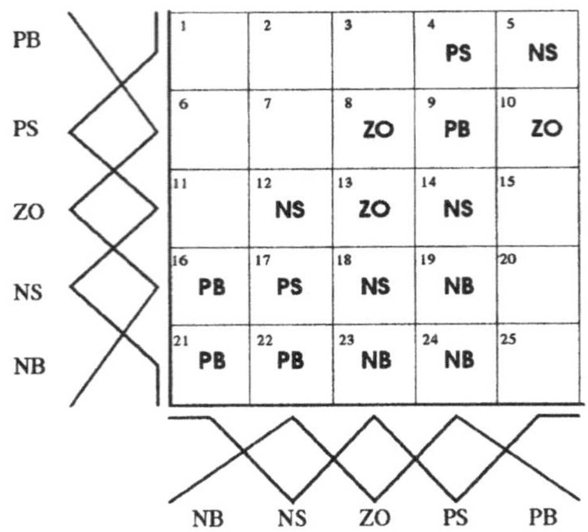

Fig. 2.18. A rule base in matrix structure. Each of the 25 cells can represent a rule. There are only 16 rules in this rule base.

(see Pedrycz [164]). We will briefly discuss these three properties with the use of the FKBC as given in Fig. 2.18. The general form of an if-then rule in this table is given by

$$\textit{if } e \text{ is } LE^{(k)} \textit{ and } \dot{e} \text{ is } L\dot{E}^{(k)} \textit{ then } u \text{ is } LU^{(k)}, \quad k = 1, \ldots, n.$$

Let us denote the output fuzzy set for crisp input values e^* and $\dot{e}^*$ by $\mathrm{OUT}(e^*, \dot{e}^*)$, meaning that OUT is a function of e^* and $\dot{e}^*$.

2.4.3 Completeness of a set of rules

Definition 2.78 *A set of if-then rules is complete if any combination of input values results in an appropriate output value. Formally, this means that*

$$\forall e, \dot{e} : \mathrm{hgt}(\mathrm{OUT}(e, \dot{e})) > 0, \tag{2.219}$$

where the height of a fuzzy set was defined in Def. 2.25 as $\mathrm{hgt}(A) = \sup_{u \in X} \mu_A(u)$. Hence, all input combinations $(e, \dot{e})$ result in an output fuzzy set. The FKBC as given in Fig. 2.18 is *not* complete, because input $(e^*, \dot{e}^*)$ falling in cell 1 does not produce any output.

Consider the rule

$$\textit{if } e \text{ is } PB \textit{ and } \dot{e} \text{ is } NB \textit{ then } u \text{ is } \langle \text{null} \rangle,$$

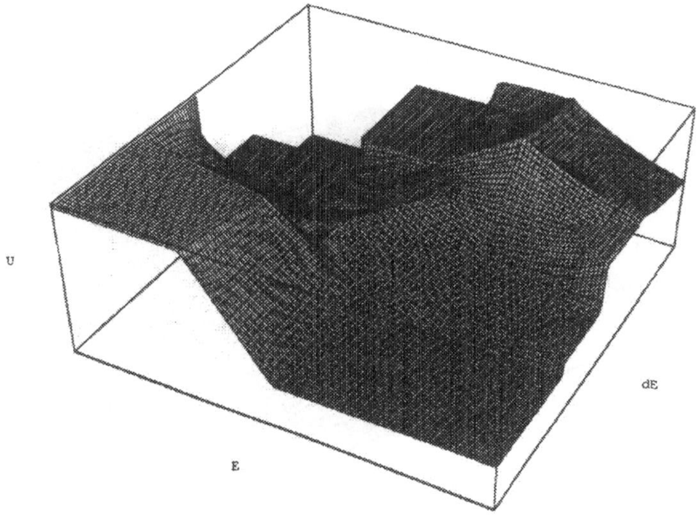

Fig. 2.19. The output behavior of the controller as described in Fig. 2.18. For each crisp input pair $(e^*, \dot{e}^*)$ the crisp output value u on the domain $\mathcal{U}$ is calculated. $\mathcal{U} = [0, 10]$. In the lower left corner, the values for u are high, because rules 16, 21 and 22 have consequent part PB. If PB is defined by the membership function $\Gamma(x; 7, 9)$, then $u = 10$ in this corner. According to the rules, and if NB is defined by the membership function $\mathrm{L}(x; 1, 3)$, in the lower right corner $u = 0$. According to the rules, in the upper left corner no output for u is defined. This is expressed by the value $u = -2$. In Fig. 2.20 the figure is rotated through 180°, such that this part of the figure can be shown.

which corresponds to cell 25. If the actual input $(e^*, \dot{e}^*)$ is such that $e^* \in PB \cap PS$ and $\dot{e}^* \in NB \cap NS$, then the rule

$$\textit{if } e \text{ is } PS \textit{ and } \dot{e} \text{ is } NB \textit{ then } u \text{ is } \langle \text{null} \rangle,$$

will fire. Thus though the cell 25 was empty, because of the fact that $\forall e : e \in PB$ implies that $e \in PS$, there was the rule specified in cell 24 which fired.

However, this is not the case for the rules specified in cells 1, 2 and 6: no matter what the actual input is, these rules will never fire and thus produce an appropriate control output.

It has to be noticed that almost no rule base in practical applications of FKBC is complete. This has to do with the fact that certain regions of the input domain (combinations of error and change-of-error) are not of interest and therefore do not have to be defined. The well known inverted pendulum problem performs optimally with only ten or twelve from twenty-five rules!

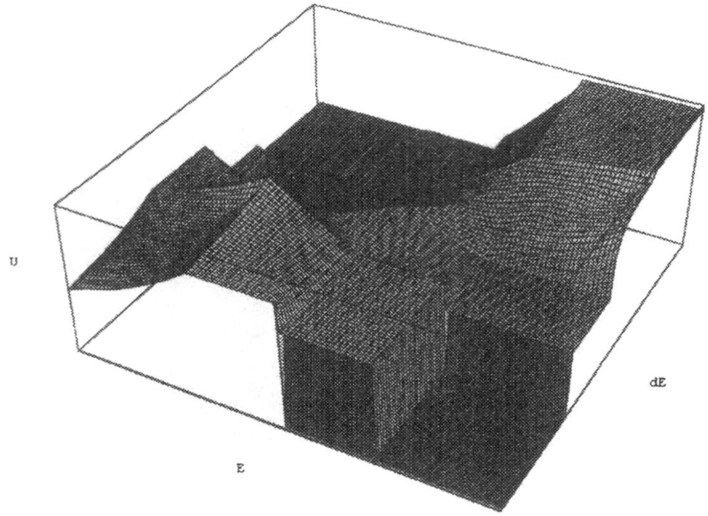

Fig. 2.20. Figure 2.19 rotated through 180°.

2.4.4 Consistency of a Set of Rules

A set of if-then rules is *consistent* if it does not contain contradictions. There exist some contradictions in the literature about consistency. Pedrycz [164] for example, defines consistency in the way we have defined continuity (see below), where others define it as following:

Definition 2.79 *A set of if-then rules is inconsistent if there are two rules with the same rule-antecedent but different rule-consequents.*

For example, a set is inconsistent if the following two rules are in the rule base:

if e is *ZO and* $\dot{e}$ is *ZO then* u is *ZO*,
if e is *ZO and* $\dot{e}$ is *ZO then* u is *NB*.

It can be questioned if the rule base would also have been inconsistent in case *NB* were replaced by *NS* or *PS*, i.e., the fuzzy sets representing their meaning overlap with $\widetilde{ZO}$. An alternative definition therefore would be to say that a rule base is inconsistent if there are two rules, rule k and rule ℓ, with the same rule-antecedents and *mutually exclusive* rule-consequents,

if e is $LE_i^{(k)}$ *and* $\dot{e}$ is $L\dot{E}_j^{(k)}$ *then* u is $LU_m^{(k)}$,
if e is $LE_i^{(k)}$ *and* $\dot{e}$ is $L\dot{E}_j^{(k)}$ *then* u is $LU_n^{(k)}$,

where $\text{hgt}(\widetilde{LU}_m^{(k)} \cap \widetilde{LU}_n^{(k)}) = 0$ (the intersection of the outputs is empty).

It should be noticed that the notion of consistency plays a minor role in fuzzy control. It seldom happens that two rules have the same antecedent. Many computer aided FKBC development systems use matrix editors like that in Fig. 2.18 or explicitly forbid the above situation.

2.4.5 Continuity of a set of rules

Consider again the matrix in Fig. 2.18. We can define the notion of 'neighboring rule' as follows: two rules are neighbors, if their cells are neighbors. So, rule 17 (*if* e is *NS and* $\dot{e}$ is *NS then* u is *PS*) has neighboring rules 12, 16, 18 and 22. It is possible to have also 11, 13, 21 and 23, but we will not do so.

Definition 2.80 *A set of if-then rules is continuous if it does not have neighboring rules with output fuzzy sets that have empty intersection.*

In this case the rule base is discontinuous, while for example (the worst case) rule 23 has output *NB* and its neighbor rule 22 has output *PB*, that clearly have an empty intersection.

2.4.6 Interaction of a set of rules

There is much confusion about *interaction* in the current literature on fuzzy logic. Usually, interaction has the meaning described in the end of Example 2.70, i.e., suppose there is a rule "*if* X is A *then* Y is B," whose meaning is represented by a fuzzy relation $\widetilde{R}$, then the composition of A and R does not deliver B:

$$\widetilde{A} \circ \widetilde{R} \neq \widetilde{B}, \tag{2.220}$$

as one should expect. However, that definition is of little use for control theory, because in fuzzy control one does not deal with fuzzy inputs.

Therefore, we will define interaction in another way. In Section 2.4.1 it was shown that there are two ways to fire a set of rules. The *first way*, composition based inference, is to first combine all the rules into one relation, e.g., the relation $\widetilde{R}_m$ in equation (2.202), and then take the composition of the fuzzified crisp input μ^* and R_m. The *second way*, individual-rule based inference, is to first fire each rule separately with a crisp input e^*, thus obtaining n clipped fuzzy sets $\widetilde{CLU}^{(1)}, \ldots, \widetilde{CLU}^{(n)}$ (see equation (2.205)) and finally combining these fuzzy sets into one (see equation (2.206)). Let us call the first output fuzzy set $\widetilde{U}_{\text{comb}}$ and the second $\widetilde{U}_{\text{indiv}}$, i.e.,

$$\widetilde{U}_{\text{comb}} = e^* \circ \widetilde{R}_m \qquad \widetilde{U}_{\text{indiv}} = \bigcup_k \widetilde{CLU}^{(k)}, \tag{2.221}$$

then we get the following definition.

Definition 2.81 *A set of if-then rules interacts if*

$$\widetilde{U}_{\mathrm{comb}} \neq \widetilde{U}_{\mathrm{indiv}}. \tag{2.222}$$

This means that FKBC which use Mamdani implication will never have interaction.

3. FKBC Design Parameters

3.1 Introduction: The Structure of a FKBC

This chapter introduces the principal design parameters of a FKBC. These include scaling factors, fuzzification and defuzzification methods, rule base and membership function construction and representation. Furthermore, we discuss the relevance of the different design parameters with respect to the performance of a FKBC. Different design options for particular parameters are presented. Choice of membership functions, defuzzification methods, inference engine, and form and meaning of rules are considered in detail. Design parameters like fuzzification, and checking the properties of the rule base are only mentioned, since these were already described in Chapter 2. Other design parameters like scaling factors, derivation of rules, tuning of the FKBC, and stability analysis, are only informally discussed or just mentioned. Thus, this chapter gives a general view of the FKBC design problem and prepares the reader for the formal treatment and/or presentation of systematic techniques for rule derivation and scaling factors determination (Chapter 4), FKBC tuning and adaptation (Chapter 5), and stability analysis (Chapter 6).

The principal structure of a FKBC, as illustrated in Fig. 3.1 consists of the following components.

3.1.1 Fuzzification Module

The fuzzification module (FM) performs the following functions:

- FM-F1: Performs a scale transformation (i.e., an input normalization) which maps the physical values of the current process state variables into a *normalized* universe of discourse (normalized domain). It also maps the normalized value of the control output variable onto its physical domain (output denormalization). When a non-normalized domain is used then there is no need for FM-F1.

- FM-F2: Performs the so-called fuzzification which converts a point-wise (crisp), current value of a process state variable into a fuzzy set, in order to make it compatible with the fuzzy set representation of the process state variable in the rule-antecedent.

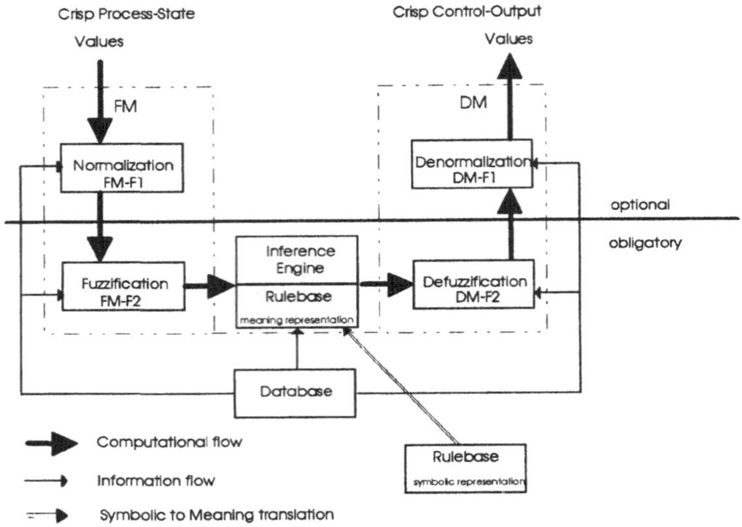

Fig. 3.1. The structure of a FKBC.

The design parameter of the fuzzification module is

* Choice of fuzzification strategy.

The above choice is determined by the type of the inference engine or rule firing employed in the particular application of a FKBC. Thus there are only two choices available: (1) fuzzification in the case when inference is composition based and (2) fuzzification in the case when inference is individual-rule-firing based (see Section 2.4.1).

3.1.2 Knowledge Base

The knowledge base of a FKBC consists of a data base and a rule base.

The basic function of the *data base* is to provide the necessary information for the proper functioning of the fuzzification module, the rule base, and the defuzzification module. This information includes:

- Fuzzy sets (membership functions) representing the meaning of the linguistic values of the process state and control output variables.
- Physical domains and their normalized counterparts together with the normalization/denormalization (scaling) factors.

If the continuous domains of the process state and control output variables have been discretized then the data base also contains information concerning the quantization (discretization) look-up tables defining the discretization policy. For the prevailing case of continuous, normalized domains the design parameters of the data base include:

* Choice of membership functions,
* Choice of scaling factors.

The basic function of the *rule base* is to represent in a structured way the control policy of an experienced process operator and/or control engineer in the form of a set of production rules such as

if ⟨process state⟩ *then* ⟨control output⟩

The *if*-part of such a rule is called the rule-antecedent and is a description of a process state in terms of a logical combination of atomic fuzzy propositions

X is LX according to the notation introduced in Section 2.3.2, or
x is LX according to the notation accepted in the literature on fuzzy control

where x is a linguistic variable and LX is its linguistic value describing a property of x. Sections 2.3.2 and 2.3.3 are devoted to the detailed treatment of linguistic variables and fuzzy propositions. The *then*-part of the rule is called the rule-consequent and is again a description of the control output in terms of a logical combinations of fuzzy propositions. These propositions state the linguistic values which the control output variables take whenever the current process state matches (at least to a certain degree) the process state description in the rule-antecedent. The design parameters involved in the construction of the rule base include:

* Choice of process state and control output variables,
* Choice of the contents of the rule-antecedent and the rule-consequent,
* Choice of term-sets (ranges of linguistic values) for the process state and control output variables,
* Derivation of the set of rules.

3.1.3 Inference Engine

There are two basic types of approaches employed in the design of the inference engine of a FKBC: (1) composition based inference (firing) and (2) individual-rule based inference (firing). Both types of inference are described in detail in Section 2.4.1 and their equivalence with respect to the control output produced is shown. In the sequel we will consider only the second type of inference because of its predominant use in applications of FKBC. The basic function of the inference engine of the second type is to compute the overall value of the control output variable based on the individual contributions of each rule in the rule base. Each such individual contribution represents the values of the control output variables as computed by a single rule. The output of the fuzzification module, representing the current, crisp values of the process state variables, is matched to each rule-antecedent, and a degree of match for each rule is established. This degree of match represents the degree of satisfaction of a fuzzy proposition and was described in Section 2.3.3. Based on this degree of match, the value of the control output variable in the rule-antecedent is modified, i.e., the "clipped" fuzzy set representing the fuzzy value of the control output variable is determined (see Sections 2.3.7 and 2.4.1). The set of all clipped control output values of the matched rules represents the overall fuzzy value of the control output. In this context the design parameters for the inference engine are as follows:

* Choice of representing the meaning of a single production rule,
* Choice of representing the meaning of the set of rules,
* Choice of inference engine,
* Testing the set of rules for consistency and completeness.

The first choice determines uniquely the second one, i.e., once the representation of the meaning of a single rule is decided upon, then there is a unique way of aggregating the individual representations into a global one. The two basic approaches to the representation of the meaning of a set of rules, are based on the Mamdani and Gödel type of representation of the meaning of a single rule and were described in Section 2.4.1. Furthermore, we will not discuss here the testing of a set of rules for consistency and completeness since these were already presented in Section 2.4.2.

3.1.4 Defuzzification Module

The functions of the defuzzification module (DM) are as follows:

- DM-F1: Performs the so-called *defuzzification* which converts the set of modified control output values into a single point-wise value.
- DM-F2: Performs an output *denormalization* which maps the point-wise value of the control output onto its physical domain. DM-F2 is not needed if non-normalized domains are used.

The design parameter of DM is

* Choice of defuzzification operators.

In the rest of this chapter we will present the available choices for the design parameters of a FKBC. We will also discuss the relevance of some of these choices for the performance of the controller. Other design parameters, like normalization and denormalization factors, tuning of membership functions, and the set of rules, are crucial with respect to performance and stability and will be discussed in detail in the next chapters.

3.2 Rule Base

As already said, the design parameters of the rule base include:

* Choice of process state and control output variables,
* Choice of the content of the rule-antecedent and the rule-consequent,
* Choice of term-sets for the process state and control output variables,
* Derivation of the set of rules.

3.2.1 Choice of Variables and Content of Rules

If one has made the choice of designing a P-, PD-, PI- or PID-like FKBC this already implies the choice of process state and control output variables, as well as the content of the rule-antecedent and the rule-consequent for each of the rules. The process state variables representing the contents of the rule-antecedent (if-part of a rule) are selected amongst

- error, denoted by e,
- change-of-error, denoted by Δe or $\dot{e}$,
- sum-of-errors, denoted by δe.

The control output (process input) variables representing the contents of the rule-consequent (then-part of the rule) are selected amongst

- change-of-control output, denoted by Δu or $\dot{u}$,
- control output, denoted by u.

Furthermore, by analogy with a conventional controller, we have that

- $e(k) = y_{sp} - y(k)$,
- $\Delta e(k) = e(k) - e(k-1)$,
- $\Delta u(k) = u(k) - u(k-1)$.

In the above expressions, y_{sp} stays for desired process output or set-point: y is the process output variable (control variable); k is the sampling time.

PD-like FKBC

The equation giving a conventional PD-controller is

$$u = K_P \cdot e + K_D \cdot \dot{e}, \tag{3.1}$$

where K_P and K_D are the proportional and the differential gain coefficients. Then a PD-like FKBC consists of rules, the symbolic description of each rule given as

> *if* $e(k)$ is ⟨property symbol⟩ *and* $\Delta e(k)$ is ⟨property symbol⟩
> *then* $u(k)$ is ⟨property symbol⟩

where ⟨property symbol⟩ is the symbolic name of a linguistic value. The natural language equivalent of the above symbolic description reads as follows:

> For each sampling time k,
> *if* the value of error has the property of being ⟨linguistic value⟩ and the value of change-of-error has the property of being ⟨linguistic value⟩
> *then* the value of control output has the property of being ⟨linguistic value⟩

For the sake of simplicity, we will omit the explicit reference to sampling time k since such a rule expresses a causal relationship between process state and control output variables which holds for any sampling time k. Thus the final symbolic representation of the above rule is

> *if* e is ⟨property symbol⟩ *and* Δe is ⟨property symbol⟩
> *then* u is ⟨property symbol⟩.

PI-like FKBC

The equation giving a conventional PI-controller is

$$u = K_P \cdot e + K_I \cdot \int e \, \mathrm{d}t, \tag{3.2}$$

where K_P and K_I are the proportional and the integral gain coefficients. When the derivative, with respect to time, of the above expression is taken, it is transformed into an equivalent expression

$$\dot{u} = K_P \cdot \dot{e} + K_I \cdot e. \tag{3.3}$$

The PI-like FKBC consists of rules of the form

> *if* e is ⟨property symbol⟩ *and* Δe is ⟨property symbol⟩
> *then* Δu is ⟨property symbol⟩.

In this case, to obtain the value of the control output variable $u(k)$, the change-of-control output $\Delta u(k)$ is added to $u(k-1)$. It is to be stressed here that this takes place outside the PI-like FKBC, and is not reflected in the rules themselves.

P-like FKBC

The symbolic representation of a rule for a P-like FKBC is given as

if e is ⟨property symbol⟩
then u is ⟨property symbol⟩.

PID-like FKBC

The equation describing a conventional PID-controller is

$$u = K_P \cdot e + K_D \cdot \dot{e} + K_I \cdot \int e \, dt. \tag{3.4}$$

Thus, in the discrete case of a PID-like FKBC one has an additional process state variable, namely

- sum-of-errors, denoted δe and computed as

$$\delta e(k) = \sum_{i=1}^{k-1} e(i). \tag{3.5}$$

Then the symbolic expression for a rule of a PID-like FKBC is

if e is ⟨property symbol⟩ *and* Δe is ⟨property symbol⟩ *and* δe is ⟨property symbol⟩
then u is ⟨property symbol⟩

In some cases, when knowledge about the parameter estimation of the process and its structure is available, one may not want to confine himself to the use of error, change-of-error, and sum-of-errors as process state variables, but rather use the actual process state variables. The symbolic expression for a rule in the case of multiple inputs and a single output (MISO) is as follows

if x_1 is ⟨property symbol⟩ *and* ... *and* x_n is ⟨property symbol⟩
then u is ⟨property symbol⟩

Rules of this if-then type are usually derived from a fuzzy process model. Details concerning the derivation of a FKBC employing such rules are given in Section 5.3.4. Another, principally different type of rule is symbolically represented as

if x_1 is ⟨property symbol⟩ *and* ... *and* x_n is ⟨property symbol⟩
then $u = f(x_1, \ldots, x_n)$

where f is a linear function of the process state variables x_i ($i = 1, ..., n$). The type of a FKBC employing the latter type of rules is called a Sugeno controller and is considered in detail in (Section 4.3.3). Here we will only briefly describe the problem of a FKBC design via the use of a fuzzy process model. We consider the multiple-input single-output case without any loss of generality.

Let $\mathbf{x} = (x_1, ..., x_n)$ be a vector of process state variables, y be the process output variable, and u be the process input variable (control variable). Furthermore, let $\mathcal{X} = \mathcal{X}_1 \times \ldots \times \mathcal{X}_n$, $\mathcal{Y}$, and $\mathcal{U}$ be the domains of $\mathbf{x}$, y, and u respectively. The conventional closed-loop model, when linearized around the set-point, is given as

$$\mathbf{x}(k+1) = A \cdot \mathbf{x}(k) + \mathbf{b}^T \cdot u(k) \tag{3.6}$$

$$y(k) = \mathbf{c}^T \cdot \mathbf{x}(k) \tag{3.7}$$

$$u(k) = k \cdot y(k), \tag{3.8}$$

where A is the process matrix, $\mathbf{b}$ and $\mathbf{c}$ are vectors, and k is a scalar. These state equations can be rewritten in the following form

$$\mathbf{x}(k+1) = A \cdot \mathbf{x}(k) + \mathbf{b}^T \cdot u(k) \tag{3.9}$$

$$u(k) = k \cdot \mathbf{c}^T \cdot \mathbf{x}. \tag{3.10}$$

Now the fuzzy counterpart of the above model is defined as follows. Let the linguistic variable (see Section 2.3.2) x_i have the term set $\mathcal{L}X_i$. An arbitrary element of this term-set will be denoted by LX_i. Furthermore, the meaning of a linguistic value LX_i is given by a membership function (fuzzy set) $\widetilde{LX}_i$. Thus, the linguistically defined process state vector is denoted by, $\widetilde{LX} = (\widetilde{LX}_1, ..., \widetilde{LX}_n)$. In a similar manner, y takes linguistic values $LY \in \mathcal{L}Y$ with corresponding membership functions $\widetilde{LY}$, and u takes linguistic values $LU \in \mathcal{L}U$. The meaning of the latter linguistic values is represented by membership functions $\widetilde{LU}$. Then the following equations describe the fuzzy model of the closed-loop system

$$\widetilde{LX}(k+1) = \left(\widetilde{LX}(k) \times (\widetilde{LU}(k)\right) \circ \widetilde{A} \tag{3.11}$$

$$\widetilde{LU}(k) = \widetilde{LX}(k) \circ \widetilde{K}, \tag{3.12}$$

where

- $\widetilde{A}$ is a fuzzy relation on $\mathcal{X} \times \mathcal{U} \times \mathcal{X}$ and is called the fuzzy process transition map. $\widetilde{A}$ can be obtained in an explicit form by on/off-line identification as in [224], or is the fuzzy relation giving the overall meaning of a set of production rules, each rule having the following symbolic representation:

  if $x_1(k)$ is LX_1 *and* ... *and* $x_n(k)$ is LX_k *and* $u(k)$ is LU
  then $x_1(k+1)$ is LX_1 *and* ... *and* $x_n(k+1)$ is LX_k

- $\widetilde{K}$ is the controller which is a fuzzy relation on $\mathcal{X} \times \mathcal{Y}$ representing the meaning of a set of *if-then* rules, each rule having the symbolic representation

  if $x_1(k)$ is LX_1 *and* ... *and* $x_n(k)$ is LX_k *then* $u(k)$ is LU

- $\circ$ is the composition operation as defined in Section 2.2.3.

If we replace $\widetilde{LU}(k)$ by $\widetilde{LX}(k) \circ \widetilde{K}$ we obtain that

$$\widetilde{LX}(k+1) = \left(\widetilde{LX}(k) \times (\widetilde{LU}(k)\right) \circ \tilde{A} = \left(\widetilde{LX}(k) \times \left(\widetilde{LX}(k) \circ \widetilde{K}\right)\right) \circ \tilde{A} \quad (3.13)$$

Now using the following lemma: If $\widetilde{LA}$ and $\widetilde{LB}$ are fuzzy sets on the domains $\mathcal{A}$ and $\mathcal{B}$ respectively, and $\tilde{R}$ is a fuzzy relation on $\mathcal{A} \times \mathcal{B} \times \mathcal{A}$, then

$$(\widetilde{LA} \times \widetilde{LB}) \circ \tilde{R} = \widetilde{LA} \circ (\widetilde{LB} \circ \tilde{R}) = \widetilde{LB} \circ (\widetilde{LA} \circ \tilde{R}), \quad (3.14)$$

we can rewrite the above equation as

$$\widetilde{LX}(k+1) = (\widetilde{LX}(k) \circ \widetilde{K}) \circ ((\widetilde{LX}(k) \circ \tilde{A}), \quad (3.15)$$

since $\widetilde{LX}(k)$ is a membership function defined on $\mathcal{X}$; the result of $\widetilde{LX}(k) \circ \widetilde{K}$ is a membership function on $\mathcal{U}$ because $\widetilde{LX}(k)$ is defined on $\mathcal{X}$ and $\widetilde{K}$ on $\mathcal{X} \times \mathcal{U}$; $\tilde{A}$ is a fuzzy relation defined on $\mathcal{X} \times \mathcal{U} \times \mathcal{X}$. Now let us rewrite the last equation as

$$\widetilde{LX}(k+1) = (\widetilde{LX}(k) \times \widetilde{LX}(k)) \circ \widetilde{K} \circ \tilde{A} = (\widetilde{LX}(k) \times \widetilde{LX}(k)) \circ (\widetilde{K} \circ \tilde{A}). \quad (3.16)$$

Then the FKBC design problem is stated as follows. Let $\tilde{T}$ be a fuzzy relation on $\mathcal{X} \times \mathcal{U} \times \mathcal{X}$ such that

$$\widetilde{LX}^{\text{desired}}(k+1) = (\widetilde{LX}(k) \times \widetilde{LX}(k)) \circ \tilde{T}. \quad (3.17)$$

In other words, $\tilde{T}$ helps to achieve the next, desired process state $\widetilde{LX}^{\text{desired}}(k+1)$. Thus, if we can construct such a $\tilde{T}$ (see [224] for a possible approach to $\tilde{T}$'s construction) we have to find a $\widetilde{K}$ such that

$$\widetilde{K} \circ \tilde{A} = \tilde{T}, \quad (3.18)$$

where $\widetilde{K}$ is the unknown fuzzy relation representing the FKBC. In Sections 5.3.3 and 5.3.4 the derivation of K from a fuzzy process model is described.

3.2.2 Choice of a Term Set

In Section 2.3.2 the term set $\mathcal{L}X$ of a linguistic variable X was described as consisting of a finite number of verbally (linguistically) expressed values which X can take. The linguistic values, members of the term set, are expressed as tuples of the form ⟨value sign, value magnitude⟩, e.g., ⟨positive big⟩, ⟨negative small⟩, etc. The value sign component of such a tuple takes on either one of the following two values: positive or negative. The value magnitude component can take on any number of linguistically expressed magnitudes, e.g., {zero, small, medium, big}, or {zero, small, big}, or {zero, very small, small, medium, big, very big}, etc. The meaning of a tuple, in the case of a PI-like FKBC, can be summarized as follows:

- Linguistic values of e with a negative sign mean that the current process output y has a value below the set-point y_{sp} since $e(k) = y_{sp} - y(k) < 0$. The magnitude of a negative value describes the magnitude of the difference $y_{sp} - y$. On the other hand, linguistic values of e with a positive sign mean that the current value of y is above the set-point. The magnitude of such a positive value is the magnitude of the difference $y_{sp} - y$.

- Linguistic values of Δe with a negative sign mean that the current process output $y(k)$ has increased when compared with its previous value $y(k-1)$ since $\Delta e(k) = -(y(k) - y(k-1)) < 0$. The magnitude of such a negative value is given by the magnitude of this increase. Linguistic values of $\Delta e(k)$ with a positive sign mean that $y(k)$ has decreased its value when compared to $y(k-1)$. The magnitude of such a value is the magnitude of the decrease.

- A linguistic value of "zero" for e means that the current process output is at the set-point. A "zero" for Δe means that the current process output has not changed from its previous value, i.e., $-(y(k) - y(k-1)) = 0$.

- Linguistic values of $\Delta u(k)$ with a positive sign mean that the value of the control output $u(k-1)$ has to be increased to obtain the value of the control output for the current sampling time k. A value with a negative sign means a decrease in the value of $u(k-1)$. The reason for this is that $u(k) = u(k-1) + \Delta u(k)$. The magnitude of a value is the magnitude of increase/decrease of the value of $u(k-1)$.

Suppose now that the term sets of e, Δe, and Δu all have been chosen equal in size and contain the same linguistic expressions for the magnitude part of the linguistic values, i.e., $\mathcal{L}E = \mathcal{L}\Delta E = \mathcal{L}\Delta U = \{NB, NM, NS, ZE, PS, PM, PB\}$ where, for example, *NB* reads "negative big" and *PS* reads "positive small". Let us return now to the rule base of a PI-like FKBC and describe the operational meaning of the if-then rules. We will present the rule base in the table format shown in Fig. 3.2. The cell defined by the intersection of the first row and the first column represents a rule such as,

if $e(k)$ is *NB* and $\Delta e(k)$ is *NB*
then $\Delta u(k)$ is *NB*

Now the set of rules can be divided in the following five groups:

- **Group 0**: In this group of rules both e and Δe are (positive or negative) small or zero. This means that the current value of the process output variable y has deviated from the set-point but is still close to it. The amount of change $\Delta u(k)$, in the previous control output $u(k-1)$ prescribed by these rules is also small or zero in magnitude and is intended to correct small deviations from the set-point. Therefore, the rules in this group are related to the steady-state behavior of the process.

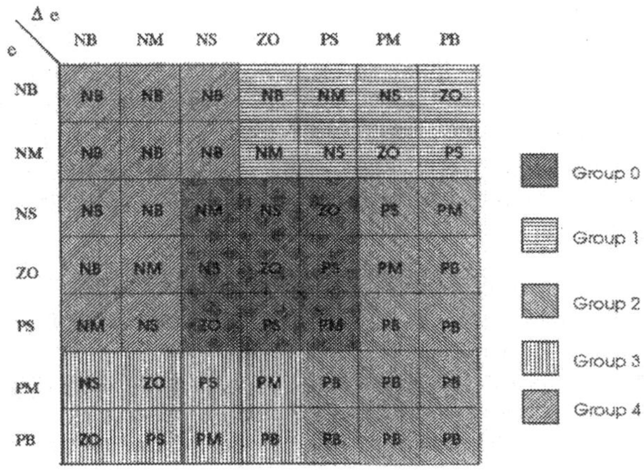

Fig. 3.2. The rule base of a PI-like FKBC in tabular form: the five groups of rules.

- **Group 1**: For this group of rules e is negative big or medium which implies that $y(k)$ is significantly above the set-point. At the same time, since $\Delta e(k)$ is positive, this means that y is moving towards the set-point. The amount of change Δu which the rules of this group introduce to the previous control output $u(k-1)$ is intended to either speed up or slow down the approach to the set-point. For example, if $y(k)$ is much above the set-point ($e(k)$ is NB) and it is moving towards the set-point with a small step ($\Delta e(k)$ is PS) then the magnitude of this step has to be significantly increased ($\Delta u(k)$ is NM).

- **Group 2**: For this group of rules e is either close to the set-point (PS, ZO, NS) or is significantly below it (PM, PB). At the same time, since Δe is positive, y is moving away from the set-point. Thus a positive change $\Delta u(k)$ in the previous control output $u(k-1)$ is intended to reverse this trend and make y, instead of moving away from the set-point, to start moving towards it.

- **Group 3**: For this group of rules e is positive medium or big which means that $y(k)$ is significantly below the set-point. At the same time, since Δe is negative, y is moving towards the set-point. The amount of change Δu which the rules of this group introduce to the previous control output $u(k-1)$ is intended to either speed up or slow down the approach to the set-point. For example, if $y(k)$ is much below the set-point ($e(k)$ is PB) and it is moving towards the set-point with a somewhat large step ($\Delta e(k)$

is *NM*) then the magnitude of this step has to be only slightly enlarged ($\Delta u(k)$ is *PS*).

- **Group 4**: For this group of rules e is either close to the set-point (*PS*, *ZO*, *NS*) or is significantly above it (*NM*, *NB*). At the same time, since Δe is negative, y is moving away from the set-point. Thus a negative change $\Delta u(k)$ in the previous control output $u(k-1)$ is intended to reverse this trend and make y, instead of moving away from the set-point, start moving towards it.

In the context of the above five groups of rules, we can see that the size of the term set determines the granularity of the control action of the FKBC. If one desires better control resolution around the set-point then one can consider a larger range of linguistic values for *NS*, *ZO*, and *PS*, e.g., *PZO* for "positive zero," *NZO* for "negative zero", *PVS* for "positive very small," and *NVS* for "negative very small." However, one should be aware of the fact that the use of term sets with larger size leads to an increase in the number of rules. For example, if each of the term sets of e and Δe has 10 elements then the maximum number of rules in the rule base is 100.

3.2.3 Derivation of Rules

There are three major approaches to the derivation of the rules of a FKBC. These three approaches complement each other and it seems that a combination of them would be necessary to construct an effective approach to the derivation of rules.

- **Approach 1**: This approach is the one that is most widely used today. It is based on the derivation of rules from the experience-based knowledge of the process operator and/or control engineer. The approach is realized using two types of techniques.

  1. An introspective verbalization of experience-based knowledge.
  2. Interrogation of a process operator and/or control engineer using a carefully organized questionnaire.

  Both techniques help in providing an initial prototype version of the rule base, specific for a particular application domain. Consequent tuning of membership functions and rules is a necessary next step. For example, in the case of a PI-like FKBC, an initial set of rules, selected using either one of the above two techniques, is implemented on the process under control. The subsequent response of the closed-loop system to non-aperiodic disturbance is monitored in the phase plane. The latter is composed of error and change of error, and the obtained trajectory is used to update the set of initial rules. This procedure is then repeated as often as required until an acceptable response is obtained. Sections 5.3.1–5.3.3 are devoted to rule base and membership function tuning.

- **Approach 2**: This approach, still in its early stages, uses a linguistic description, viewed as a fuzzy model of the process under control, to derive the set of rules of a FKBC. The general controller design ideas using such a fuzzy process model were already described in the previous subsection. Based on the fuzzy process model, either the set of rules or an explicit fuzzy relation describing the FKBC is derived. In Section 4.3.3 the derivation of a set of rules from a Sugeno-type fuzzy process model is described. In Section 5.3.4 a different approach to the derivation of a FKBC from a fuzzy process model is presented. Work on the identification of a fuzzy process model is to be found in [163, 162].

- **Approach 3**: This approach, also in its early development stages, relies on the existence of a conventional process model, usually a nonlinear one. A well developed formal technique which uses a "fuzzy" version of the so-called Sliding Mode Control is presented in detail in Chapter 4.

3.3 Data Base

The two design parameters involved in the construction of the data base are

* Choice of membership functions,
* Choice of scaling factors.

3.3.1 Choice of Membership Functions

The linguistic values taken by the variables in the rule-antecedent and rule-consequent, and the symbolic representation of the rules are good enough to allow some qualitative analysis concerning the stability of the closed-loop system [30]. However, for the needs of a quantitative description of the behavior of the closed-loop system, involving the computation of the quantitative control output, one needs a quantitative interpretation of the meaning of the linguistic values (see Section 6.2).

Let the physical domains of e, Δe, and Δu be $\mathcal{E}$, $\Delta\mathcal{E}$, and $\Delta\mathcal{U}$. respectively. The elements of these domains will be denoted by e, Δe, and Δu. As already described in Section 2.3.2 the meaning or interpretation of a particular linguistic value LX of a linguistic variable x is given by a fuzzy set $\widetilde{LX}$ or μ_{LX} defined on the domain (universe of discourse) $\mathcal{X}$ of x as

$$\widetilde{LX} = \mu_{LX} = \int_{\mathcal{X}} \mu_{LX}(x)/x \tag{3.19}$$

Now suppose that $\mathcal{L}E = \mathcal{L}\Delta E = \mathcal{L}\Delta U = \{NB, NM, NS, ZO, PS, PM, PB\}$, i.e., the term sets containing the linguistic values for the three linguistic variables are the same. In this case we have to define twenty-one membership functions representing the meaning of each linguistic value from the above term

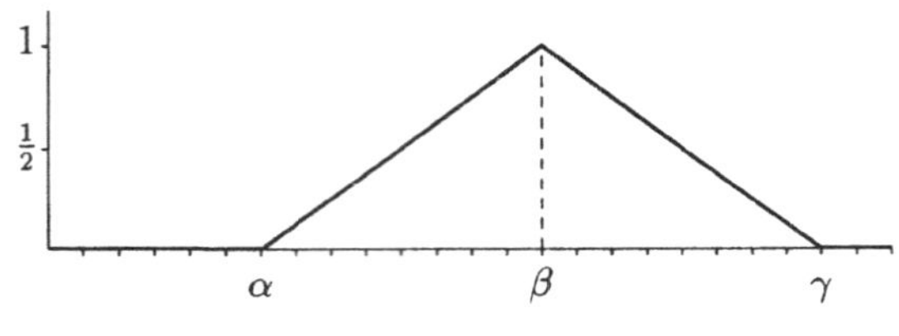

Fig. 3.3. A triangular membership function $\Lambda(u; \alpha, \beta, \gamma)$ (see Section 2.1.4).

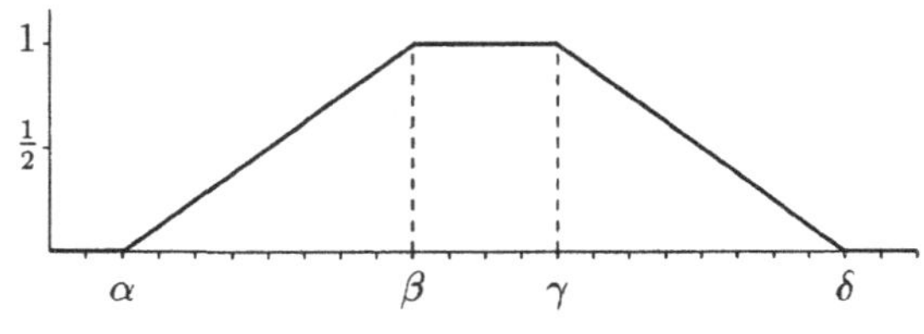

Fig. 3.4. A trapezoidal membership function $\Pi(u; \alpha, \beta, \gamma, \delta)$ (see Section 2.1.4).

set on the respective domains $\mathcal{E}$, $\Delta\mathcal{E}$, and $\Delta\mathcal{U}$. For computational efficiency, efficient use of memory, and performance analysis needs, a uniform representation of the membership functions is required. This uniform representation can be achieved by employing membership functions with uniform shape and parametric, functional definition.

The most popular choices for the shape of the membership function include triangular-, trapezoidal-, and bell-shaped functions. These three choices can be explained by the ease with which a parametric, functional description of the membership function can be obtained, stored with minimal use of memory, and manipulated efficiently, in terms of real-time requirements, by the inference engine. Figures 3.3–3.5 depict the above three choices for the membership functions.

It is easy to see that the parametric, functional description of the triangular-shaped membership function is the most economic one. This explains the predominant use of this type of membership functions.

Once the shape of the membership function is selected, one has to map each element of the term set on the domain of the corresponding linguistic variable. For example, this mapping in the case of e and $\mathcal{E} = [-6, +6]$ would be as shown in Fig. 3.6.

The mapping of the linguistic values of a variable on its domain can affect the performance of the FKBC in a number of ways. Before we discuss this in detail, let us first introduce some parameters which characterize a mem-

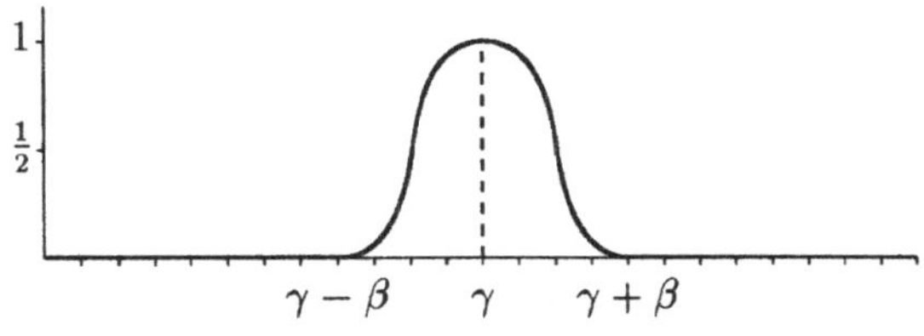

Fig. 3.5. A bell-shaped membership function $\pi(u;\beta,\gamma)$ (see Section 2.1.4).

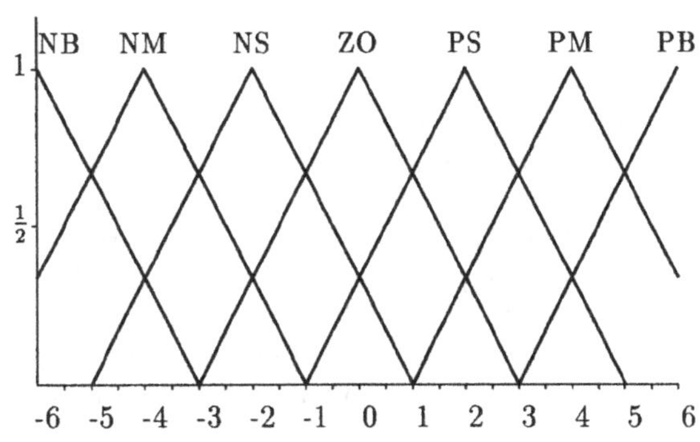

Fig. 3.6. The mapping of the term set of error on a normalized domain $[-6,6]$.

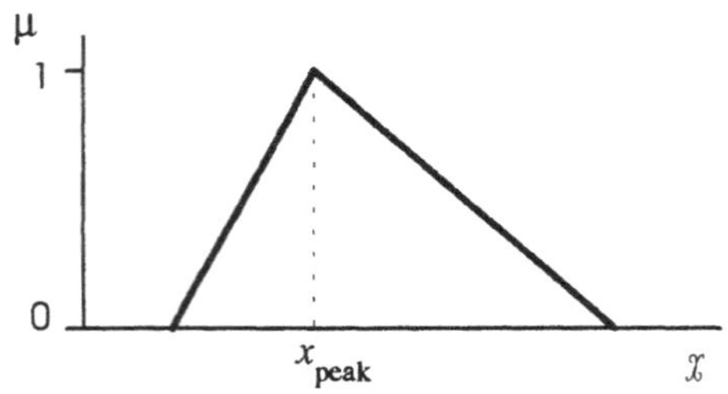

Fig. 3.7. The peak value of a triangular membership function.

bership function. We will consider here only triangular membership functions, but the parameters described and discussed with respect to how they influence the FKBC performance are relevant for any membership function (see Section 2.1.5).

Peak Value

Let X be a linguistic variable and let the meaning of its linguistic value LX be represented by a membership function $\mu_{LX} : \mathcal{X} \to [0,1]$ where $\mathcal{X}$ is the domain of X. The peak value of μ_{LX} is then this value from the domain $\mathcal{X}$ whose degree of membership is 1, i.e., $\mu_{LX}(x_{\text{peak}}) = 1$. Figure 3.7 illustrates the peak value of a triangular membership function. In the case of trapezoidal-shaped membership functions the peak value is an interval.

Left and Right Width

The left width of μ_{LX} is the length of the interval from the peak value to that value in $\mathcal{X}$ which is situated to the left of the peak value and whose degree of membership is zero. Similarly, the right width is the length of the interval from the peak value to that value in $\mathcal{X}$ situated to the right of the peak value and whose degree of membership is zero. The sum of the left and right widths defines the length of the interval, which is called the support of μ_{LX}. If the left width is equal to the right width then the membership function is symmetric, otherwise it is asymmetric. Figure 3.8 illustrates the above notions for a triangular membership function.

Crosspoints

Let μ_{LX^1} and μ_{LX^2} be two membership functions representing the meaning of two different linguistic values of X which are elements of the term set $\mathcal{L}X$. A cross-point between μ_{LX^1} and μ_{LX^2} is that value x_{cross} in $\mathcal{X}$ such that $\mu_{LX^1}(x_{\text{cross}}) = \mu_{LX^2}(x_{\text{cross}}) > 0$. Furthermore, a cross-point level is defined by the degree of membership of x_{cross}, i.e., $\mu_{LX^1}(x_{\text{cross}})$ which, by the definition of

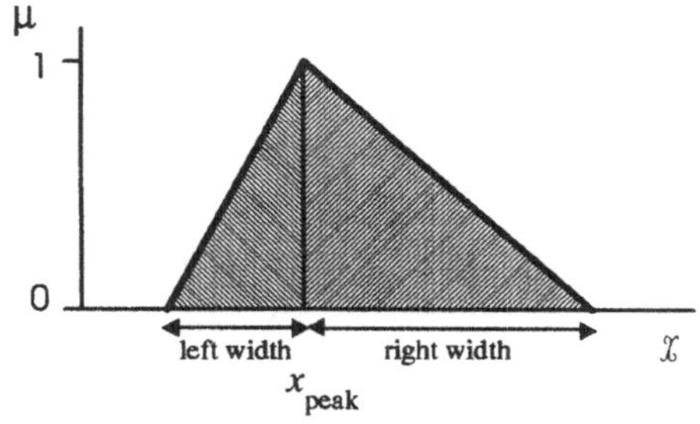

Fig. 3.8. The left and right width of a triangular membership function.

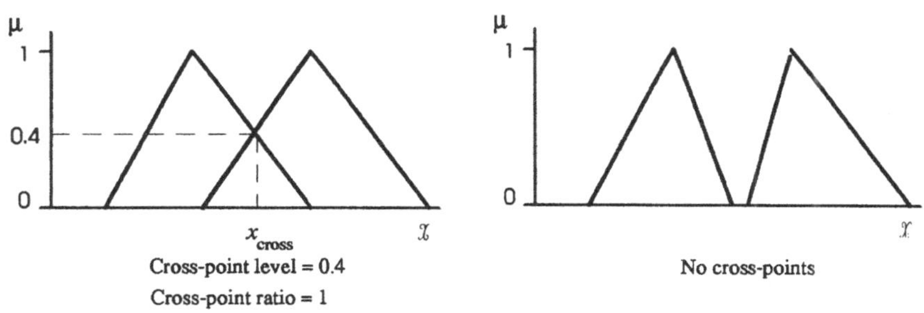

Fig. 3.9. Crosspoints and cross-point levels of triangular membership functions.

the cross-point, is the same as $\mu_{LX^2}(x_{\text{cross}})$. It may be the case that two membership functions defining the meaning of two different linguistic values may have more than one cross-point. The number of cross-points between two membership functions is called the cross-point ratio. Figure 3.9 illustrates the above notions in the case of triangular membership functions.

Now we are ready to discuss how the above described membership function parameters influence the performance of a FKBC.

Influence of Crosspoint Level

First, the mapping from a term set, say $\mathcal{LE}$, to membership functions defined on the domain of error $\mathcal{E}$ has to be such that the cross-point level for every two membership functions is greater than zero. This means that every crisp value of error belongs to at least one membership function with degree of membership strictly greater than zero. If this is not the case then there will be a crisp, input value of error which cannot be matched, during the fuzzification phase, to a rule-antecedent. Thus, none of the rules will fire and consequently, no value for the control output will be computed. This in turn leads to discontinuities

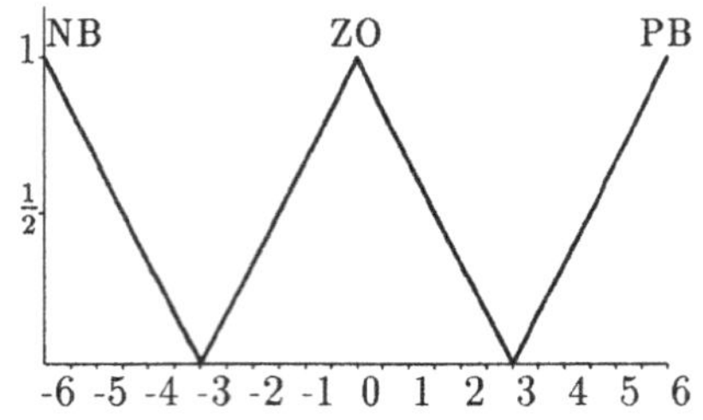

Fig. 3.10. The cross-point ratio between membership functions is zero.

in the control output. Furthermore, if the cross-point ratio between every two membership functions is zero then only one rule at a time can fire. This situation is depicted in Fig. 3.10.

Second, as shown in Boverie et al. [27], for linear systems up to third order and for symmetrical membership functions there are some "optimal" values for the cross-point level and ratio. If each two adjacent membership functions have a cross-point level of 0.5 and a cross-point ratio of 1, then this provides for significantly less overshoot, faster rise-time, and less undershoot. The shape of the membership functions does not play a significant role, but trapezoidal functions are responsible for a slower rise time. Though these results have an empirical character, the choice of cross-point level of 0.5 and a cross-point ratio of 1 is the usual choice reported elsewhere in the literature.

Influence of Symmetry and Width

Inference performed with a single rule (single rule firing) is based on the following idea. For the sake of simplicity consider a P-like FKBC, that is, the process state variable is error e and the control output variable is u. Let the current process state e^* match to some degree α the rule-antecedent μ_{LE}. Then the actual value of the control output should match to the same degree the rule-consequent, i.e., μ_{LU}. That is why, as already described in Section 2.4.1, the membership function of the control output from the rule-consequent is modified in proportion to α. The modified μ_{LU} is called the clipped μ_{LU}, denoted by μ_{CLU}, and is such that $\mu_{CLU} : \mathcal{E} \to [0, \alpha]$, where α is the degree of match of the rule-antecedent with the current value of e.

Now let the n-th rule of the FKBC have a rule-consequent such as "u is NM" and its meaning represented by μ_{NM} as given in Fig. 3.11.

Now if the rule-antecedent is satisfied or matched to, say, a degree of 0.7 then there are two values in the domain of u which match to degree 0.7 the membership function μ_{NM}, i.e., u_1 situated to the left of the peak value $u_{\mathbf{peak}}$ and u_2 situated to the right of $u_{\mathbf{peak}}$. Since only a single value of u can be passed on to compute the next process input, there is a need to compute this single value from u_1 and u_2. One way to do this is to simply take the average of u_1

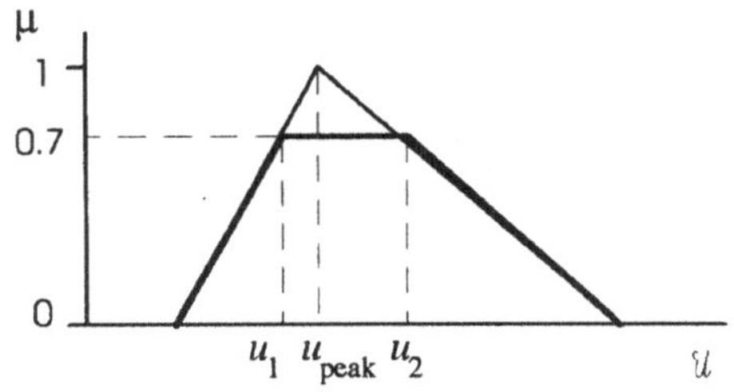

Fig. 3.11. The membership function of "negative medium" and its clipped version. The "clipping" is done with respect to a 0.7 degree of match.

and u_2, i.e.,

$$u = \frac{u_1 + u_2}{2}, \tag{3.20}$$

The so-called Center-of-Gravity/Area method usually used to obtain a single value of the control output expands the above averaging operation by including every element of $\mathcal{U}$ and its corresponding degree of membership to μ_{CNM}, where $\mu_{CNM} : \mathcal{U} \to [0, 0.7]$. Thus we have that

$$u_{\text{cog}} = \frac{\int_{\mathcal{U}} \mu_{CNM}(u) \cdot u \, du}{\int_{\mathcal{U}} \mu_{CNM}(u) \, du}. \tag{3.21}$$

Consider the case when the degree of match of the rule-antecedent is 1 as illustrated in Fig. 3.12. Then since μ_{CNM} has a triangular shape, there is only one element of $\mathcal{U}$ which is satisfied to degree 1, namely u_{peak}. Thus it makes sense to take u_{peak} as the actual value of u. On the other hand, for the degree of match 1, we have that $\mu_{CNM} = \mu_{NM}$. Now if we take the Center-of-Gravity of μ_{CNM} we will obtain that $u_{\text{cog}} \neq u_{\text{peak}}$ as shown in Fig. 3.12.

This clash between the underlying idea of inference with a single rule and the result produced by the Center-of-Gravity method can be resolved if μ_{NM} is chosen to be symmetrical around its peak value. Then we would obtain that $u_{\text{cog}} = u_{\text{peak}}$. This is the main reason why it is recommended that the membership functions describing the meaning of the linguistic values in the rule-consequent should be symmetrical.

To illustrate the influence of the width of the membership functions employed for representing the meaning of the linguistic values in the rule antecedent, let us consider the following example of a P-like FKBC with two rules:

if e is *PM* *then* u is *PB*
if e is *PS* *then* u is *PM*

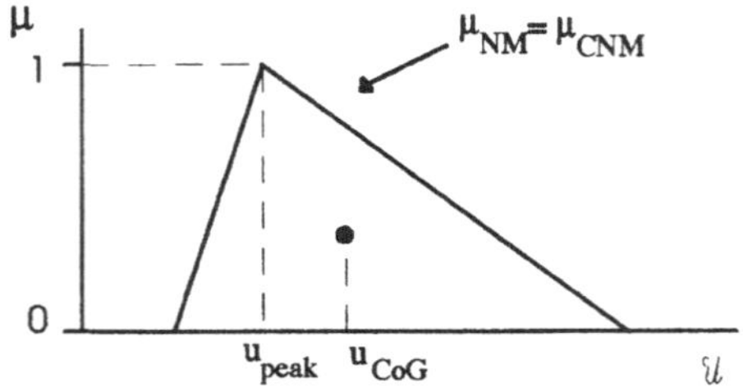

Fig. 3.12. When the degree of match is 1 then the original membership function and its "clipped" version are the same. In the case of asymmetric membership functions the peak value is different from the Center-of-Gravity.

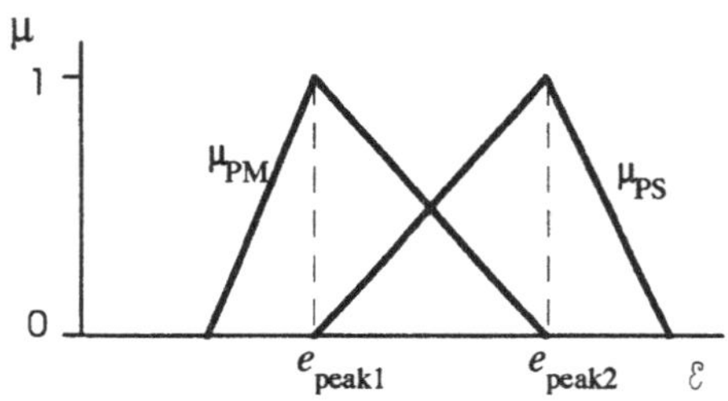

Fig. 3.13. Version 1 of the membership functions representing the meaning of the rule-antecedents e is *PM* and e is *PS*.

and let *PM* and *PS* belong to the term set $\mathcal{L}E$ of e, and *PB* and *PM* belong to the term set $\mathcal{L}\Delta U$ of u. Furthermore, let the membership functions representing the meaning of the linguistic values *PM* and *PS* in the rule-antecedents be denoted by μ_{PM} and μ_{PS} and given graphically in two different versions as in Fig. 3.13 and Fig. 3.14.

The only difference between version 1 and version 2, representing two different versions of μ_{PM} and μ_{PS}, is that in version 1 the following condition is satisfied:

- **condition width**: The left-width of μ_{PS} is equal to the right-width of μ_{PM} and they are both equal to the length of the interval between the peak values of the two adjacent membership functions.

In version 2 the above condition is not satisfied. In the case of version 1 when e changes smoothly from $e_{\mathbf{peak1}}$ to $e_{\mathbf{peak2}}$, and after inference and consequent application of the Center-of-Gravity method, one obtains that u also changes

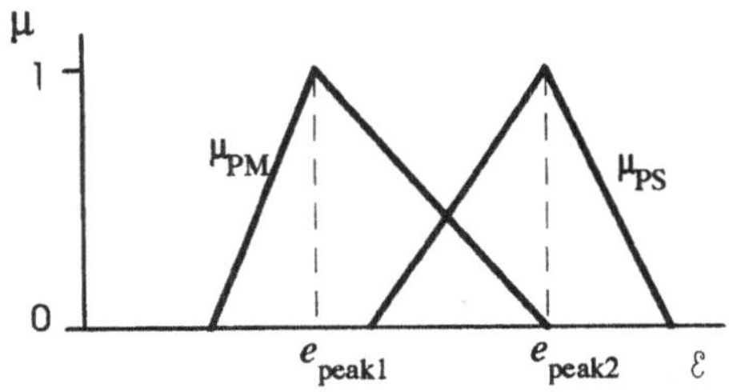

Fig. 3.14. Version 2 of the membership functions representing the meaning of the rule-antecedents *e* is *PM* and *e* is *PS*.

smoothly from u_{peak1} to u_{peak2} as illustrated in Fig. 3.15. In the case of version 2, u changes step-wise from u_{peak1} to u_{peak2}. This is illustrated in Fig. 3.16.

This is the reason why the above mentioned condition is required to hold for any two adjacent membership functions describing the meaning of the linguistic values in the rule-antecedent.

Noise

The actual values of e, Δe and Δu may be subject to external disturbances in the form of noisy signals (bias, sine wave, random noise, etc.). The experience has shown that in such a case the membership functions representing the meaning of the linguistic values of e, Δe and Δu should not be constructed without reference to the probability distribution of the noise. However, this problem has not so far been studied in any satisfactory formal manner. There are some rather heuristic guidelines presented in [29].

Let $e(t)$ be a stationary and ergodic (Gaussian) process with mean

$$E[e(t)] = \lim_{t\to\infty} \frac{1}{2T} \cdot \int_{-T}^{T} e(t)\, dt, \tag{3.22}$$

and variance

$$\sigma^2[e(t)] = \frac{1}{2T} \cdot \int_{-T}^{T} (e(t) - E[e(t)])^2\, dt. \tag{3.23}$$

The above two notions are represented graphically in Fig. 3.17. Let also $I = [-a, a]$ be the normalized domain on which the membership functions describing the linguistic values of error are defined.

Now $E[e(t)]$ should be placed in the middle of the interval $[-a, a]$; thus the width of the domain of error $[-a, a]$ is

$$\text{width} = a - (-a) = 4 \cdot \sigma, \tag{3.24}$$

because the probability that a measurement e^* falls into the domain $[-a, a]$ is $\text{Prob}(e^*) = 0.95$.

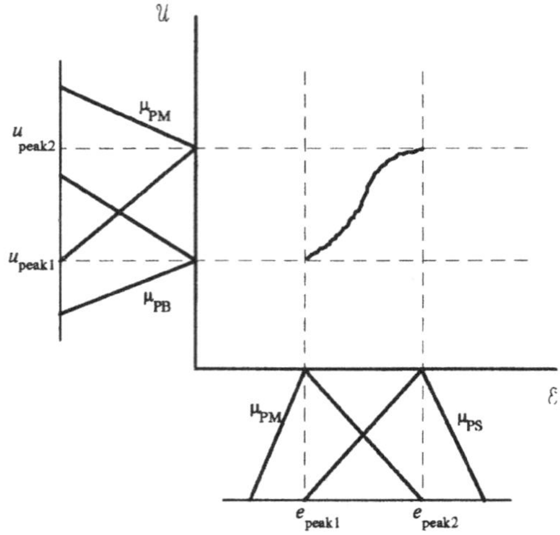

Fig. 3.15. The smooth (continuous) change in control output when condition width is satisfied.

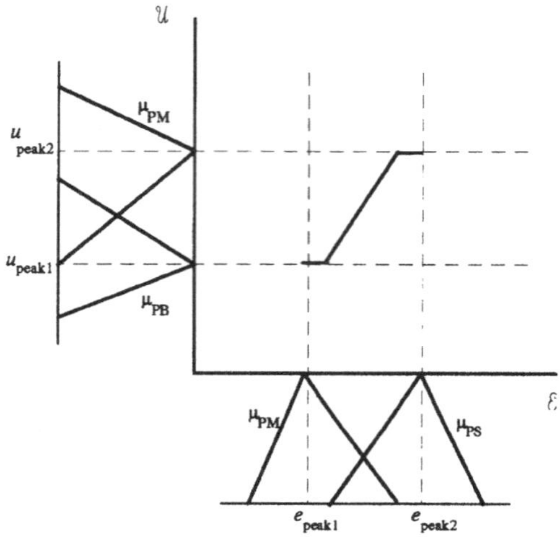

Fig. 3.16. The step-wise (discontinuous) change in control output when condition width is not satisfied.

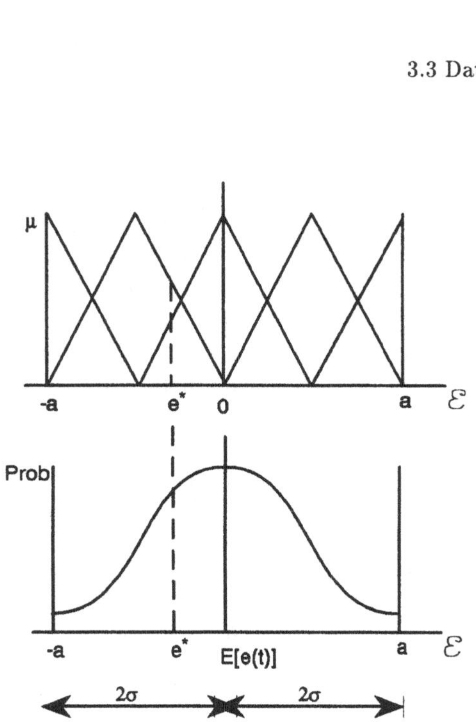

Fig. 3.17. The mean and variance of noise on the domain over which the membership functions for the linguistic values of error are defined.

Influence of Discretized Domains

So far our discussion about the proper choice of membership functions was about membership functions defined on continuous domains. The process of discretization of a given continuous domain is called "quantization" and is done in a number of steps:

1. The continuous domain is quantized into a finite number of segments called "quantization levels". Each segment is labeled as a generic element and the set of all generic elements forms a discrete universe of discourse.

2. Given a linguistic value from a certain term set the fuzzy set defining the meaning of this linguistic value is constructed by assigning a degree of membership to each generic element. This is done for every linguistic value in a term set.

Table 3.1, called a *look-up table* in the literature, gives an example of discretizing the continuous domain $[-3.2, +3.2]$ into thirteen segments, then labeling each segment with a label from -6 to $+6$ and finally assigning the fuzzy sets representing the meaning of the linguistic values from the term set $\{NB, NM, NS, ZO, PS, PM, PB\}$. The number of quantization levels, as observed in [27], influences significantly the overshoot and damping effects. A

Table 3.1. A look-up table for a discretized domain

Segment Label	Segment	NB	NM	NS	ZO	PS	PM	PB
-6	$e \le -3.2$	1.0	0.3	0.0	0.0	0.0	0.0	0.0
-5	$-3.2 \le e \le -1.6$	0.7	0.0	0.0	0.0	0.0	0.0	0.0
-4	$-1.6 \le e \le -0.8$	0.3	1.0	0.3	0.0	0.0	0.0	0.0
-3	$-0.8 \le e \le -0.4$	0.0	0.7	0.7	0.0	0.0	0.0	0.0
-2	$-0.4 \le e \le -0.2$	0.0	0.3	1.0	0.3	0.0	0.0	0.0
-1	$-0.2 \le e \le -0.1$	0.0	0.0	0.7	0.7	0.0	0.0	0.0
0	$-0.1 \le e \le 0.1$	0.0	0.0	0.3	1.0	0.3	0.0	0.0
1	$0.1 \le e \le 0.2$	0.0	0.0	0.0	0.7	0.7	0.0	0.0
2	$0.2 \le e \le 0.4$	0.0	0.0	0.0	0.3	1.0	0.3	0.0
3	$0.4 \le e \le 0.8$	0.0	0.0	0.0	0.0	0.7	0.7	0.0
4	$0.8 \le e \le 1.6$	0.0	0.0	0.0	0.0	0.3	1.0	0.3
5	$1.6 \le e \le 3.2$	0.0	0.0	0.0	0.0	0.0	0.7	0.7
6	$3.2 \le e$	0.0	0.0	0.0	0.0	0.0	0.3	1.0

coarse quantization for large errors and finer quantization for small errors is the usual choice in the case of discretized continuous domains. However, these findings have a purely empirical nature and so far no formal analysis tools exist for studying how the quantization affects controller performance. This explains the preference for continuous domains since, as pointed out in [27], quantization is the source of instability and oscillation problems.

3.3.2 Choice of Scaling Factors

The use of normalized domains (universes of discourse) requires a scale transformation which maps the physical values of the process state variables into a normalized domain. This is called input normalization. Furthermore, output denormalization maps the normalized value of the control output variables into their respective physical domains. Such scale transformations are required both for discrete and continuous domains.

The scaling factors which describe the particular input normalization and output denormalization play a role similar to that of the gain coefficients in a conventional controller. In other words, they are of utmost importance with respect to controller performance and stability related issues, i.e., they are the source of possible instabilities, oscillation problems, and deteriorated damping effects.

For example, a PI-like FKBC can be represented as

$$N_u \cdot \Delta u(k) = F(N_e \cdot e(k), N_{\Delta e} \cdot \Delta e(k)), \tag{3.25}$$

where N_e, $N_{\Delta e}$, and N_u are the scaling factors for e, Δe, and Δu respectively, and F is a nonlinear function representing the FKBC. In this context, one can clearly see the analogy to the gain coefficients K_P and K_I of a conventional PI-controller in which F is a linear function of e, Δe, and Δu.

There are basically two major approaches to the determination of the scaling factors: (1) heuristic, and (2) formal (analytic). The first approach has a trial-and-error nature and a good example of it is the work of Daugherity [45] which we will outline in brief. The FKBC considered is a PI-like FKBC and the performance criteria are defined as follows:

- Desired value of overshoot (OV_d),
- Desired rise-time (RT_d),
- Desired amplitude of oscillations (OSC_d).

Thus, the performance measures used are

$$\Delta OV = OV - OV_d \tag{3.26}$$

$$\Delta RT = RT - RT_d \tag{3.27}$$

$$\Delta OSC = OSC - OSC_d, \tag{3.28}$$

where OV, RT, and OSC are the actual values of the variables describing the above performance criteria.

The heuristic production rules used for adjusting the scaling factors have the form:

if $\langle$value of performance variable$\rangle$ = $\langle$value$\rangle$
then $\Delta N_e = \langle$value$\rangle$

if $\langle$value of performance variable$\rangle$ = $\langle$value$\rangle$
then $\Delta N_{\Delta e} = \langle$value$\rangle$

In the above rules $\langle$value$\rangle$ can be a linguistic value or a crisp value. Finally, the FKBC is connected to the actual process, thus closing the control loop, and the scaling factors are updated after each iteration as follows:

$$N_e(i+1) = N_e(i) + \Delta N_e \tag{3.29}$$

$$N_{\Delta e}(i+1) = N_{\Delta e}(i) + \Delta N_{\Delta e}. \tag{3.30}$$

The iterations are continued until the desired performance values are obtained. The heuristic rules can be explained as follows. By changing the scaling factors for each one of e and Δe we are in fact changing the weights given to this particular process state variable. For example, if the process response is slower than the desired one, i.e., ΔRT is positive, then we need to increase the effect of error on the process, hence N_e is increased. Similarly, if the overshoot or amplitude of oscillation is higher than the desired one, then we need to increase the effect of change-of-error, hence $N_{\Delta e}$ is increased.

The second approach to the derivation of the scaling factors aims at establishing an analytic relationship between the values of the scaling factors and the closed-loop behavior of the controlled process. In this case it is assumed that a conventional model of the process under control exists and the FKBC is considered as a nonlinear transfer element (TE). This approach is described in detail in Chapter 4.

3.4 Inference Engine

The inference engine or rule firing, as already described in detail in Section 2.4.1, can be of two basic types.

- **Composition based inference**: In this case, the fuzzy relations representing the meaning of each individual rule are aggregated into one fuzzy relation describing the meaning of the overall set of rules. Then inference or firing with this fuzzy relation is performed via the operation composition between the fuzzified crisp input and the fuzzy relation representing the meaning of the overall set of rules. As a result of the composition one obtains the fuzzy set describing the fuzzy value of the overall control output.

- **Individual-rule based inference**: In this case, first each single rule is fired. This firing can be simply described by: (1) computing the degree of match between the crisp input and the fuzzy sets describing the meaning of the rule-antecedent and (2) "clipping" the fuzzy set describing the meaning of the rule-consequent to the degree to which the rule-antecedent has been matched by the crisp input. Finally, the "clipped" values for the control output of each rule are aggregated, thus forming the value of the overall control output.

Usually, the second type of inference is preferred since it is computationally very efficient and saves a lot of memory. The FKBC described and analyzed in the subsequent chapters all use the individual-rule type of inference unless stated otherwise.

To assist the reader, we will reintroduce here the basic notions in the representation of rules and inference. We will consider the particular case of a PD-like FKBC, based on Mamdani implication. The formalism presented can be easily extended to other types of FKBC.

Representation of a Single Rule

The symbolic expression of the k-th rule of a PD-like FKBC is

$$\textit{if } e \text{ is } LE^{(k)} \textit{ and } \Delta e \text{ is } L\Delta E^{(k)} \textit{ then } u \text{ is } LU^{(k)}$$

where $LE^{(k)}$, $L\Delta E^{(k)}$ and $LU^{(k)}$ are linguistic values from the term-sets $\mathcal{L}E$, $\mathcal{L}\Delta E$ and $\mathcal{L}U$, respectively. The meaning of these linguistic values is represented by the membership functions or fuzzy sets $\mu_{LE^{(k)}}$, $\mu_{L\Delta E^{(k)}}$ and $\mu_{LU^{(k)}}$, defined on the domains $\mathcal{E}$, $\Delta\mathcal{E}$ and $\mathcal{U}$. The meaning of the above rule in terms of Mamdani-type implication is given as a fuzzy relation $R^{(k)}$ defined on $\mathcal{E} \times \Delta\mathcal{E} \times \mathcal{U}$:

$$\forall e, \Delta e, u : \mu_{R^{(k)}}(e, \Delta e, u) = \min\left(\mu_{LE^{(k)}}(e), \mu_{L\Delta E^{(k)}}(\Delta e), \mu_{LU^{(k)}}(u)\right). \tag{3.31}$$

Representation of a Set of Rules

The meaning of the whole set of rules is simply the union of all individual-rule meanings, $\mu_{R^{(k)}}$, $k = 1, \ldots, n$. Thus we have that the fuzzy relation representing the set of rules is

$$\forall e, \Delta e, u : \mu_R(e, \Delta e, u) = \max\left(\mu_{R^{(1)}}(e, \Delta e, u), \ldots, \mu_{R^{(n)}}(e, \Delta e, u)\right), \tag{3.32}$$

where each of $\mu_{R^{(k)}}$ is defined in equation (3.31).

Representation of Input Values

Suppose that the actual point-wise (crisp) values of e and Δe are e^* and Δe^*. These are provided as an input to the fuzzification module, which produces their set-wise representations as the following two membership functions:

$$\forall e : \mu_E^*(e) = \begin{cases} 1 & \text{for } e = e^*, \\ 0 & \text{otherwise}, \end{cases} \tag{3.33}$$

$$\forall \Delta e : \mu_{\Delta E}^*(\Delta e) = \begin{cases} 1 & \text{for } \Delta e = \Delta e^*, \\ 0 & \text{otherwise}. \end{cases} \tag{3.34}$$

The above transformation from point-wise to set-wise representation of the input is called *fuzzification.*

Representation of the Rule-Antecedent

At this stage, μ_E^* and $\mu_{\Delta E}^*$ are combined into one membership function, μ_{ant}, representing the meaning of the rule-antecedent as follows

$$\forall e, \Delta e : \mu_{ant}(e, \Delta e) = \min_{\mathcal{E} \times \Delta\mathcal{E}}(\mu_E^*(e), \mu_{\Delta E}^*(\Delta e)). \tag{3.35}$$

Inference (Firing) with a Set of Rules

The overall value of the control output variable is obtained by composition of μ_{ant} and μ_R. That is,

$$\forall u : \mu_U(u) = \max_{e, \Delta e} \min\left(\mu_{ant}(e, \Delta e), \mu_R(e, \Delta e, u)\right). \tag{3.36}$$

The disadvantage of this type of composition based inference is in the computational cost of computing μ_{ant} and performing the operation composition plus the memory required to store the whole of μ_R. Fortunately, this composition based inference is equivalent to individual-rule based inference. In what follows we will show this.

We had that

$$\forall e, \Delta e : \mu_{ant}(e, \Delta e) = \min_{\mathcal{E}\times\Delta\mathcal{E}}(\mu^*_E(e), \mu^*_{\Delta E}(\Delta e)). \tag{3.37}$$

According to the way in which μ^*_E and $\mu^*_{\Delta E}$ were defined and the nature of the minimum operation, we obtain that

$$\forall e, \Delta e : \mu_{ant}(e, \Delta e) = \begin{cases} 1 & \text{if } e = e^* \text{ and } \Delta e = \Delta e^* \\ 0 & \text{otherwise.} \end{cases} \tag{3.38}$$

Now let us consider the membership function of the control output as obtained by composition based inference

$$\forall u : \mu_U(u) = \max_{x} \min_{e,\Delta e} \left(\mu_{ant}(e, \Delta e), \mu_R(e, \Delta e, u)\right). \tag{3.39}$$

Looking at this expression we can see that $\mu_U(u) = 0$ whenever $e \neq e^*$ and/or $\Delta e \neq \Delta e^*$, since $\mu_{ant}(e, \Delta e) = 0$ for exactly these cases. Thus,

$$\forall u : \mu_U(u) = \max_{x} \min_{\substack{e\neq e^* \\ \Delta e\neq\Delta e^*}} (0, \mu_R(e, \Delta e, u) = 0.) \tag{3.40}$$

Furthermore, $\mu_U(u)$ is equal to $\mu_R(e^*, \Delta e^*, u)$ when $e = e^*$ and/or $\Delta e = \Delta e^*$, since $\mu_{ant}(e, \Delta e) = 1$ for exactly these cases. In other words,

$$\forall u : \mu_U(u) = \max_{x} \min_{\substack{e=e^* \\ \Delta e=\Delta e^*}} (1, \mu_R(e, \Delta e, u)) = \mu_R(e^*, \Delta e^*, u). \tag{3.41}$$

Thus the membership function of the overall control output obtained via *composition based inference* looks like

$$\forall u : \mu_U(u) = \mu_R(e^*, \Delta e^*, u). \tag{3.42}$$

Now let us consider μ_R from the above definition of μ_U. We had that

$$\forall e, \Delta e, u : \mu_R(e, \Delta e, u) = \max\left(\mu_{R^{(1)}}(e, \Delta e, u), \ldots, \mu_{R^{(n)}}(e, \Delta e, u)\right), \tag{3.43}$$

and replacing e with e^* and Δe with Δe^* we obtain

$$\forall u : \mu_R(e^*, \Delta e^*, u) = \max\left(\mu_{R^{(1)}}(e^*, \Delta e^*, u), \ldots, \mu_{R^{(n)}}(e^*, \Delta e^*, u)\right). \tag{3.44}$$

Let us consider a particular $\mu_{R^{(k)}}$:

$$\begin{aligned} \forall u : \mu_{R^{(k)}}(e^*, \Delta e^*, u) &= \min\left(\mu_{LE^{(k)}}(e^*), \mu_{L\Delta E^{(k)}}(\Delta e^*), \mu_{LU^{(k)}}(u)\right) \\ &= \min\left(\min(\mu_{LE^{(k)}}(e^*), \mu_{L\Delta E^{(k)}}(\Delta e^*)), \mu_{LU^{(k)}}(u)\right) \end{aligned} \tag{3.45}$$

where

- $\mu_{LE^{(k)}}(e^*)$ is the degree of membership of the crisp input e^* in $\mu_{LE^{(k)}}$,
- $\mu_{L\Delta E^{(k)}}(\Delta e^*)$ is the degree of membership of the crisp input Δe^* in $\mu_{L\Delta E^{(k)}}$.

Let us denote $\min(\mu_{LE^{(k)}}(e^*), \mu_{L\Delta E^{(k)}}(\Delta e^*))$ as $\mu^{(k)*}$. Thus

$$\forall u : \mu_{R^{(k)}}(e^*, \Delta e^*, u) = \min\left(\mu^{(k)*}, \mu_{LU^{(k)}}(u)\right). \tag{3.46}$$

Taking into account the nature of the minimum operation, we see that

$$\forall u : \mu_{R^{(k)}}(e^*, \Delta e^*, u) = \begin{cases} \mu^{(k)*} & \text{if } \mu^{(k)*} \leq \mu_{LU^{(k)}}(u), \\ \mu_{LU^{(k)}}(u) & \text{otherwise,} \end{cases} \tag{3.47}$$

or the maximal degree of membership in $\mu_{R^{(k)}}$ is equal to $\mu^{(k)*}$. This is exactly the definition of the *clipped* value of the control output in the case of individual-rule based inference as described in Section 2.4 and denoted by $\mu_{CLU^{(k)}}$. Thus we have that

$$\forall u : \mu_{R^{(k)}}(e^*, \Delta e^*, u) = \mu_{CLU^{(k)}}(u). \tag{3.48}$$

In the case of this *individual-rule based inference* the overall control output was obtained as

$$\begin{aligned} \forall u : \mu_U(u) &= \max\left(\mu_{CLU^{(1)}}(u), \ldots, \mu_{CLU^{(n)}}(u)\right) \\ &= \max\left(\mu_{R^{(1)}}(e^*, \Delta e^*, u), \ldots, \mu_{R^{(n)}}(e^*, \Delta e^*, u)\right) \\ &= \mu_R(e^*, \Delta e^*, u). \end{aligned} \tag{3.49}$$

The latter shows the equivalence between composition based inference and individual-rule based inference in the case of Mamdani-type implication used to represent the meaning of the individual rules.

3.5 Choice of Fuzzification Procedure

As already said, there are two basic types of inference, namely

- Composition based inference,
- Individual-rule based inference.

When the first type of inference is chosen, the fuzzification procedure is defined as follows. We will consider here the case of a PI-like FKBC without any loss of generality. Let the crisp input values of e and Δe be $e^\star$ and $\Delta e^\star$. It is these two values that have to be fuzzified, i.e., their crisp (point-wise) representation is transformed into a fuzzy (set-wise) representation defined as:

$$\mu_{e^\star} = \begin{cases} 1 & \text{if } e = e^\star \text{ and } e \in \mathcal{E}, \\ 0 & \text{otherwise,} \end{cases} \tag{3.50}$$

$$\mu_{\Delta e^\star} = \begin{cases} 1 & \text{if } \Delta e = \Delta e^\star \text{ and } \Delta e \in \Delta\mathcal{E}, \\ 0 & \text{otherwise.} \end{cases} \tag{3.51}$$

In the case of the second type of inference the result of defuzzification is obtained as follows. Let us consider the k-th rule of the PI-like FKBC,

$$\textit{if } e \text{ is } LE^{(k)} \textit{ and } \Delta e \text{ is } L\Delta E^{(k)} \textit{ then } \Delta u \text{ is } L\Delta U^{(k)},$$

where $LE^{(k)}$, $L\Delta E^{(k)}$, and $L\Delta U^{(k)}$ are the linguistic values taken by e, Δe, and Δu in the k-th rule. The meaning of these linguistic values is represented by membership functions $\mu_{LE^{(k)}}$, $\mu_{L\Delta E^{(k)}}$, and $\mu_{L\Delta U^{(k)}}$ defined on the domains $\mathcal{E}$, $\Delta\mathcal{E}$, and $\Delta\mathcal{U}$ respectively. Now let $e^\star$ and $\Delta e^\star$ be two crisp inputs. Then the fuzzified version of $e^\star$ is its degree of membership in $\mu_{LE^{(k)}}$, i.e., $\mu_{LE^{(k)}}(e^\star)$, and the fuzzified version of $\Delta e^\star$ is $\mu_{L\Delta E^{(k)}}(\Delta e^\star)$.

3.6 Choice of Defuzzification Procedure

In what follows we will cover the six most often used defuzzification methods:

- Center-of-Area/Gravity defuzzification, ,
- Center-of-Sums defuzzification,
- Center-of-Largest-Area defuzzification,
- First-of-Maxima defuzzification,
- Middle-of-Maxima defuzzification,
- Height defuzzification.

Let us mention here that the influence of the defuzzification method on the controller performance has been paid very little attention so far. That is why we will confine ourselves just to the presentation of the above six defuzzification methods. In the literature on fuzzy control, the terms used in describing the different defuzzification methods vary from author to author. That is why the reader, when comparing other literature sources with the presentation below, should pay attention to the formal definitions of the defuzzification methods rather than to their names.

To facilitate the presentation, we will reintroduce some of the basic notions already presented in Sections 2.1.5 and 2.3.2 dealing with linguistic variables and properties of fuzzy sets. A linguistic variable is defined by

$$\langle X, \mathcal{L}X, \mathcal{X}, M_X \rangle . \qquad (3.52)$$

Here X denotes the symbolic name of a linguistic variable. $\mathcal{L}X$ is the set of *linguistic values* that X can take on. A linguistic value denotes a symbol for a particular property of X. We denote an arbitrary element of $\mathcal{L}X$ by LX. $\mathcal{X}$ is the actual physical domain over which the meaning of the linguistic values. $\mathcal{X}$ can be discrete or continuous. M_X is a semantic function which gives a 'meaning' (interpretation) of a linguistic value in terms of the quantitative elements of $\mathcal{X}$, i.e.,

$$M_X : LX \to \widetilde{LX}, \qquad (3.53)$$

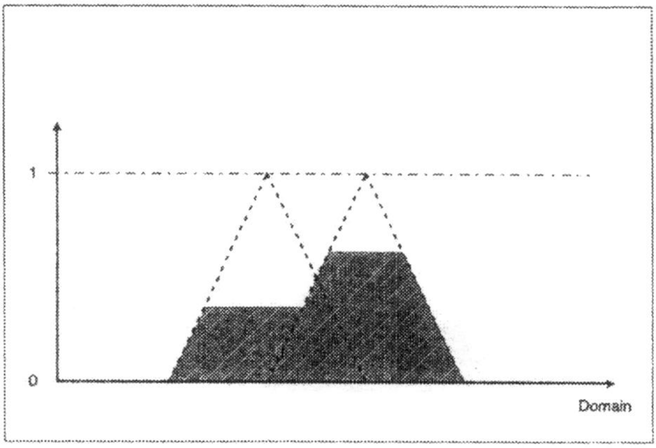

Fig. 3.18. Two clipped fuzzy sets.

where $\widetilde{LX}$ is a denotation for a fuzzy set or a membership function defined over $\mathcal{X}$, i.e.,

$$\widetilde{LX} = \sum_{\mathcal{X}} \mu_{LX}(x)/x \quad \text{in the case of discrete } \mathcal{X}, \tag{3.54}$$

$$\widetilde{LX} = \int_{\mathcal{X}} \mu_{LX}(x)/x \quad \text{in the case of continuous } \mathcal{X}. \tag{3.55}$$

In other words, M_X is a function which takes a symbol as its argument and returns the 'meaning' of this symbol in terms of a fuzzy set. Instead of $\widetilde{LX}$ we will also use μ_{LX}, i.e., the membership function without an argument.

We can now define a set of m rules as

if x_1 is $LX^{(k)}$ *and* ... *and* x_n is $LX_n^{(k)}$ *then* u is $LU^{(k)}$, $k = 1, \ldots, m$.

The result of firing these rules with physical, crisp input values $x_1^*, \ldots, x_n^*$ will either results in m clipped fuzzy sets denoted by

$$\widetilde{CLU}^{(1)}, \ldots \widetilde{CLU}^{(m)} \tag{3.56}$$

or m scaled fuzzy sets denoted by

$$\widetilde{SLU}^{(1)}, \ldots \widetilde{SLU}^{(m)}. \tag{3.57}$$

The way a scaled fuzzy set is obtained as the result of firing an individual rule was described in Section 2.3.7. In the formulas realizing the different defuzzification methods it does not make any difference whether one uses clipped or scaled fuzzy sets. That is why we will use CLU as the general notation for the value for the control output after firing a rule. Figures 3.18 and 3.19 give examples of

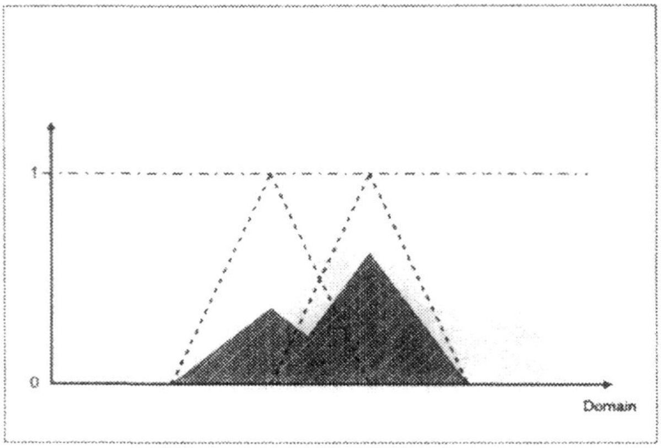

Fig. 3.19. Two scaled fuzzy sets.

two clipped and two scaled fuzzy sets. Finally, the overall control output $\widetilde{U}$ or μ_U is obtained as the *union* of the clipped or scaled control outputs

$$\widetilde{U} = \bigcup_{k=1}^{m} \widetilde{CLU}^{(k)}; \tag{3.58}$$

$\widetilde{U}$ may be a convex fuzzy set (Section 2.5.1) or may be a non-convex fuzzy set which consists of a number of convex fuzzy subsets. The latter case is illustrated by Fig. 3.23. The *crisp value* of the control output resulting from a particular defuzzification output will be denoted by u^*.

The *area* of the fuzzy set $\widetilde{U}$ is defined as

$$\int_{\mathcal{U}} \mu_U(u)\, \mathrm{d}u, \tag{3.59}$$

where $\int$ is a classical integral operation.

The *height* of $\widetilde{CLU}^{(k)}$ is equal to the degree of match of the k-th rule-antecedent, and will be denoted by f_k. The *peak value* of $\widetilde{CLU}^{(k)}$ is equal to the peak value of its unclipped version $\widetilde{LU}^{(k)}$ or $\mu_{LU}^{(k)}$. If $\widetilde{LU}^{(k)}$ is a triangular membership function, then its peak value is that domain element on $\mathcal{U}$ which has degree of membership 1. If $\widetilde{LU}^{(k)}$ is a trapezoidal membership function, then its peak value is an interval. Instead of taking the whole interval, we take its middle point as the peak value of $\widetilde{LU}^{(k)}$.

3.6.1 Center-of-Area/Gravity

The Center-of-Area method (in the literature also referred to as Center-of-Gravity method) is the best well-known defuzzification method.

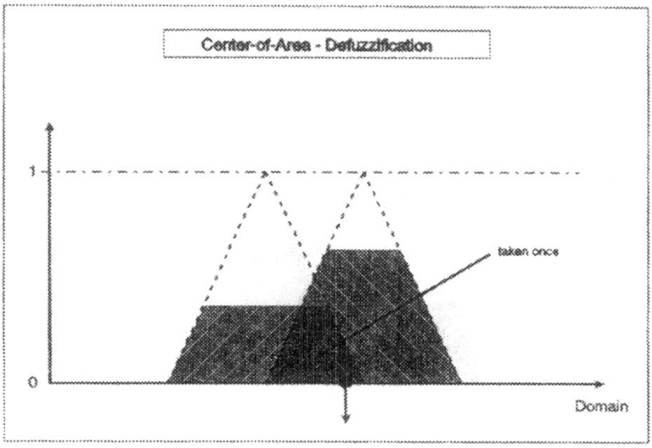

Fig. 3.20. A graphical representation of the Center-of-Area defuzzification method.

In the discrete case ($\mathcal{U} = \{u_1, \ldots, u_\ell\}$) this results in

$$u^* = \frac{\sum_{i=1}^{\ell} u_i \cdot \mu_U(u_i)}{\sum_{i=1}^{\ell} \mu_U(u_i)} = \frac{\sum_{i=1}^{\ell} u_i \cdot \max_k \mu_{CLU^{(k)}}(u_i)}{\sum_{i=1}^{\ell} \max_k \mu_{CLU^{(k)}}(u_i)}. \tag{3.60}$$

In the continuous case we obtain

$$u^* = \frac{\int_{\mathcal{U}} u \cdot \mu_U(u)\, du}{\int_{\mathcal{U}} \mu_U(u)\, du} = \frac{\int_{\mathcal{U}} u \cdot \max_k \mu_{CLU^{(k)}}(u)\, du}{\int_{\mathcal{U}} \max_k \mu_{CLU^{(k)}}(u)\, du}, \tag{3.61}$$

where $\int$ is the classical integral. So this method determines the center of the area below the combined membership function. Figure 3.20 shows this operation in a graphical way. It can be seen that this defuzzification method takes into account the area of $\widetilde{U}$ as a whole. Thus if the areas of two clipped fuzzy sets constituting $\widetilde{U}$ overlap (see Fig. 3.20), then the overlapping area is not reflected in the above formula. This operation is computationally rather complex and therefore results in quite slow inference cycles.

3.6.2 Center-of-Sums

A similar but faster defuzzification method is Center-of-Sums. The motivation for using this method is to avoid the computation of $\widetilde{U}$. The idea is to consider the contribution of the area of each $\widetilde{CLU}^{(k)}$ individually. Mathematically, the Center-of-Area/Gravity method builds $\widetilde{U}$ by taking the union of all $\widetilde{CLU}^{(k)}$. Center-of-Sums, however, takes the sum of the $\widetilde{CLU}^{(k)}$. Thus overlapping areas, if such exist, are reflected more than once by this method (see Fig. 3.21). The

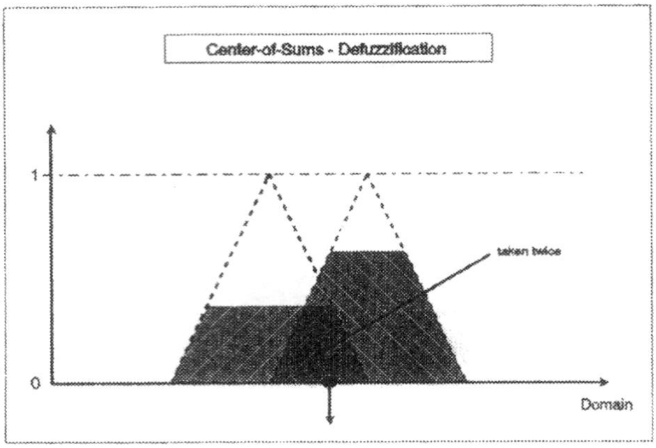

Fig. 3.21. A graphical representation of the Center-of-Sums defuzzification method.

faster algorithm for this defuzzification method is one reason why most FKBC use this method. There are some fuzzy systems which claim that the defuzzification method used is Center-of-Gravity, but actually use Center-of-Sums. In the discrete case Center-of-Sums is formally given by

$$u^* = \frac{\sum_{i=1}^{\ell} u_i \cdot \sum_{k=1}^{n} \mu_{CLU^{(k)}}(u_i)}{\sum_{i=1}^{\ell} \sum_{k=1}^{n} \mu_{CLU^{(k)}}(u_i)}. \tag{3.62}$$

In the continuous case we obtain

$$u^* = \frac{\int_{\mathcal{U}} u \cdot \sum_{k=1}^{n} \mu_{CLU^{(k)}}(u)\, du}{\int_{\mathcal{U}} \sum_{k=1}^{n} \mu_{CLU^{(k)}}(u)\, du}. \tag{3.63}$$

3.6.3 Height

Height defuzzification is a method which instead of using $\widetilde{U}$ uses the individual clipped or scaled control outputs. This method takes the peak value of each $\widetilde{CLU}^{(k)}$ and builds the weighted (with respect to the height f_k of $\widetilde{CLU}^{(k)}$) sum of these peak values (Fig. 3.22). Thus neither the support or shape of $\widetilde{CLU}^{(k)}$ play a role in the computation of u^*. The Height method is both a very simple and very quick method. Let $c^{(k)}$ be the peak value of $\widetilde{LU}$, and f_k is the height of $\widetilde{CLU}^{(k)}$. Then the Height defuzzification method in a system of m rules is formally given by

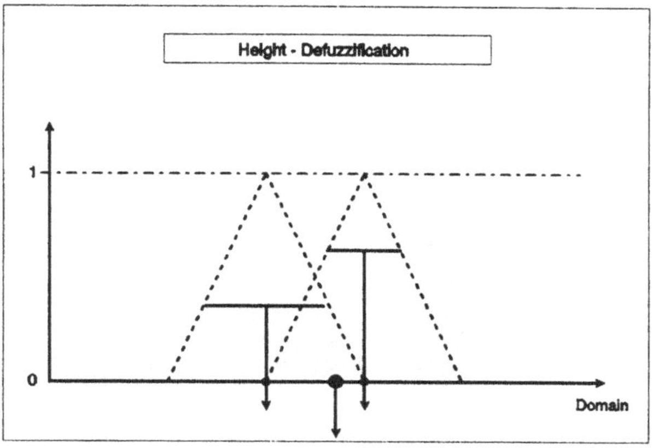

Fig. 3.22. A graphical representation of the Height defuzzification method.

$$u^* = \frac{\sum_{k=1}^{m} c^{(k)} \cdot f_k}{\sum_{k=1}^{n} f_k}. \tag{3.64}$$

3.6.4 Center-of-Largest-Area

The Center-of-Largest-Area is used in the case when $\widetilde{U}$ is non-convex, i.e., it consists of at least two convex fuzzy subsets. Then the method determines the convex fuzzy subset with the largest area and defines the crisp output value u^* to be the Center-of-Area of this particular fuzzy subset (Fig. 3.23). It is difficult to represent this defuzzification method formally, because it involves first finding the convex fuzzy subsets, then computing their areas, etc. Figure 3.23 illustrates this method.

3.6.5 First-of-Maxima

First-of-Maxima uses $\widetilde{U}$ and takes the smallest value of the domain $\mathcal{U}$ with maximal membership degree in $\widetilde{U}$. This is realized formally in three steps: let

$$\text{hgt}(U) = \sup_{u \in \mathcal{U}} \mu_U(u) \tag{3.65}$$

be the highest membership degree of $\widetilde{U}$, and let

$$\{u \in \mathcal{U} \mid \mu_U(u) = \text{hgt}(U)\} \tag{3.66}$$

be the set of domain elements with degree of membership equal to hgt(U). Then u^* is given by

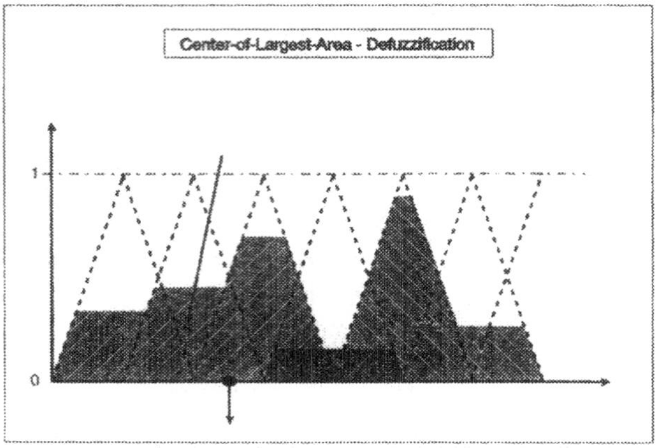

Fig. 3.23. A graphical representation of the Center-of-Largest-Area defuzzification method. The dark shaded area represents the overlap of the two convex subset.

$$u^* = \inf_{u \in \mathcal{U}} \{u \in \mathcal{U} \mid \mu_U(u) = \text{hgt}(U)\}. \tag{3.67}$$

The alternative version of this method is called Last-of-Maxima and is given as

$$u^* = \sup_{u \in \mathcal{U}} \{u \in \mathcal{U} \mid \mu_U(u) = \text{hgt}(U)\}. \tag{3.68}$$

3.6.6 Middle-of-Maxima

Middle-of-Maxima is very similar to First-of-Maxima or Last-of-Maxima. Instead of determining u^* to be the first or last from all values where $\mathcal{U}$ has maximal membership degree, this method takes the average of these two values (see Fig. 3.25). Formally,

$$u^* = \frac{\inf_{u \in \mathcal{U}} \{u \in \mathcal{X} \mid \mu_U(u) = \text{hgt}(U)\} + \sup_{u \in \mathcal{U}} \{u \in \mathcal{U} \mid \mu_U(u) = \text{hgt}(U)\}}{2} \tag{3.69}$$

3.6.7 Description of Two Examples

In this section we will present two examples showing some possible problems in the application of the two most widely used defuzzification methods. The first example is from Pfluger, Yen and Langari [167] concerning a mobile robot path planning problem. The robot has to move from point A to point B without colliding into walls or an obstacle and following the shortest possible path. In Fig. 3.26 the angle ϕ of the desired movement direction is stated in the form of a fuzzy set $\widetilde{D}$, defined over the interval $[-90, 0]$. The fuzzy set $\widetilde{A}$ contains

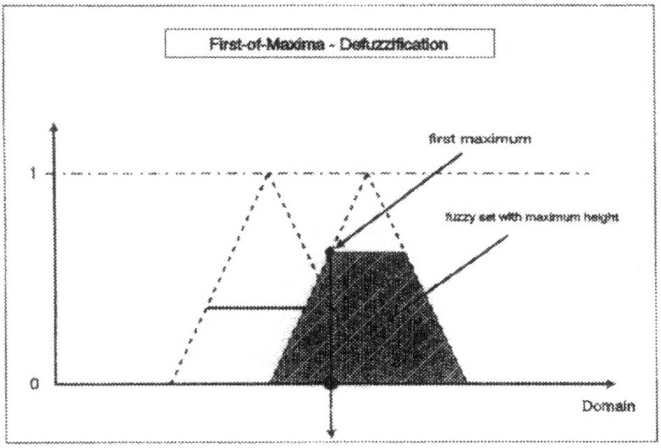

Fig. 3.24. A graphical representation of the First Maximum defuzzification method.

the so-called non-prohibitive information which defines which sections of the movement space are allowed for the robot to move in. This fuzzy set is defined on the intervals $[-90, 75]$ and $[-25, 90]$. When these two fuzzy sets are combined using the intersection operation, a fuzzy set with two "peaks" is the result, where each "peak" points out a possible movement direction.

The *Middle-of-Maxima* defuzzification method results in an angle $\phi = -25°$. If the robot follows the direction given by this angle, it will pass the table via its right side. The *Center-of-Area* method results in an angle ϕ of almost $-45°$. In this case the robot will move towards the right side table with the possibility of colliding with it. If this orientation process is repeated often enough, so that the robot does not collide with the table, this will result in the following. Suppose the robot chooses the $-45°$ direction. Then just before colliding with the table, it constructs new versions $\widetilde{A}'$ and $\widetilde{D}'$ and their intersection. The Center-of-Area of $\widetilde{A}' \cap \widetilde{B}'$ will be close to $-75°$ direction, which finally results in the robot going to the left around the table.

The second example is about the inverted pendulum control problem. The problem is to balance a pole on a mobile platform that can move in only two directions, to the left or to the right. In Yamakawa [229] the mathematical model of this system is given as follows:

$$I \cdot \ddot{\theta} = V \cdot L \cdot \sin\theta - H \cdot L \cdot \cos\theta, \tag{3.70}$$

$$V - mg = -mL(\ddot{\theta}\sin\theta + \dot{\theta}^2\cos\theta), \tag{3.71}$$

$$H = m\ddot{y} + mL(\ddot{\theta}\cos\theta + \dot{\theta}^2\sin\theta), \tag{3.72}$$

$$U - H = M\ddot{y}, \tag{3.73}$$

where θ is the angle between the platform and the pendulum, $2L$ the length of the pendulum, y the position of the platform bearing the pendulum, m the mass of the pendulum, M the mass of the platform, H the horizontal force at the

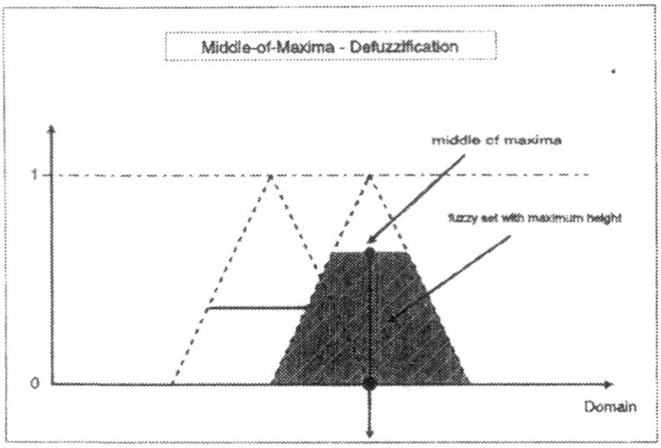

Fig. 3.25. A graphical representation of the Middle-of-Maxima defuzzification method.

pivot, V the vertical force at the pivot, U the driving force given to the platform, and $I = \frac{1}{3}mL^2$ the moment of inertia. These formulas can be simplified and linearized for small angles $|\theta| \leq 10°$, but even then it is a complicated model for a relatively simple process. That is why a conventional control strategy cannot be easily obtained. Yamakawa describes the rules of the FKBC for this same process as follows:

1. *if* θ is *PM and* $\dot{\theta}$ is *ZO*, *then* $\dot{y}$ is *PM*,
2. *if* θ is *PS and* $\dot{\theta}$ is *PS*, *then* $\dot{y}$ is *PS*,
3. *if* θ is *PS and* $\dot{\theta}$ is *NS*, *then* $\dot{y}$ is *ZO*,
4. *if* θ is *NM and* $\dot{\theta}$ is *ZO*, *then* $\dot{y}$ is *NM*,
5. *if* θ is *NS and* $\dot{\theta}$ is *NS*, *then* $\dot{y}$ is *NS*,
6. *if* θ is *NS and* $\dot{\theta}$ is *PS*, *then* $\dot{y}$ is *ZO*,
7. *if* θ is *ZO and* $\dot{\theta}$ is *ZO*, *then* $\dot{y}$ is *ZO*.

where $\dot{y}$ is the velocity of the platform. Consider now the so-called scaled inference as described in Example 2.15. It should be noticed that almost all existing software based FKBC development tools use this composition rule, because it is the computationally fastest method. Suppose now that the actual input θ^* is somewhere between *PS* and *PM* and the actual input $\dot{\theta}^*$ is somewhere between *ZO* and *PS*, then only rules 1 and 2 can fire. Thus the aggregated value of the control output $\tilde{Y}$ is the union of the scaled versions of *PM* (rule 1) and *PS* (rule 2). The result is, for example, one of the three figures given in Fig. 3.27 dependent on the input values of θ and $\dot{\theta}$. Consider now the case where θ^* and $\dot{\theta}^*$ change slightly in such a way, that the three cases in Fig. 3.27 take place one after another in time. If one now uses the Center-of-Area method, then the crisp output $\dot{y}^*$ of the system will change continuously, from somewhere right of α to somewhere left of β. If one uses the Middle-of-Maxima defuzzification

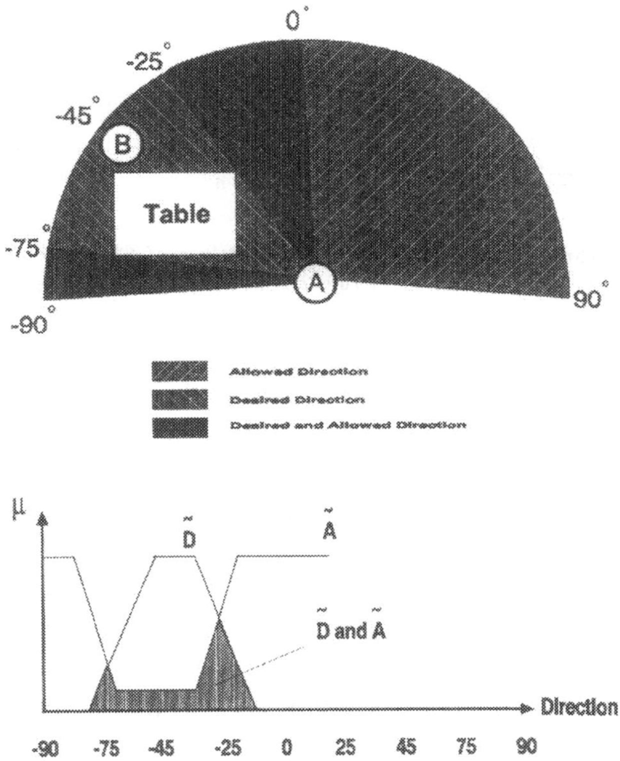

Fig. 3.26. The robot path planning problem.

method, then the result will be α in the first case, $(\alpha + \beta)/2$ in the second case and β in the third case. This means that a small change in the input will cause a big change in the output values.

3.6.8 Comparison and Evaluation of Defuzzification Methods

In this paragraph we suggest some criteria which the 'ideal' defuzzification method should satisfy. Now it must be stated in advance that none of our defuzzification methods satisfies all criteria listed below, i.e., one has to weight these criteria for the particular application to be able to make the right choice of defuzzification method.

Some criteria for defuzzification methods are:

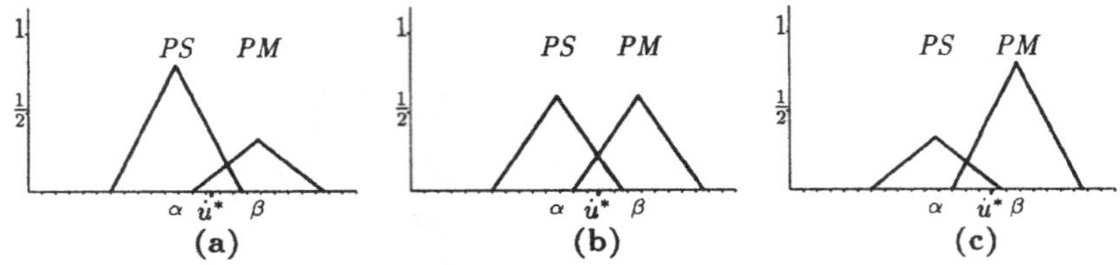

Fig. 3.27. Three output states for different values of θ^* and $\dot{\theta}^*$. Case (a) takes place when the rule-antecedent of rule 2 is matched to a high degree ($> .5$), while this for rule 1 is matched to a low degree ($> .5$). Case (b) takes place when both rule-antecedents are matched to the same high degree. Case (c) is the inverse of case (a). $\dot{y}^*$ denotes the Center-of-Area, α and β denote peak values.

1. **Continuity:** A small change in the input of the FKBC should not result in a large change in the output, for example, in the case of a two-input, one-output FKBC, when two inputs $(e_1^*, \dot{e}_1^*)$ and $(e_2^*, \dot{e}_2^*)$ differ slightly, then the corresponding output values u_1^* and u_2^* should differ slightly too, i.e.,

$$\forall \epsilon > 0 \; \exists \delta > 0 : |e_1^* - e_2^*| < \delta \text{ and } |\dot{e}_1^* - \dot{e}_2^*| < \delta \text{ then } |u_1^* - u_2^*| < \epsilon \tag{3.74}$$

2. **Disambiguity:** In Fig. 3.28 there are two equally large areas covered by the two fuzzy convex subsets constituting the overall control output for both max-min composition based inference and scaled inference. Thus, the

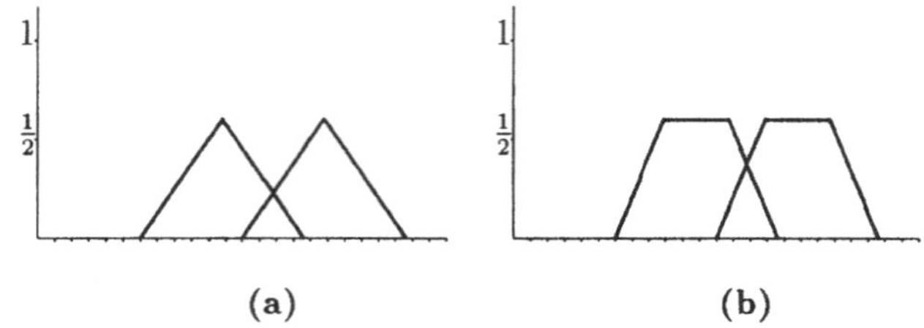

Fig. 3.28. A case where the Center-of-Largest-Area defuzzification method will deliver ambiguous results for the scaled inference (a) and max-min inference method (b).

defuzzification method cannot choose between these two areas. In other words it is ambiguous. This criterion is not satisfied by the Center-of-Largest-Area defuzzification method.

3. **Plausibility:** Every defuzzified control output has a horizontal component $u^* \in \mathcal{U}$, and a vertical component $\mu_U(u^*) \in [0, 1]$. We define u^* to

be plausible if it lies approximately in the middle of the support of $\widetilde{U}$ and has a high degree of membership in $\widetilde{U}$. The Center-of-Area method, for example, applied to Fig. 3.27b does not satisfy these properties: although the Center-of-Area lies in the middle of the support set, its membership degree is one of the lowest possible.

4. **Computational complexity:** This criterion is particularly important in practical applications of FKBC. The Height method, together with the Middle- and First-of-Maxima are fast methods, whereas the Center-of-Area method is slower. The computational complexity of Center-of-Sums depends on the shape of the output membership functions and whether max-min composition based inference or scaled inference is chosen. The Middle-of-Maxima method, for example, is faster with scaled inference. There is also the issue of the representation of the fuzzy sets. In case of the Center-of-Area or Center-of-Sums defuzzification, a tabular representation of the fuzzy sets in combination with clipped fuzzy sets (max-min composition) makes the defuzzification procedure very slow.

A fifth criterion is the so-called weighting of the output fuzzy sets which constitutes the difference between the Center-of-Area and Center-of-Sums method. This criterion is handled separately from the former four criteria, because it is hard to say whether it is to be preferred or not. We consider it as a positive property.

5. **Weight counting:** Suppose there are three rules and all have PB in the rule-consequent, The first is matched to 0.2 resulting in $\widetilde{PB}^{(1)}$, the second to 0.4 resulting in $\widetilde{PB}^{(2)}$, and the third to 0.6 resulting in $\widetilde{PB}^{(3)}$. Suppose there is a fourth rule with a rule-consequent PS, and it is matched to 0.7 resulting in $\widetilde{PS}^{(4)}$. The question now is how to compute u^*. The aggregated value of the control output is

$$\widetilde{U} = \widetilde{PB}^{(1)} \cup \widetilde{PB}^{(2)} \cup \widetilde{PB}^{(3)} \cup \widetilde{PS}^{(4)} = \widetilde{PB}^{(3)} \cup \widetilde{PS}^{(4)}. \tag{3.75}$$

If we defuzzify the $\widetilde{U}$ so obtained with Center-of-Area, First-of-Maxima, Middle-of-Maxima and Center-of-Largest-Area, we will get u^* having a higher degree of membership in $\widetilde{PS}$ than to $\widetilde{PB}$. This means that there were three rules producing clipped versions of $\widetilde{PB}$ and only one rule producing a clipped version of $\widetilde{PS}$. It is desirable to somehow take this fact into account, which is done by *weight counting* given, for example, in the Height defuzzification method as follows: $f_1 = 0.2$, $f_2 = 0.4$, $f_3 = 0.6$, $f_4 = 0.7$, and let $u_{1\mathbf{peak}}$ and $u_{2\mathbf{peak}}$ be the peak values of $\widetilde{PB}$ and $\widetilde{PB}$ respectively. Then

$$u^* = \frac{0.2 \cdot u_{1\mathbf{peak}} + 0.4 \cdot u_{1\mathbf{peak}} + 0.6 \cdot u_{1\mathbf{peak}} + 0.7 \cdot u_{2\mathbf{peak}}}{0.2 + 0.4 + 0.6 + 0.7} \tag{3.76}$$

has a higher degree of membership in $\widetilde{PB}$ than to $\widetilde{PS}$. Center-of-Sums and Height defuzzification are weight counting methods.

Table 3.2. The defuzzification methods Center-of-Area (CoA), Center-of-Sums (CoS), Middle-of-Maxima (MoM), First-of-Maxima (FoM), Height Method (HM) and Center-of-Largest-Area (CLA) and their criteria. (*) no only in the case of scaled inference.

	CoA	CoS	MoM	FoM	HM	CLA
Continuity	yes	yes	no	no	yes	no
Disambiguity	yes	yes	yes	yes	yes	no
Plausibility	yes	yes	no*	no	yes	yes
Comp. complexity	bad	good	good	good	good	bad
Weight counting	no	yes	no	no	yes	no

Table 3.2 gives an overview of the six defuzzification methods and their performance with respect to these five criteria.

4. Nonlinear Fuzzy Control

4.1 Introduction

The analytic functions employed in models of linear and nonlinear systems (processes) operate on the domain of crisp (point-wise) reals. In addition, we have the class of fuzzy systems whose models, in general, are algebraic mappings from the domain of crisp reals into a prespecified domain of fuzzy (set-wise defined) reals.

Likewise, the class of controllers can be divided into linear, nonlinear, and fuzzy knowledge based controllers (FKBC). The class of FKBC can be considered both as an independent class and as a variety of nonlinear controllers. The subsequent sections focus on the latter point of view because FKBC are also applied to systems which are described in a crisp manner. Figure 4.1 shows an "open scheme" for different types of systems and controllers.

The controllers which we describe are used in systems with crisp inputs and crisp outputs. The controller design uses two major knowledge sources:

> The process operator's or control engineer's heuristic knowledge about the process and/or controller. In this case the model of the process/controller is described in terms of production rules or if-then rules only.
>
> The non-fuzzy model of the process (e.g., the phase plane of a second order system).

With respect to these two ways of formally describing the process and the controller, the tuple "crisp process – FKBC" is, from the point of view of systems theory, a hybrid system. However, the general control law design principles are the same as in the case of crisp linear and nonlinear systems:

1. Stability analysis,
2. Performance analysis according to selected criteria,
3. Robustness analysis concerning parameter fluctuations, model uncertainties, and disturbances.

In this sense, a constructive approach which enables the designer to analyze, at least in qualitative terms, the behavior of the tuple "crisp process – FKBC"

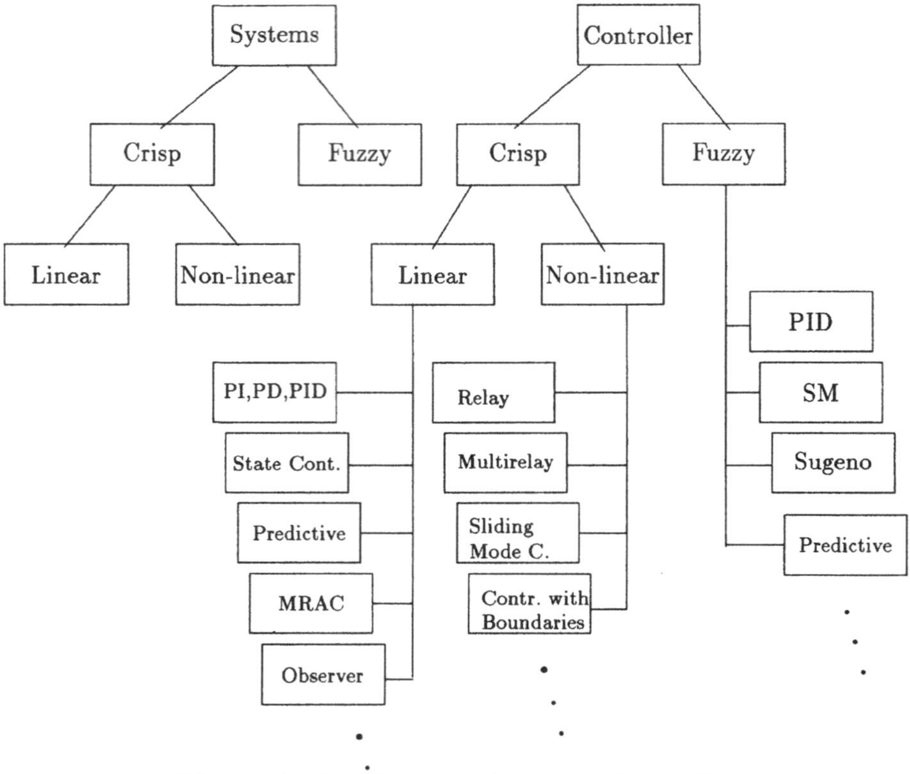

Fig. 4.1. An "open" scheme of systems and controllers.

requires a unification of the crisp and fuzzy ways of looking at the components of this tuple. One method is the transformation of the process model into a symbolic representation, e.g., linguistic variables, if-then rules, etc. This method is helpful in the case when the process cannot be described satisfactorily by conventional means (differential equations, balance equations, etc.). Yet, using this method, stability, performance, and robustness can be analyzed only qualitatively so that a comparison with other methods requires quantitative representation in any case.

On the other hand, if there exists a good enough non-fuzzy (crisp) model of the process, we can adopt methods from linear and nonlinear control theory for the FKBC design. Then a FKBC becomes a symbolic nonlinear controller. Looking at it as a nonlinear controller, the system of if-then rules is transformed into a nonlinear transfer element with scaling factors yet to be determined in a crisp manner. Thus qualitative or symbolic control design is completed by a quantitative (numerical) design phase. General design rules which can be derived from the basic structure of a FKBC (see Section 4.2.1) are the following:

1. Qualitative (symbolic) design of if-then rules. This includes the following steps:

   Defining the linguistic term-sets for the process state and control output variables and the corresponding membership functions describing the meaning of the elements of these term-sets.

   Formulation of the *set of if-then rules.*

   Testing the set of if-then rules with respect to their *completeness and consistency.*

2. Quantitative design of scaling or normalization factors. This includes the following steps:

   Testing the system to be controlled with respect to *controllability* and *observability.*[1]

   Analysis of the *operating points* and *operation areas* of the crisp process states, process outputs, and control variables.

   Transformation of the FKBC into, from system theoretical aspect, a controllable, *nonlinear transfer element* whose properties are well known.

   Utilization of design methods with origins in *nonlinear system theory.*

Section 4.2 deals with the computational structure of a FKBC and its formal description as a nonlinear transfer element. Furthermore, the connection between conventional and rule based transfer elements is established showing the compatibility of conventional control and FKBC. Finally, the most frequently encountered types of FKBC are discussed with respect to the design principles mentioned previously. These include PID-like FKBC, sliding mode based FKBC and Sugeno-Takagi FKBC.

4.2 The Control Problem

The simplest and most general multivariable control problem can be represented by the following system of state equations:

$$\begin{aligned} \mathbf{f}(\mathbf{x}, \dot{\mathbf{x}}, \mathbf{u}, \mathbf{z}) &= \mathbf{0} \\ \mathbf{y} &= \mathbf{g}(\mathbf{x}, \mathbf{u}) \\ \mathbf{u} &= \mathbf{h}(\mathbf{w}, \mathbf{y}), \end{aligned} \tag{4.1}$$

where

[1]For nonlinear systems: *local controllability* and *local observability.*

$\mathbf{x} = (x_1, x_2, \ldots, x_l)^T$:	$(1 \times l)$-state vector
$\dot{\mathbf{x}} = (\dot{x}_1, \dot{x}_2, \ldots, \dot{x}_l)^T$:	derivative of $\dot{\mathbf{x}}$
$\mathbf{u} = (u_1, u_2, \ldots, u_k)^T$:	$(1 \times k)$-control vector
$\mathbf{y} = (y_1, y_2, \ldots, y_m)^T$:	$(1 \times m)$-output vector
$\mathbf{w} = (w_1, w_2, \ldots, w_m)^T$:	$(1 \times m)$-desired input vector
$\mathbf{z} = (z_1, z_2, \ldots, z_k)^T$:	$(1 \times k)$-vector of disturbances.

$\mathbf{f}, \mathbf{g}, \mathbf{h}$ are vectors whose elements are functions of $\mathbf{x}, \dot{\mathbf{x}}, \mathbf{u}, \mathbf{y}, \mathbf{w}$ and $\mathbf{z}$ which in turn are functions of the time parameter t. For simplicity, in the following we do not write the time parameter t as an argument of states or functions of states if there is no other reason to mention the time t explicitly. The dependence of

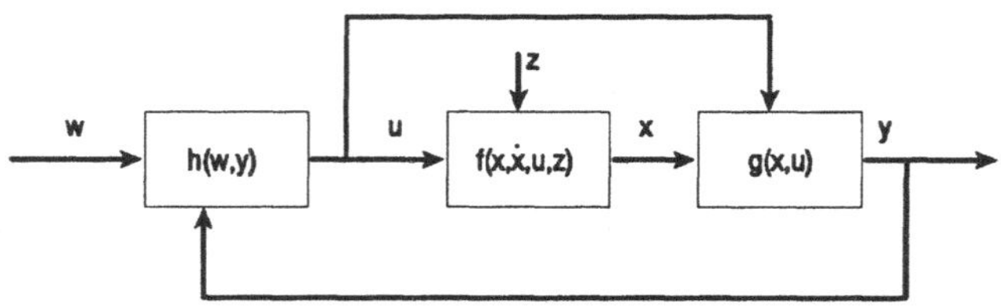

Fig. 4.2. General control scheme.

$\mathbf{y}$ on $\mathbf{u}$ in $\mathbf{g}(\mathbf{x}, \mathbf{u})$ only exists for systems with step behavior. For systems with low-pass behavior this dependence does not hold. The corresponding control block scheme is shown in Fig. 4.2.

For a large class of systems (e.g., mechanical systems) equation (4.1) can be rewritten in explicit form where the derivative $\dot{\mathbf{x}}$ of the state vector $\mathbf{x}$ can be separated:

$$\begin{aligned} \dot{\mathbf{x}} &= \tilde{\mathbf{f}}(\mathbf{x}) + \mathbf{B} \cdot \mathbf{u} + \mathbf{z} \\ \mathbf{y} &= \mathbf{C} \cdot \mathbf{x} + \mathbf{D} \cdot \mathbf{u} \\ \mathbf{u} &= \mathbf{h}(\mathbf{w}, \mathbf{y}), \end{aligned} \tag{4.2}$$

where $\tilde{\mathbf{f}}(\mathbf{x})$ is a vector whose elements are nonlinear functions of $\mathbf{x}$ and where $\mathbf{B}, \mathbf{C}$, and $\mathbf{D}$ are constant matrices. For systems with low-pass characteristics we obtain $\mathbf{D} = \mathbf{0}$.

The control law, represented by $\mathbf{h}(\mathbf{w}, \mathbf{y})$, contains the FKBC. It is determined in such a way so that the output vector $\mathbf{y}$ follows the desired input vector $\mathbf{w}$.

A convenient representation of a second order dynamic system is in terms of its phase plane. Figure 4.3 shows the phase plane of a double integrator using constant control values $u = \pm K$, where K is a positive constant scalar value.

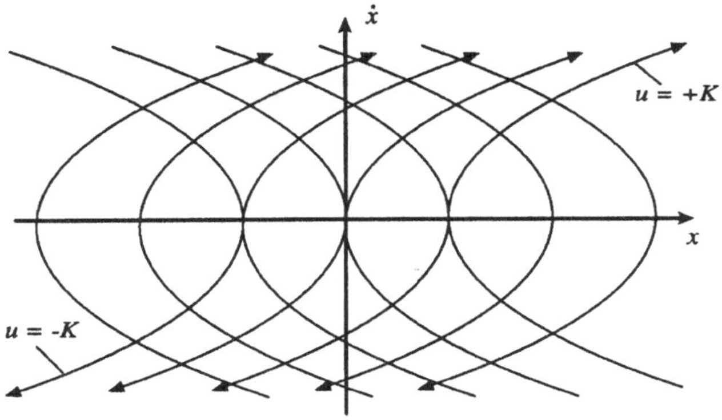

Fig. 4.3. Phase plane of a double integrator.

This way of representing the system's behavior is essential because it shows the system's trajectories in a compact and clear manner. In the first place, the FKBC is a nonlinear controller for controlling nonlinear processes. The following section deals with the construction of an FKBC as a nonlinear transfer element (TE).

4.3 The FKBC as a Nonlinear Transfer Element

In principle, all operations with respect to time, e.g., derivation, integration, etc., are not included in the FKBC. This means that the rule based representation of a FKBC does not include any dynamics. This makes a FKBC a static transfer element (TE), like a state controller. In addition to that, a FKBC has nonlinear transfer characteristic which derives from the nonlinear properties of the computational structure of the controller (see Section 4.3.2).

4.3.1 FKBC Computational Structure

The computational structure of a FKBC in the closed loop (Fig. 4.2) consists of a number of computational steps as presented in Fig. 4.4. There are five such computational steps

1. Input scaling (normalization),
2. Fuzzification of inputs,
3. Inference or rule firing,
4. Defuzzification of outputs,

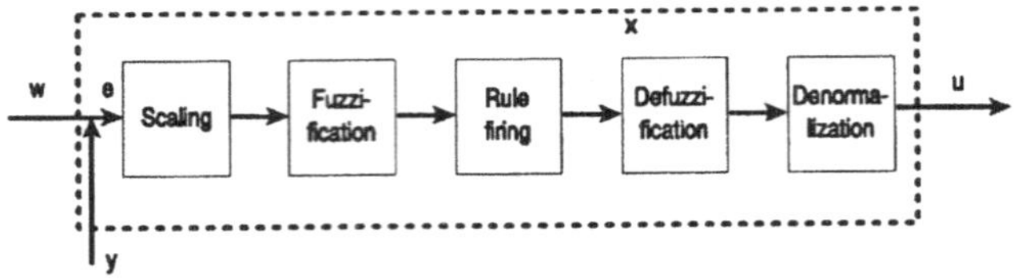

Fig. 4.4. FKBC computational structure.

5. Output-denormalization.

In what follows we consider each of the above computational steps for the case of a multiple-input single-output (MISO) FKBC. The generalization to the case of multiple-input multiple-output (MIMO), in which there are k control outputs $u_1, \ldots, u_k$, can be made easily.

Input Scaling

There are two principle cases in the context of input scaling:

1. The membership functions for the inputs (process state variables) and outputs (control output variables) of the controller are defined off-line on their physical domains. In this case the inputs and outputs of the controller are processed only using *fuzzification*, *rule firing* and *defuzzification*.

2. The membership functions for both inputs and outputs of the controller are defined off-line, on a common normalized domain. This means that the physical values of the actual inputs and outputs of the controller are mapped on the predetermined normalized domain. This mapping, called *normalization*, is done by the so-called *normalization factors*. *Scaling* is the multiplication of the physical input value with a *normalization factor* so that it is mapped onto the normalized input domain. *Denormalization* is the multiplication of the normalized output value with a *denormalization factor* so that it maps onto the physical output domain.
   The advantage of the second case is that *fuzzification*, *rule firing* and *defuzzification* can be designed independently of the physical domains of the inputs and output.

In the input scaling (normalization) procedure the error-vector $\mathbf{e} = \mathbf{x} - \mathbf{w}$ of physical signals is normalized with the help of a matrix $\mathbf{N_e}$ containing predetermined normalization factors for each component of $\mathbf{e}$:

$$\mathbf{e_N} = \mathbf{N_e} \cdot \mathbf{e}, \tag{4.3}$$

with

$$\mathbf{N_e} = \begin{pmatrix} N_{e_1} & 0 & \dots & 0 \\ 0 & N_{e_2} & \dots & 0 \\ \vdots & \vdots & \ddots & \vdots \\ 0 & 0 & \dots & N_{e_k} \end{pmatrix}, \tag{4.4}$$

where N_{e_i} are real numbers and the normalized domain for $\mathbf{e}$ is, say, $[-a, +a]$.

Example 4.1 Let the vector of errors be $\mathbf{e} = (e_1, e_2) = (e, \dot{e})$ with

$$e = x - w; \qquad \dot{e} = \dot{x} - \dot{w}. \tag{4.5}$$

Then, scaling (normalization) of e into e_N and $\dot{e}$ into $\dot{e}_N$ yields

$$e_N = N_e \cdot e; \qquad \dot{e}_N = N_{\dot{e}} \cdot \dot{e}, \tag{4.6}$$

where N_e and $N_{\dot{e}}$ are normalization factors. Using the phase plane of the system to be controlled, the normalization affects the angle of the so-called switching line which divides the plane into two semiplanes (see Section 4.4).

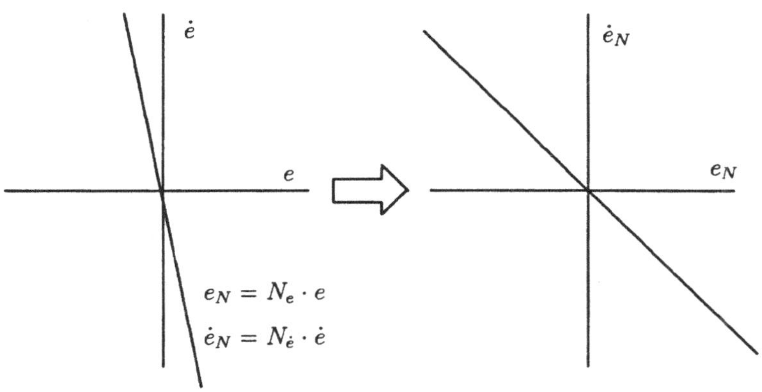

Fig. 4.5. Normalization of the phase plane.

In addition to that, we can see how the universes of discourse (domains of membership functions) change because of normalization (see Figs. 4.6. and 4.7.)

Fuzzification

In the next paragraphs, we consider only normalized values unless stated otherwise. Let $LE_1^{(i)}, \dots, LE_m^{(i)}$ be the linguistic values taken by the process state

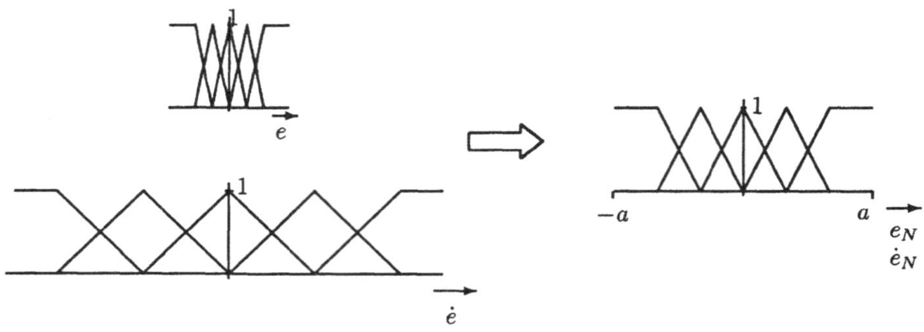

Fig. 4.6. Changing of supports in the presence of normalization.

variables e, $\dot{e}$ ..., $e^{(m)}$ in the rule-antecedent of the i-th rule. The meaning of each linguistic value $LE_k^{(i)}$ is represented by a membership function defined on the normalized domain $\mathcal{E}_k$ of the process state variable e_k. Thus the meaning of $LE_k^{(i)}$ is given by $\mu_{LE_k^{(i)}} : \mathcal{E}_k \to [0,1]$. Let us consider a normalized input vector

$$\mathbf{e}^* = (e_1^*, \ldots, e_m^*), \tag{4.7}$$

where each of e_k^* is the current normalized measurement. The *fuzzification* then consists of finding the membership degree of e_k^* in $\mu_{LE_k^{(i)}}$. This is done for every element of the $\mathbf{e}^*$.

Example 4.2 Suppose the rule

if e is *PS* *and* $\dot{e}$ is *NM* *then* u is *PM*.

In this example we have $\mathbf{e} = (e_1, e_2) = (e, \dot{e})$. Furthermore, let $e^* = a_1 = -0.5 \cdot a$ and $\dot{e}^* = a_2 = 0.1 \cdot a$. Then from Fig. 4.7 we obtain the degrees of membership $\mu_{NM}(a_1) = 0.65$ and $\mu_{PS}(a_2) = 0.3$.

Rule Firing

For a multi-input/single-output case (MISO) the s-th rule of the set of if-then rules has the form

if e_1 is $LE_1^{(s)}$ *and* ... *and* e_m is $LE_m^{(s)}$
then u is $LU^{(s)}$,

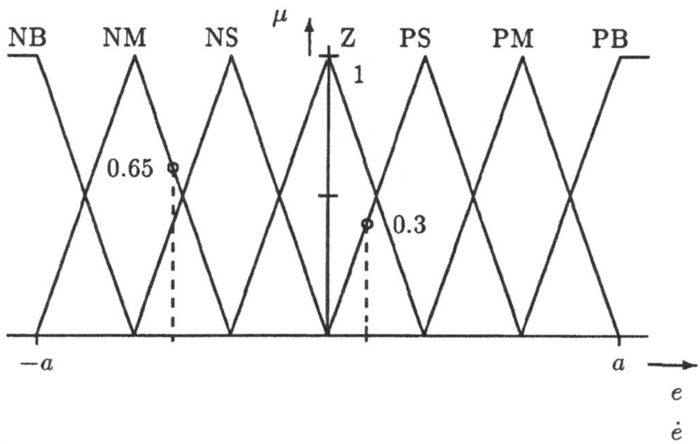

Fig. 4.7. Fuzzification of crisp values e^* and $\dot{e}^*$.

where $e_i = x_i - w_i$ is the error with respect to the i-th process state variable x_i, and u is the control output variable, which take linguistic values $LE_i^{(s)}$ and $LU^{(s)}$ respectively. The membership functions of these two linguistic values are denoted by $\mu_{LE_i^{(s)}}$ and $\mu_{LU^{(s)}}$ defined on a common normalized domain, i.e.,

$$\mathcal{E}_1 = \ldots = \mathcal{E}_m = \mathcal{U} \tag{4.8}$$

Given an input vector consisting of the normalized measurements of $e_1^*, \ldots, e_m^*$, the output of the s-th rule is $\mu_{CLU^{(s)}}$ i.e., the clipped membership function for the control output variable u (see Section 2.3.7). The working principle corresponds to the block scheme of Fig. 4.8.

In the next step, the clipped output fuzzy sets for each rule are combined in the following way:

$$\forall u : \mu_U(u) = \max\left(\mu_{CLU^{(1)}}, \ldots, \mu_{CLU^{(s)}}\right), \tag{4.9}$$

where μ_U is the membership function representing the value of the overall control output.

Defuzzification

The result of the rule firing is a fuzzy set μ_U as defined in (4.9). The purpose of defuzzification is to obtain a scalar value u from μ_U. This is done using the Center-of-Area/Gravity. In the continuous case we have

$$u = \frac{\int\limits_{u \in \mathcal{U}} \mu_U(u) \cdot u \, du}{\int\limits_{u \in \mathcal{U}} \mu_U(u) \, du} \tag{4.10}$$

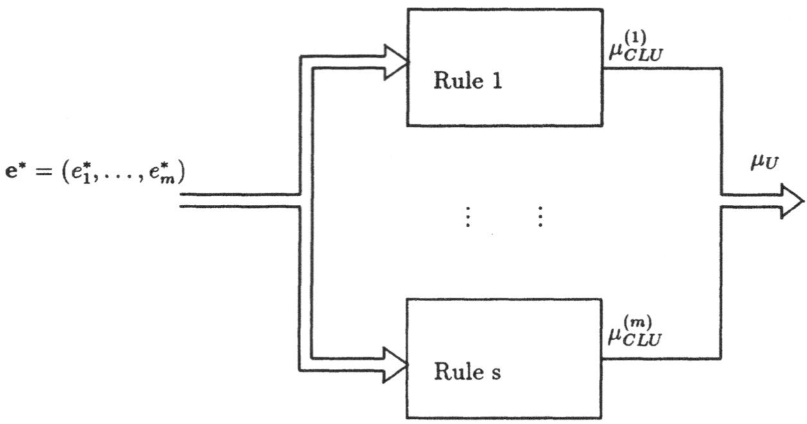

Fig. 4.8. A scheme of the rule firing.

and for the discrete case

$$u = \frac{\sum_{u \in \mathcal{U}} \mu_U(u) \cdot u}{\sum_{u \in \mathcal{U}} \mu_U(u)}. \tag{4.11}$$

Example 4.3 Let the common normalized domain $\mathcal{U} = \{1, 2, \dots, 8\}$ and

$$\mu_U = (0/1, 0.2/2, 0.5/3, 0.8/4, 1/5, 0.5/6, 0.2/7, 0/8). \tag{4.12}$$

Then

$$u = \frac{0 \cdot 1 + 0.2 \cdot 2 + 0.5 \cdot 3 + 0.8 \cdot 4 + 1 \cdot 5 + 0.5 \cdot 6 + 0.2 \cdot 7 + 0 \cdot 8}{0 + 0.2 + 0.5 + 0.8 + 1 + 0.5 + 0.2 + 0} = 4.53 \tag{4.13}$$

(see Fig. 4.9).

Denormalization

In the denormalization procedure, the control value u_N obtained after defuzzification is denormalized with the help of an off-line predetermined scalar denormalization factor N_u. Let the normalization procedure be

$$u_N = N_u \cdot u. \tag{4.14}$$

Then the denormalization procedure is simply the inverse method

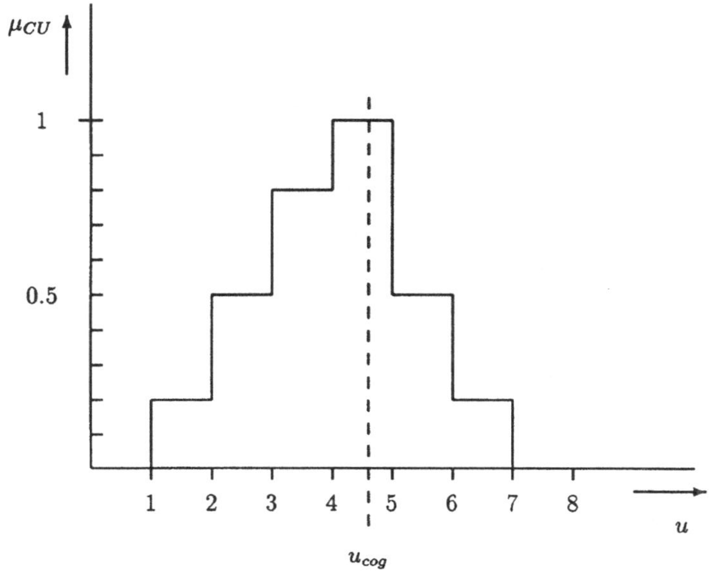

Fig. 4.9. Defuzzification of a fuzzy set.

$$u = N_u^{-1} \cdot u_N. \tag{4.15}$$

As we see in Section 4.4.2, the choice of N_U essentially determines, together with the scaling factors, the stability of the system to be controlled.

4.3.2 The Nonlinearity of the Controller

System theory distinguishes between two types of systems: linear and nonlinear. The linearity of a system is characterized by the following two properties:

1. Additivity property or superposition property:

   Let

   $$y_1 = f(x) \qquad \text{and} \qquad y_2 = f(z)$$

   Then for the additivity property to be obeyed it is required that

   $$y_1 + y_2 = f(x + z). \tag{4.16}$$

   Hence

   $$f(x) + f(z) = f(x + z). \tag{4.17}$$

2. Scaling property or homogenity property:
   Let
   $$y = f(x). \tag{4.18}$$
   Then for the scaling property to be obeyed it is required that
   $$\alpha \cdot y = f(\alpha \cdot x)$$
   $$\alpha \cdot f(x) = f(\alpha \cdot x).$$

All systems which do not have these properties are nonlinear systems. Because of *fuzzification* and *defuzzification*, a FKBC is a crisp TE. The type of the membership functions and the defuzzification procedure lead to a nonlinear transfer characteristic. The argument for this is that if one element within the computational structure of the controller is nonlinear then the whole transfer characteristic is nonlinear too. Looking at the *additivity property* and *scaling property* of a linear system we check each step of the computation of the control output with respect to these properties:

Let

if x is LX *then* u is LU

be the only control rule where LX and LU are the linguistic values taken by the process state variable x and the control output variable u. The meaning of these two linguistic values is given by the membership functions $\mu_{LX} : \mathcal{X} \to [0,1]$ and $\mu_{LU} : \mathcal{U} \to [0,1]$.

Furthermore, let x_1 and x_2 be two crisp inputs and u_1 and u_2 be their respective crisp outputs. Then the sources of nonlinearity are as follows.

Scaling

Scaling is a linear element because it simply multiplies the inputs with a scalar N_x:
$$N_x \cdot x_1 + N_x \cdot x_2 = N_x \cdot (x_1 + x_2) \tag{4.19}$$
and
$$\alpha \cdot N_x \cdot x_1 = N_x \cdot (\alpha \cdot x_1). \tag{4.20}$$

Denormalization

Likewise, normalization and denormalization are also linear elements.

Fuzzification

Let the membership function μ_{LX} of the linguistic value LX be, in general, a nonlinear function. Then the fuzzification of x_1 and x_2 result in finding $\mu_{LX}(x_1)$ and $\mu_{LX}(x_2)$.

Linearity requires
$$\mu_{LX}(x_1) + \mu_{LX}(x_2) = \mu_{LX}(x_1 + x_2). \tag{4.21}$$
which, however, cannot be fulfilled because of the nonlinear characteristic of μ_{LX}.

Rule Firing
Let the membership function μ_{LU} of the linguistic value LU be, in general, a nonlinear function. Then we obtain as the result of firing the rule for input x_1

$$\forall u : \mu'_{CLU}(u) = \mu_{LX}(x_1) \wedge \mu_{LU}(u), \tag{4.22}$$

and for input x_2

$$\forall u : \mu''_{CLU}(u) = \mu_{LX}(x_2) \wedge \mu_{LU}(u). \tag{4.23}$$

Linearity requires

$$\forall u : \mu'_{CLU}(u) + \mu''_{CLU}(u) = \mu_{LX}(x_1 + x_2) \wedge \mu_{LU}(u). \tag{4.24}$$

This equation cannot be fulfilled because:

μ_{LU} is a nonlinear function,

μ'_{CLU} and μ''_{CLU} are nonlinear functions,

The operation "$\wedge$" is nonlinear for $\wedge$=min.

Defuzzification
Let the defuzzification procedure be performed with the help of the Center-of-Area method (Section 3.6). Furthermore, let u_1 and u_2 be the defuzzification results obtained by

$$u_1 = \frac{\sum\limits_{U} \mu'_{CLU}(u) \cdot u}{\sum\limits_{U} \mu'_{CLU}(u)} \tag{4.25}$$

and

$$u_2 = \frac{\sum\limits_{U} \mu''_{CLU}(u) \cdot u}{\sum\limits_{U} \mu''_{CLU}(u)}. \tag{4.26}$$

Linearity requires

$$u_1 + u_2 = \frac{\sum\limits_{U} (\mu'_{CLU}(u) + \mu''_{CLU}(u)) \cdot u}{\sum\limits_{U} (\mu'_{CLU}(u) + \mu''_{CLU}(u))}. \tag{4.27}$$

which cannot be fulfilled. Instead of this equation we have

$$u_1 + u_2 = \frac{\sum\limits_{U} \mu'_{CLU}(u) \cdot u}{\sum\limits_{U} \mu'_{CLU}(u)} + \frac{\sum\limits_{U} \mu''_{CLU}(u) \cdot u}{\sum\limits_{U} \mu''_{CLU}(u)}, \tag{4.28}$$

which shows that the nonlinearity comes from the normalization with regard to the sums $\sum_U \mu'_{CLU}$ and $\sum_U \mu''_{CLU}$, respectively.

From this explanation it is readily seen that a FKBC is a TE whose sources of nonlinearity are *the nonlinearity of membership functions, rule firing* and *defuzzification*. Nevertheless, FKBC can be linearized piecewise, so that they locally work as linear TE (see paragraph "Fuzzy Sliding Mode" in Section 4.4.2).

Discontinuities in the Output

Because of special properties of the membership functions, discontinuities in the transfer characteristic can be the result which affects the control performance negatively. Discontinuities in the output mainly arise at discontinuity points in the membership functions. This can be explained by a SISO-system with x as a crisp input and u as crisp (defuzzified) output.
Assume the two rules

$R^{(1)}$: *if* x is S *then* u is B
$R^{(2)}$: *if* x is B *then* u is S

Figure 4.10 shows the membership functions assumed for the linguistic values of x and u. Figure 4.11 shows the corresponding transfer characteristic between x and u. The points x_1 and x_3 are discontinuity points of the membership functions for x, and the u_i are the defuzzified control outputs for x_i.

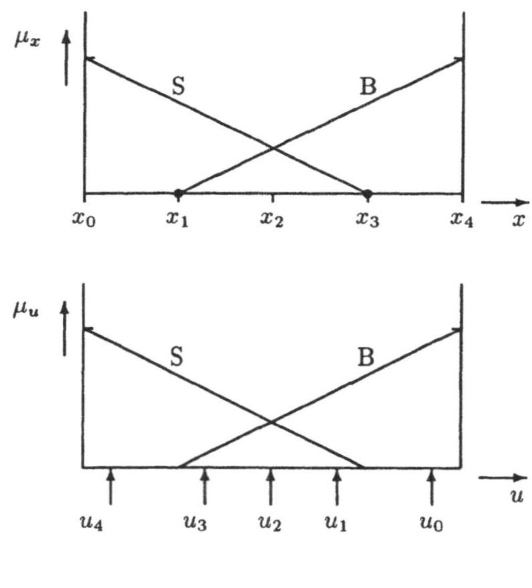

Fig. 4.10. Membership functions for S and B.

The commonly used functions with triangular shape and trapezoidal functions have discontinuous points. Figure 4.12 shows the output of a FKBC where the discontinuities can be recognized easily at the marked edges. However, if discontinuities in the output actually lead to worse control behavior, they can be avoided by either

- smoothing out the membership functions, or

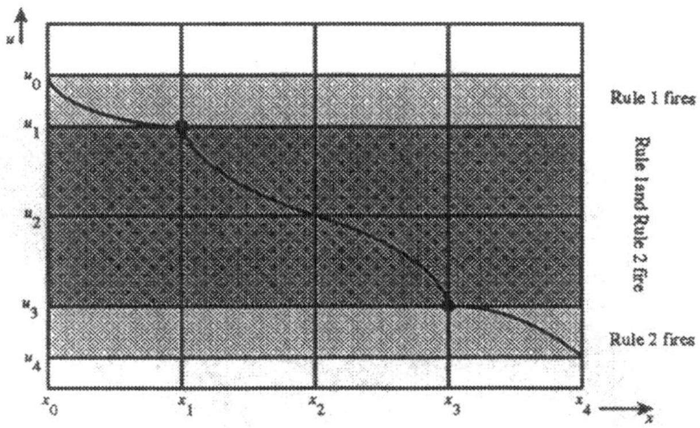

Fig. 4.11. Discontinuities in the transfer characteristic of a FKBC.

- lowpass filtering of the output signal.

4.3.3 Rule Based Representation of Conventional TE

This section deals with the connection between conventional TE (Transfer Elements) and rule based TE. The transition between these two concepts reveals similarities between different types of FKBC (Mamdani FKBC and Sugeno FKBC, see [129, 200, 204]). In a Mamdani FKBC the meaning of each rule is represented by Mamdani implication.

Conventional TE can be classified into linear and nonlinear TE. TE are dynamical systems in so far that they are represented by a *functional* of an input vector $\mathbf{x} = (x_1, x_2, \ldots, x_n)^T$. Linear TE provide a linear relationship between $\mathbf{x}$ and u corresponding to

$$u = f(\mathbf{x}) = a_1 \cdot x_1 + a_2 \cdot x_2 + \ldots + a_n \cdot x_n. \tag{4.29}$$

A nonlinear description starts from the nonlinear combination of the components of $\mathbf{x}$:

Example 4.4 From

$$\frac{1}{u} = \frac{b_1}{x_1} + \frac{b_2}{x_2} \tag{4.30}$$

we get the nonlinear function

$$u = \frac{x_1 \cdot x_2}{b_1 \cdot x_1 + b_2 \cdot x_2}. \tag{4.31}$$

Let the domain of x_i be $\mathcal{X}_i$, $i = 1, \ldots, n$. Furthermore, each $\mathcal{X}_i$ is divided into m disjoint intervals

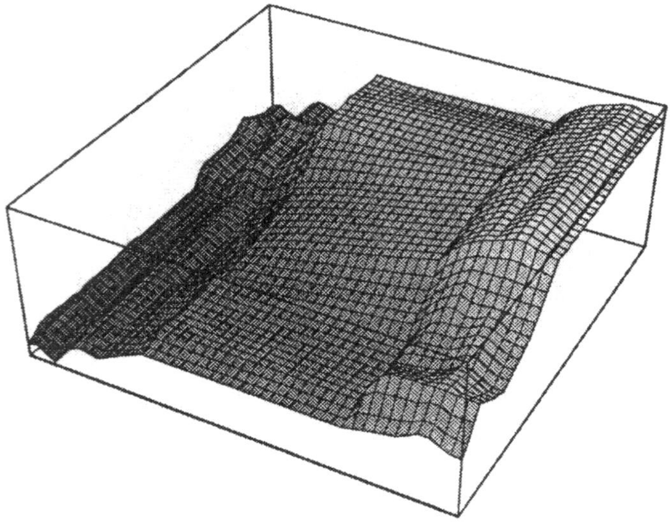

Fig. 4.12. Transfer characteristic of a FKBC.

$$I_{i1}, \ldots, I_{im}, \quad \text{where } \forall j, \ell : I_{ij} \cap I_{i\ell} = \emptyset. \tag{4.32}$$

Then the k-th rule describing the relationship between process state variables and control output is given as

$$R^{(k)}\text{: } \textit{if } \mathbf{x} \in (I_{1k} \times \ldots \times I_{nk}) \textit{ then } u = u_k,$$

or

$$R^{(k)}\text{: } \textit{if } x_1 \in I_{1k} \textit{ and } \ldots \textit{ and } x_n \in I_{nk} \textit{ then } u = u_k,$$

where $u_k \in \mathcal{U}$ and I_{jk} is the interval describing the interval-value of x_j in the k-th rule.

This means that in the case of m intervals for each element of $\mathbf{x}$ we obtain $n \cdot m$ cells in the input space each of which corresponds to one rule $R^{(k)}$ (For a 2-dimensional state-vector see Fig. 4.13). In terms of membership functions, each interval I_{jk} of $\mathbf{x}$ can be represented as a fuzzy set with membership degrees either 0 or 1 (see Fig. 4.14).

For the k-th fuzzy rule of a second order TE in which $x_1 = x$ and $x_2 = \dot{x}$ we then obtain

$$R^{(k)}\text{: } \textit{if } x \text{ is } I_{xk} \textit{ and } \dot{x} \text{ is } I_{\dot{x}k} \textit{ then } u = u_k.$$

The membership functions of I_{xk}, $I_{\dot{x}k}$ and u_k are given by

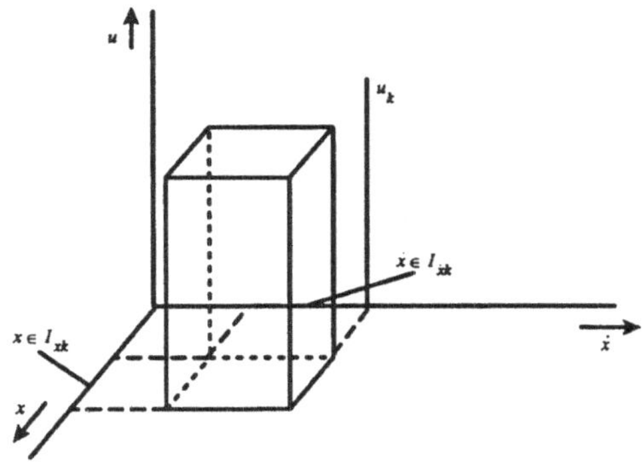

Fig. 4.13. Discrete cells in the state space.

$$\forall x : \mu_{I_{xk}}(x) = \begin{cases} 1 & \text{if } x \in I_{xk} \\ 0 & \text{otherwise.} \end{cases} \tag{4.33}$$

$$\forall \dot{x} : \mu_{I_{\dot{x}k}}(x) = \begin{cases} 1 & \text{if } \dot{x} \in I_{\dot{x}k} \\ 0 & \text{otherwise.} \end{cases} \tag{4.34}$$

$$\forall u : \mu_{u_k}(u) = \begin{cases} 1 & \text{if } u = u_k \\ 0 & \text{otherwise.} \end{cases} \tag{4.35}$$

The way of computing the output is the following. In case of measurements x^* and $\dot{x}^*$, which belong to I_{xk} and $I_{\dot{x}k}$ respectively, we obtain

- From a measured value x^* we obtain $\mu_{I_{xk}}(x^*) = 1$, and for all $r \neq k$, $\mu_{I_{xk}}(x^*) = 0$ because $I_{xk} \cap I_{xr} = \emptyset$.
- From a measured value $\dot{x}^*$ we obtain $\mu_{I_{\dot{x}k}}(\dot{x}^*) = 1$ and for all $r \neq k$, $\mu_{I_{\dot{x}k}}(\dot{x}^*) = 0$.
- The conjunction of the previous two yields $\mu_{I_{xk}}(x^*) \wedge \mu_{I_{\dot{x}k}}(\dot{x}^*) = 1$. Thus one obtains that only the control output of rule $R^{(k)}$ is u_k since $\mu_{u_k}(u) = \mu_{I_{xk}}(x^*) \wedge \mu_{I_{\dot{x}k}}(\dot{x}^*) = 1$ and the only u which has degree of membership 1 to μ_{U_k} is u_k.
- Finally, the result of the defuzzification is

$$u = \frac{\sum_{\mathcal{U}} \mu_{U_k}(u_k) \cdot u_k}{\sum_{\mathcal{U}} \mu_{U_k}(u_k)} = \frac{0 + 0 + \ldots + 1 \cdot u_k + \ldots + 0}{0 + 0 + \ldots + 1 + \ldots + 0} = u_k. \tag{4.36}$$

The change into a rule base employing fuzzy values can be made in three steps (see also Fig. 4.15):

1. Enlarging the intervals within the universe of discourse of the input space but keeping them disjoint,

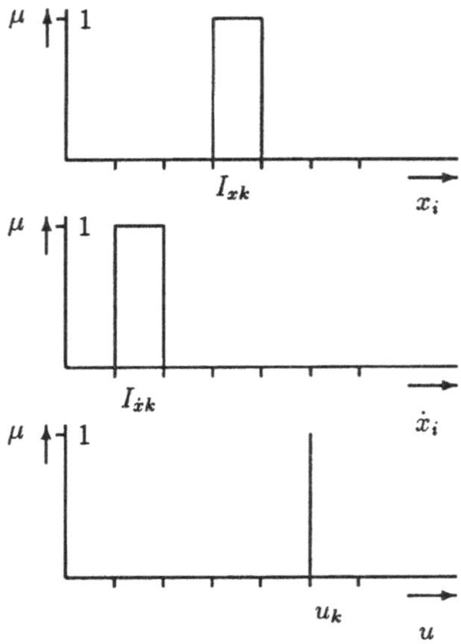

Fig. 4.14. Membership functions in the discrete case.

2. Making the intervals overlap.
3. Fuzzification of the membership functions.

A different way of describing conventional TE by means of rules is to present the u as a crisp real function of the input values, i.e., $u = f(x_1, \ldots, x_n)$ (see also the Sugeno FKBC in Section 4.3.3). The input domain for each variable x_i is divided again into m disjoint intervals as in the previous case. Then the form of a rule is given as

$$R^{(k)}\text{: } \textit{if } \mathbf{x} \in (I_{1k} \times \ldots \times I_{nk}) \textit{ then } u = f_k(x_1, \ldots, x_n),$$

where

$$\mu_{\mathbf{x}} = \begin{cases} 1 & \text{if } \mathbf{x} \in (I_{1k} \times \ldots \times I_{nk}), \\ 0 & \text{otherwise.} \end{cases} \tag{4.37}$$

which means that given a particular input vector $\mathbf{x}^*$ only *one* rule can fire. Furthermore, for each k and r, $f_k(x_1, \ldots, x_n)$ is different from $f_r(x_1, \ldots, x_n)$.

Example 4.5 The rules for a second order system are (see Fig. 4,16)

$$R^{(1)}: \quad \textit{if } \mathbf{x} \in (I_{x1} \times I_{\dot{x}1}) \textit{ then } u_1 = f_1(x, \dot{x}), \tag{4.38}$$

$$R^{(2)}: \quad \textit{if } \mathbf{x} \in (I_{x2} \times I_{\dot{x}2}) \textit{ then } u_2 = f_2(x, \dot{x}), \tag{4.39}$$

where

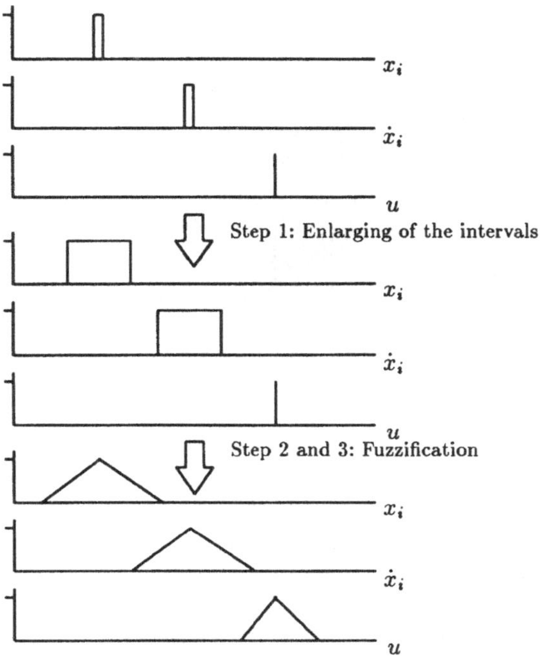

Fig. 4.15. Change from crisp to fuzzy values of variables.

$$u_1 = f_1(x, \dot{x}) = a_{01} + a_{11} \cdot x + a_{21} \cdot \dot{x}; \quad x \in I_{x1}, \quad \dot{x} \in I_{\dot{x}1} \tag{4.40}$$

$$u_2 = f_2(x, \dot{x}) = a_{02} + a_{12} \cdot x + a_{22} \cdot \dot{x}; \quad x \in I_{x2}, \quad \dot{x} \in I_{\dot{x}2}. \tag{4.41}$$

Overlapping and fuzzification leads to the Sugeno FKBC of Section 4.4.3.

4.4 Types of FKBC

Using fuzzy control methods one can fall back on well known control structures which, however, are varied, extended, and even mixed with conventional techniques. For a classification of systems and controllers see Fig. 4.1. In this section the following basic FKBC structures with some variants are:

1. PID-like FKBC with the variants

   PD-like FKBC,
   PI-like FKBC.

2. Sliding mode FKBC or SMFC with the variants

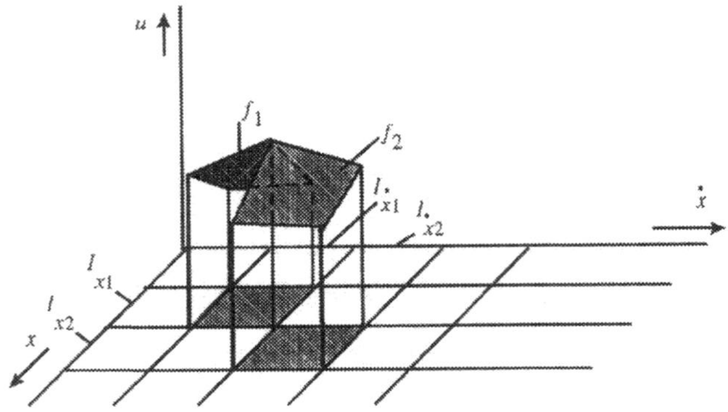

Fig. 4.16. Graphical representation of crisp rules with analytical output.

SMFC with boundary layer,

SMFC with boundary layer and compensation term,

SMFC of higher order.

3. FKBC according to Sugeno and Takagi.

It has to be mentioned that PID-like FKBC and SMFC are Mamdani FKBC (see Section 2.4.1). The PID-like FKBC is widely used for second order systems with both linear and nonlinear characteristics. The SMFC and its variants are extensions of the PID-like FKBC. Their field of application mainly deals with nonlinear systems of higher order with considerable model uncertainties and disturbances. The Sugeno FKBC also deals with highly nonlinear plants where control actions can be locally described by linear or nonlinear control consequences.

4.4.1 PID-like FKBC

The basic model of the process to be controlled by a PID-like FKBC is a system of n-th order (linear or nonlinear) with the state equation

$$\begin{aligned} x^{(n)} &= f(x, \dot{x}, \ldots, x^{(n-1)}, t, u) \\ y &= g(\mathbf{x}) \end{aligned} \tag{4.42}$$

with

$\mathbf{x}$ — process state vector $\mathbf{x} = (x_1, x_2, \ldots, x_n) = (x, \dot{x}, \ldots, x^{(n-1)})^T$
y — process output
u — control variable

t — time parameter
f, g — linear or nonlinear functions.

The basic idea of a PID-controller is to choose the control law by considering error $e = x - x_d$, change-of-error $\dot{x} - \dot{x}_d$ and integral-of-error $\delta = \frac{1}{T}\int_0^t e\,dt$:

$$u_{\text{PID}} = K_P \cdot e + K_D \cdot \dot{e} + K_I \cdot \frac{1}{T}\int_0^t e \cdot \mathrm{d}t, \tag{4.43}$$

where x_d is the desired value (set-point).

For a *linear process* the control parameters K_P, K_D and K_I are designed in such a way that the closed-loop control is stable. The corresponding analysis can be done by means of the knowledge of process parameters (e.g., mass, damper, and spring of a mechanical system) taking into account special performance criteria (e.g., sum of the squares of the error).

In the case of *nonlinear processes* which can be linearized around the operating point, conventional PID-controllers also work successfully. However, a PID-controller with constant parameters in the whole working area is robust but not optimal. In this case, tuning of the PID-parameters has to be performed. FKBC of the PID type start from the same assumptions which are decisive for a conventional PID-controller:

- The process is either linear or can be piecewise linearized,
- The process can be stabilized taking into account selected performance criteria.

The FKBC corresponding to the PID-controller

$$u = K_P \cdot e + K_D \cdot \dot{e} + K_I \cdot \delta \tag{4.44}$$

has the following rule form:

$R^{(i)}_{\text{PID}}$: *if* e is $LE^{(i)}$ *and* $\dot{e}$ is $L\dot{E}^{(i)}$ *and* δ is $L\delta^{(i)}$ *then* u is $LU^{(i)}$,

with

e — error defined on the non-normalized domain $\mathcal{E}$,
$\dot{e}$ — change-of-error defined the non-normalized domain $\dot{\mathcal{E}}$,
δ — integral-of-error defined on the non-normalized domain Δ,
u — control variable defined on the non-normalized domain $\mathcal{U}$.

$LE^{(i)}$, $L\dot{E}^{(i)}$, $L\delta^{(i)}$, $LU^{(i)}$ are the linguistic values of e, $\dot{e}$, δ, u in the i-th rule.

After normalization, the domains of $e, \dot{e}, \delta$, and u are all normalized onto the same universe of discourse:

$$\begin{aligned} e_N &= N_e \cdot e \\ \dot{e}_N &= N_{\dot{e}} \cdot \dot{e} \\ \delta_N &= N_\delta \cdot \delta \\ u_N &= N_u \cdot u. \end{aligned} \tag{4.45}$$

Hence inference and defuzzification are performed on this common universe of discourse. The corresponding normalized PID-control law is

$$u_N = K_{P_N} \cdot e_N + K_{D_N} \cdot \dot{e}_N + K_{I_N} \cdot \delta_N. \tag{4.46}$$

The connection between normalization (scaling) factors and PID-control parameters can be derived by the following procedure:

In the normalized space the PID-control parameters are set equal to one:

$$K_{P_N} = K_{D_N} = K_{I_N} = 1. \tag{4.47}$$

With this, one obtains for K_P, K_D, and K_I and the corresponding normalization factors

$$\begin{aligned} K_P &= N_e/N_u \\ K_D &= N_{\dot{e}}/N_u \\ K_I &= N_\delta/N_u. \end{aligned} \tag{4.48}$$

For

$$K_P \cdot e + K_D \cdot \dot{e} + K_I \cdot \delta < 0 \tag{4.49}$$

the PID-controller produces a negative control output (control variable) and for

$$K_P \cdot e + K_D \cdot \dot{e} + K_I \cdot \delta > 0 \tag{4.50}$$

a positive one.

The plane in which the control output u changes its sign is called a *switching surface* with its equation given as

$$s = \dot{e} + \frac{K_P}{K_D} \cdot e + \frac{K_I}{K_D} \cdot \delta = 0. \tag{4.51}$$

For stability reasons, all of K_P, K_I, $K_D < 0$ is required. The corresponding equation for the switching surface in the normalized space is (see Fig. 4.17)

$$s_N = e_N + \dot{e}_N + \delta_N = 0. \tag{4.52}$$

Constant u_N-values correspond to planes $u_N = s_N = const$ which are parallel to $s_N = 0$. Here *const* is the point where s_N intersects the ordinate δ. In the normalized phase space we obtain the result shown in Fig. 4.18.

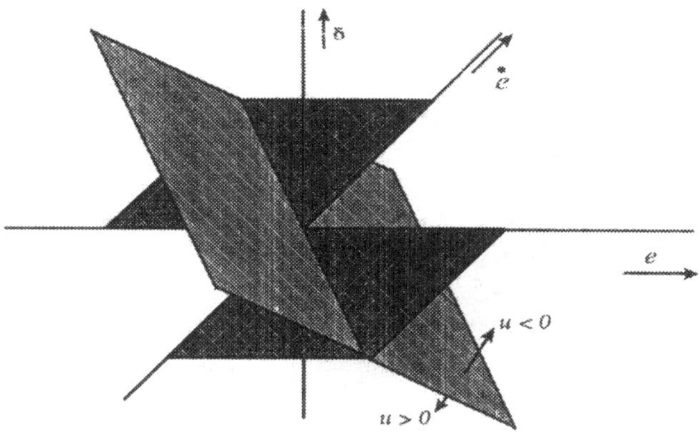

Fig. 4.17. Switching surface of a PID-controller.

Neglecting the integral term we obtain the PD-like FKBC. From equations (4.48) and (4.51) we obtain the switching lines for the non-normalized and normalized phase plane

$$s = \dot{e} + \frac{N_e}{N_{\dot{e}}} \cdot e = 0 \tag{4.53}$$

and

$$s_N = \dot{e}_N + e_N = 0, \tag{4.54}$$

respectively, at which the sign of the control variable changes (see Fig. 4.20). The corresponding rules for the PD-like FKBC within the normalized phase plane are

$R^{(i)}_{\mathrm{PD}}$: *if* e is $LE^{(i)}$ *and* $\dot{e}$ is $L\dot{E}^{(i)}$ *then* u is $LU^{(i)}$.

Changing the control output variable u_N of the previous rule into its derivative $\dot{u}_N$ and adding an integrator after the FKBC we obtain the so-called PI-like FKBC

$R^{(i)}_{\mathrm{PI}}$: *if* e is $LE^{(i)}$ *and* $\dot{e}$ is $L\dot{E}^{(i)}$ *then* $\dot{u}$ is $L\dot{U}^{(i)}$.

Figure 4.20 shows conventional PID-, PD-, and PI-controllers and their phase spaces from which the corresponding fuzzy control rules can be derived.

The following example of a PD-like FKBC is used to illustrate the design procedure for such a controller using the phase plane approach:

Example 4.6 Suppose that we have obtained the following rule base for the control variable of a PD-controller from the two-dimensional phase plane of the system in terms of e and $\dot{e}$ (see Fig. 4.21): With regard to the number of cells in the phase plane one obtains 21 rules for the positive part and the negative part, respectively. The corresponding rule for the cell 1 is given by

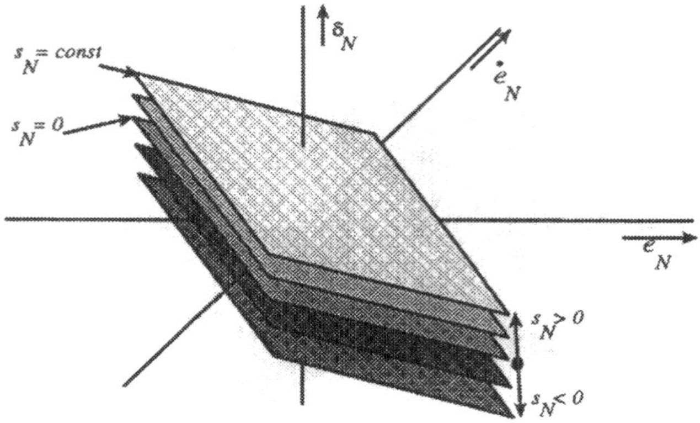

Fig. 4.18. Normalized phase space for the PID-like FKBC.

$R^{(1)}$: *if* e is *PS and* $\dot{e}$ is *PM then* u is *NM*

with *PS*, *PM*, *NM* denoting Positive Small, Positive Medium and Negative Medium respectively. Furthermore, let the membership functions representing the meaning of the above linguistic values be stored in look-up tables and defined over the common normalized universe $e_N, \dot{e}_N, u_N \in (-50, +50)$ (see Table 4.1).

Suppose that the values of e_N and $\dot{e}_N$ at the time t are $e_N^* = 27$ and $\dot{e}_N^* = 14$. Hence, the corresponding grades of membership for these two values are $\mu_{PS}(27) = 0.44$ and $\mu_{PM}(14) = 0.2$. The *and*-connection yields

$$\min(0.44, 0.2) = 0.2. \tag{4.55}$$

According to the j-th element of the control output fuzzy set of the i-th rule, the individual-rule based inference results in

$$\forall u : \mu_{CNM}(u) = \min(0.2, \mu_{NM}(u)). \tag{4.56}$$

For the first rule in our example this results in

$$\begin{aligned}&(0 \wedge 0.2, 0.2 \wedge 0.2, 0.6 \wedge 0.2, 0.8 \wedge 0.2, 1 \wedge 0.2, 0.8 \wedge 0.2, 0.6 \wedge 0.2,\\ &0.2 \wedge 0.2, 0 \wedge 0.2, 0 \wedge 0.2, \ldots, 0 \wedge 0.2)\\ &= (0, 0.2, 0.2, 0.2, 0.2, 0.2, 0.2, 0.2, 0.2, 0, 0, 0, 0, 0, 0, 0, 0, 0, 0, 0, 0).\end{aligned}$$

In this way there is one rule for each cell and one specific clipped control output value for each rule. All the resulting clipped values are combined by an *or*-operation using the maximum operator.

According to the next two cells in Table 4.1 we have the rule

$R^{(2)}$: *if* e_N is *PM and* $\dot{e}_N$ is *PS then* u_N is *NM.*

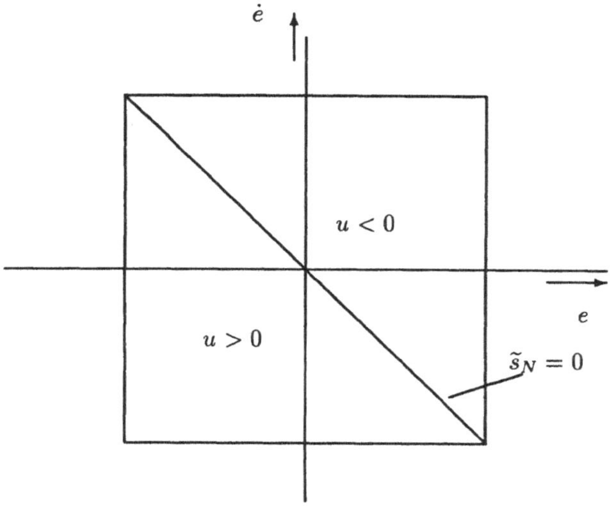

Fig. 4.19. Normalized phase plane of the PD-like FKBC.

Inference with this rule given $e_N^* = 27$ and $\dot{e}_N^* = 14$ results in the membership degrees

$$(0, 0.2, 0.6, 0.78, 0.78, 0.78, 0.6, 0.2, 0, 0, 0, 0, 0, 0, 0, 0, 0, 0, 0, 0, 0)$$

for the elements of $\mathcal{U}_N$, i.e., a clipped control output fuzzy set. Then the membership function for the combined control output is computed by taking the maximum of the two membership functions from above, and is given as

$$(0, 0.2, 0.6, 0.78, 0.78, 0.78, 0.6, 0.2, 0, 0, 0, 0, 0, 0, 0, 0, 0, 0, 0, 0, 0).$$

These membership degrees describe the degree to which the elements $u_N \in \mathcal{U}_N$ belong to the combined control output of rules $R^{(1)}$ and $R^{(2)}$. Thus we have that

$$\begin{aligned} &0.2/(-45, -40] + 0.6/(-40, -35] + 0.78/(-35, -30] + 0.78/(-30, -25] \\ &+ 0.78/(-25, -20] + 0.6/(-20, -15] + 0.2/(-15, -10]. \end{aligned} \tag{4.57}$$

This is repeated for each rule $R^{(i)}$. For example, let the final output be the fuzzy set obtained as the result of inference with $R^{(1)}$ and $R^{(2)}$. Then we obtain the Center-of-Area of this fuzzy set as follows:

$$\frac{(-42.5) \cdot 0.2 + (-37.5) \cdot 0.6 + \ldots + (-12.5) \cdot 0.2)}{0.2 + 0.6 + \ldots + 0.2} = 27.55, \tag{4.58}$$

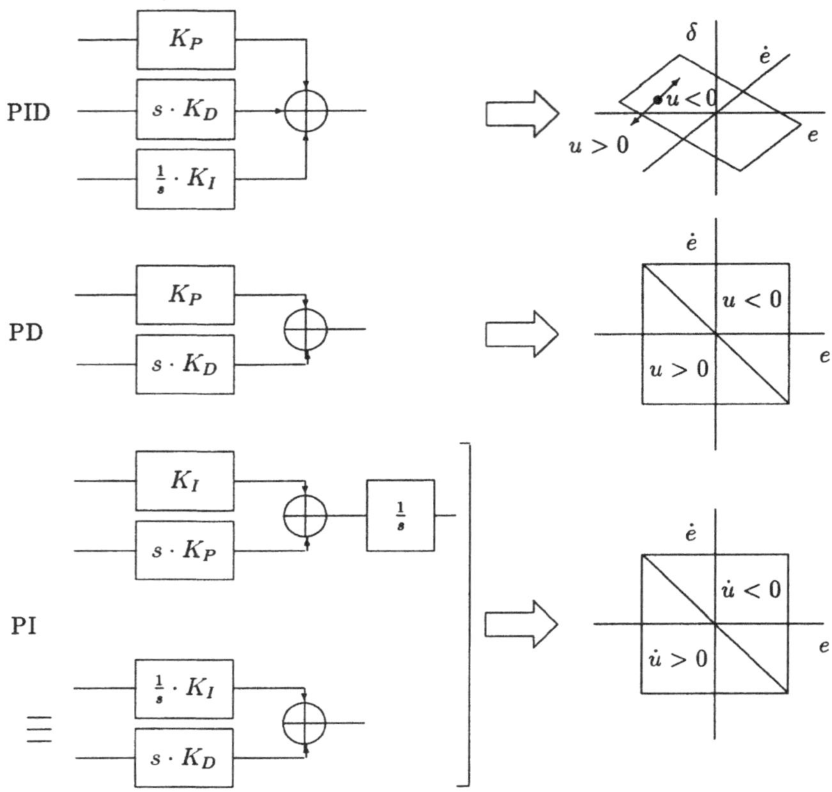

Fig. 4.20. Conventional controllers and their related phase spaces.

where, -42.5 is the center of the interval $(-45, -40]$, and so on.

In the case of the same cardinality of the term-sets for error e_N and change-of-error $\dot{e}_N$, the number of cells and, with that, the number of rules can be computed by

$$n = (\text{cardinality}(\mathcal{E}_N) + 1) \times \text{ number of cardinality}(\dot{\mathcal{E}}_N)). \tag{4.59}$$

The "1" has been added because there are diagonal elements both above and below the switching line. With a finer scaling, the number of rules increases quadratically. For a PID-like FKBC the number of rules increases as the third power of the number of fuzzy sets.

In the next section we show how to minimize the number of rules also in the case of FKBC of higher order. To the PID-like FKBC controller design also

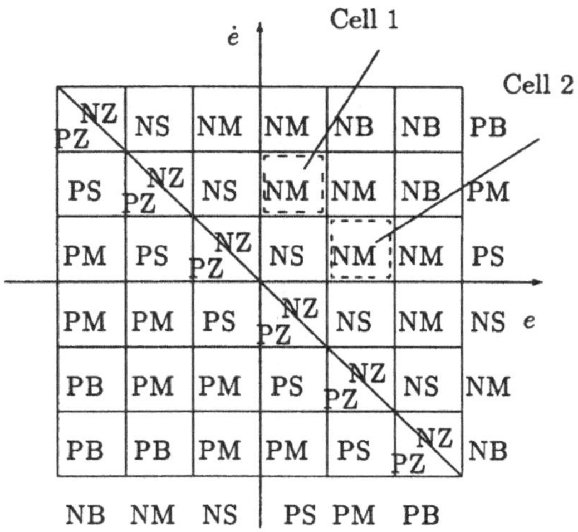

Fig. 4.21. Rule base for a PD-like FKBC.

belongs the proper choice of the normalization factors (see equation (4.45)). In the case of a linear system to be controlled, this can be done by conventional design methods adopted from linear control theory. For a large class of nonlinear systems, we show in the next section that under certain assumptions about the system under control, the normalization factors can be obtained easily.

4.4.2 Sliding Mode FKBC

As already described in the previous section, for a large class of nonlinear systems FKBC are designed with respect to phase plane determined by error e and change-of-error $\dot{e}$ with respect to the states x and $\dot{x}$ [173, 210, 223]. A fuzzy value for the control variable is determined according to fuzzy values of error and change-of-error. The general approach to control design is the division of the phase plane into two semi-planes, by means of a *switching* line. Within the semi-planes positive and negative control outputs are produced, respectively. The magnitude of the control outputs depends on the distance of the state vector from the switching line. Considering the large field of FKBC applications one is tempted to ask why the FKBC are so successful. This can be answered for a certain class of systems as follows: for a specific class of nonlinear systems there is an appropriate robust control method called *sliding mode control* [220, 195]. This control method can be applied in the presence of model uncertainties, parameter fluctuations and disturbances provided that the upper bounds of their absolute values are known. The sliding mode control is especially appropriate for the tracking control of robot manipulators and also for motors whose mechanical loads change over a wide range. The disadvantage of

Table 4.1. Discretized membership functions for $e_N, \dot{e}_N$ and u_N

interval	e_N is PS	e_N is PM	$\dot{e}_N$ is PS	$\dot{e}_N$ is PM	u_N is NM
$(-50, -45]$	0	0	0	0	0
$(-45, -40]$	0	0	0	0	0.2
$(-40, -35]$	0	0	0	0	0.6
$(-35, -30]$	0	0	0	0	0.8
$(-30, -25]$	0	0	0	0	1
$(-25, -20]$	0	0	0	0	0.8
$(-20, -15]$	0	0	0	0	0.6
$(-15, -10]$	0	0	0	0	0.2
$(-10, -5]$	0	0	0	0	0
$(-5, 0]$	0	0	0	0	0
$(0, 5]$	1	0	1	0	0
$(5, 10]$	0.88	0	0.88	0	0
$(10, 15]$	0.78	0.2	$\boxed{0.78}$	$\boxed{0.2}$	0
$(15, 20]$	0.65	0.6	0.65	0.6	0
$(20, 25]$	0.55	0.8	0.55	0.8	0
$(25, 30]$	$\boxed{0.44}$	$\boxed{1}$	0.44	1	0
$(30, 35]$	0.32	0.8	0.32	0.8	0
$(35, 40]$	0.21	0.6	0.21	0.6	0
$(40, 45]$	0.11	0.2	0.11	0.2	0
$(45, 50)$	0	0	0	0	0

this method is the drastic changes of the control variable which leads to high stress, e.g., chattering of gears for the plant to be controlled. However, this can be avoided by a small modification: a boundary layer is introduced near the switching line which smooths out the control behavior and ensures that the system states remain within this layer. Given that the upper bounds of the model uncertainties, etc., are known, stability and high performance of the controlled system are guaranteed. *In principle, FKBC work like modified sliding mode controllers.* Compared to ordinary sliding mode control, however, FKBC have the advantage of still higher robustness. In the following, the structure of a FKBC is derived from a nonlinear state equation representing a large class of physical systems. The following aspects are discussed: stability conditions, scaling (normalization) of the state vector, choice of the switching line and determination of the break frequencies of the controller. By the choice of an additional boundary layer in the phase plane, the FKBC is modified so that drastic changes of the control variable can be avoided, especially at the boundary of the normalized phase plane. In this context, the higher robustness of the modified FKBC over the modified sliding mode control is discussed. Finally, the method is extended to the n-dimensional case.

Sliding Mode Control

Let

$$x^{(n)} = f(\mathbf{x}, t) + u + d, \tag{4.60}$$

where

$$\mathbf{x} = (x, \dot{x}, \ldots, x^{(n-1)})^T$$

$\mathbf{x}$ is the state vector, d are disturbances, and u is the control variable. Furthermore, let

$$f(\mathbf{x}, t) = \hat{f}(\mathbf{x}, t) + \Delta f(\mathbf{x}, t) \tag{4.61}$$

be a nonlinear function of the state vector $\mathbf{x}$ and, explicitly, of time t, where Δf are model uncertainties, and $\hat{f}$ is an estimate of f.

Furthermore, let Δf, d and $x_d^{(n)}$ have upper bounds with known values $\tilde{F}$,D and v:

$$|\Delta f| \leq \tilde{F}(\mathbf{x}, t) \ ; \quad |d| \leq D(\mathbf{x}, t) \ ; \quad |x_d^{(n)}| \leq v. \tag{4.62}$$

The control problem is to obtain the state $\mathbf{x}$ for tracking a desired state $\mathbf{x}_d$ in the presence of model uncertainties and disturbances. With the tracking error

$$\mathbf{e} = \mathbf{x} - \mathbf{x}_d = (e, \dot{e}, \ldots, e^{(n-1)})^T, \tag{4.63}$$

a stable *sliding surface* (*switching line* for second order systems) is defined as follows:

$$s(\mathbf{x}, t) = 0 \tag{4.64}$$

$$s(\mathbf{x}, t) = (\mathrm{d}/\mathrm{d}t + \lambda)^{n-1} e; \quad \lambda \geq 0. \tag{4.65}$$

Example 4.7 For a second order system with $n = 2$ we obtain

$$s(\mathbf{x}, t) = (\mathrm{d}/\mathrm{d}t + \lambda)e = \dot{e} + \lambda \cdot e.$$

Starting from the initial conditions

$$\mathbf{e}(0) = \mathbf{0}, \tag{4.66}$$

the tracking task $\mathbf{x} \to \mathbf{x}_d$, which means that $\mathbf{x}$ has to follow $\mathbf{x}_d$ with a predefined precision, can be considered as solved if the state vector $\mathbf{e}$ remains on the sliding surface $s(\mathbf{x}, t) = 0$ for all $t \geq 0$. A sufficient condition for this behavior is to choose the control value so that

$$\frac{1}{2} \cdot \frac{\mathrm{d}}{\mathrm{d}t}(s^2(\mathbf{x}, t)) \leq -\eta \cdot |s|; \quad \eta \geq 0. \tag{4.67}$$

Considering $s^2(\mathbf{x}, t)$ as a Lyapunov-like function, the condition (4.67) ensures that the system controlled is stable. Looking at the phase plane we observe that *the system is controlled in such a way that the state vector always moves towards the sliding surface*. The sign of the control value must change at the intersection of state trajectory $\mathbf{e}(t)$ with the sliding surface. In this way, the trajectory is forced to move always towards the sliding surface (see Fig. 4.22.). A *sliding mode* along the sliding surface is thus obtained. By remaining in the sliding mode of (4.67) the system's behavior is invariant despite model uncertainties, parameter fluctuations and disturbances. However, the sliding mode causes drastic changes of the control variable, which is an evident drawback for technical systems. Returning to equation (4.67) one obtains the conventional notation for sliding mode:

$$s \cdot \dot{s} \leq -\eta \cdot |s| \quad \text{or alternatively} \quad \dot{s} \cdot \operatorname{sgn}(s) \leq -\eta. \tag{4.68}$$

In the following, without loss of generality, we focus on second order systems. Hence, from equation (4.65) it follows that

$$s = \lambda e + \dot{e} \quad \text{and} \quad \dot{s} = \lambda \dot{e} + \ddot{e} = \lambda \dot{e} + \ddot{x} - \ddot{x}_d. \tag{4.69}$$

From the above and (4.60) it follows

$$s \cdot \dot{s} = s \cdot (\lambda \dot{e} + f(\mathbf{x}, t) + u + d - \ddot{x}_d) \leq -\eta \cdot |s|. \tag{4.70}$$

Rewriting this equation we obtain

$$[f(\mathbf{x}, t) + d + \lambda \dot{e} - \ddot{x}_d] \cdot \operatorname{sgn}(s) + u \cdot \operatorname{sgn}(s) \leq -\eta. \tag{4.71}$$

To achieve the sliding mode of equation (4.68) we choose u so that

$$u = (-\hat{f} - \lambda \dot{e}) - K(\mathbf{x}, t) \cdot \operatorname{sgn}(s) \quad \text{with } K(\mathbf{x}, t) > 0 \tag{4.72}$$

where $(-\hat{f} - \lambda \dot{e})$ is a compensation term and the second term is the controller. Hence, equation (4.71) can be written as

$$(f(\mathbf{x}, t) - \hat{f}(\mathbf{x}, t) + d(t) - \ddot{x}_d) \cdot \operatorname{sgn}(s) - K(\mathbf{x}, t) \leq -\eta. \tag{4.73}$$

Installing the upper bounds of equation (4.62) in equation (4.73) one obtains finally

$$K(\mathbf{x}, t)_{\max} \geq \widetilde{F} + D + v + \eta. \tag{4.74}$$

To avoid drastic changes of the control variable mentioned above, we substitute the function $\operatorname{sgn}(s)$ by $\operatorname{sat}(s/\Phi)$ in equation (4.72), where

$$\operatorname{sat}(x) = \begin{cases} x & \text{if } |x| < 1 \\ \operatorname{sgn}(x) & \text{if } |x| \geq 1. \end{cases} \tag{4.75}$$

This substitution corresponds to the introduction of a "boundary layer" $|s| \leq \Phi$ (see Fig. 4.22.) where Φ is half of the width of the boundary layer. Thus, we have

$$u = -\hat{f} - \lambda \dot{e} - K(\mathbf{x}, t)_{\max} \cdot \operatorname{sat}(s/\Phi) \quad \text{where} \quad \Phi > 0; \quad K(\mathbf{x}, t)_{\max} > 0. \tag{4.76}$$

From equations (4.70) and (4.76) follows the filter function

$$\dot{s} + K(\mathbf{x}, t) \cdot \frac{s}{\Phi} = \Delta f + d - \ddot{x}_d \tag{4.77}$$

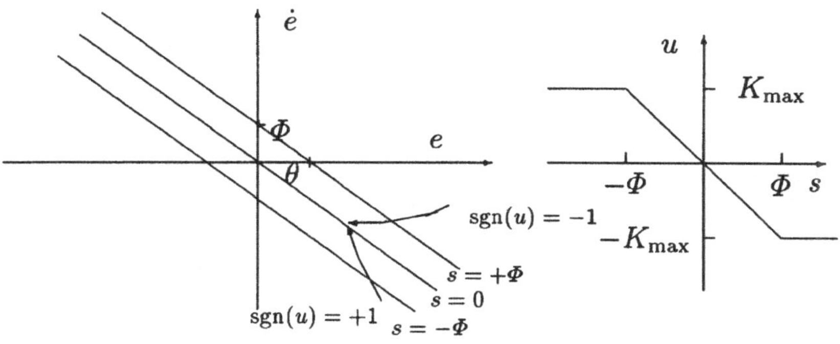

Fig. 4.22. Sliding mode principle with boundary layer.

for unmodelled disturbances d, model fluctuations Δf and the acceleration $\ddot{x}_d$ of the desired values x_d. Suppose the solution of the above filter equation is s_0. Then the distance from the state vector $\mathbf{e}$ to the switching line $s = 0$ is

$$dist = |s_0| \cdot \frac{1}{\sqrt{1+\lambda^2}}. \tag{4.78}$$

With the slope λ of the switching line $s = 0$ one obtains the guaranteed tracking precision

$$\theta = \frac{\Phi}{\lambda}. \tag{4.79}$$

The break frequency ν_Φ of filter (4.77) yields

$$\nu_\Phi = \frac{K(\mathbf{x}, t)_{\max}}{\Phi}. \tag{4.80}$$

On the other hand, even (4.65) is a filter with s as input and e as output. Hence, the break frequency λ of this filter should be, like ν_Φ, small compared to unmodelled frequencies ν_{su}:

$$\lambda \ll \nu_{su}. \tag{4.81}$$

Furthermore, ν_Φ should be less than or equal to the largest acceptable λ. From (4.80) we obtain the balance condition

$$\nu_\Phi = \frac{K(\mathbf{x}, t)_{\max}}{\Phi} = \lambda, \tag{4.82}$$

i.e., critical damping. Finally, the design rule for λ with regard to sample rate ν_{sample} and time constant τ_{plant} of the plant [195] can be denoted as

$$\lambda < \frac{\nu_{sample}}{2 \cdot (1 + \tau_{plant} \cdot \nu_{sample})}. \tag{4.83}$$

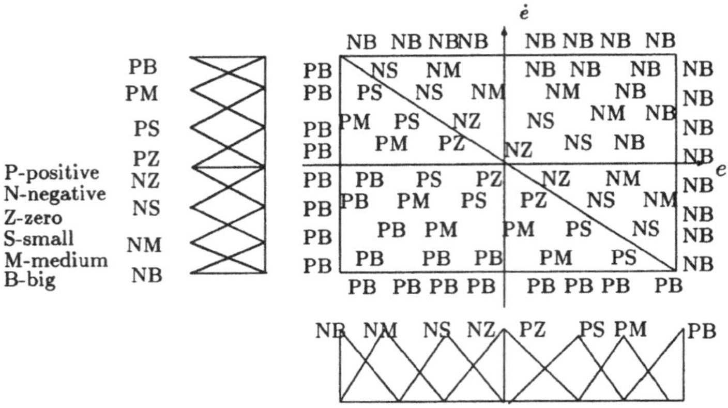

Fig. 4.23. Rules in the normalized phase plane.

Fuzzy Sliding Mode

First we will dicuss the high similarity between the sliding mode control with boundary layer and a FKBC whose rules have been derived from the phase plane.

The rules are, in general, conditioned in such a way that above the switching line $s = 0$ a negative control output is generated and a positive one below it, similar to the sliding mode control (or sub-optimal control). Let $K_{Fuzz}(e, \dot{e}, \lambda)$ be the absolute value of the crisp control output of the FKBC. Then the working principle of a FKBC can be represented by

$$u = -K_{Fuzz}(e, \dot{e}, \lambda) \cdot \operatorname{sgn}(s). \tag{4.84}$$

Figure 4.23 shows an example where close to the sliding surface (switching line) control outputs are smaller than at a longer distance from it. Neglecting the compensation term $-\hat{f} - \lambda \cdot \dot{e}$ of the sliding mode control with boundary layer in (4.76) and comparing the result with the FKBC (4.84) we come to the following conclusion: *Fuzzy control, in the above mentioned sense, is an extension of sliding mode with boundary layer.*

The Sliding Mode FKBC (SMFC) as a State Dependent Filter

The FKBC with partial compensation has the form

$$u = -\lambda \dot{e} - K_{Fuzz}(e, \dot{e}, \lambda) \cdot \operatorname{sgn}(s), \tag{4.85}$$

which is a hybrid version of a conventional compensation strategy and fuzzy control. Due to the ability of the FKBC to produce a control value depending on *dist*, one automatically obtains from (4.85) a type of boundary layer. However, this boundary layer depends strongly on the number and size of the membership

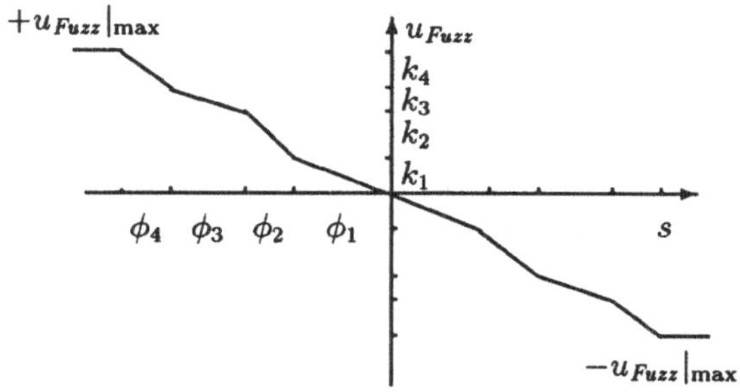

Fig. 4.24. Nonlinear operating line.

functions used in the rules. Generally, in contrast to the sliding mode control with boundary layer, the FKBC generates a piecewise linear function $u = f(s)$ (see. Fig. 4.24). Yet, with an increasing number of membership functions $u = f(s)$ becomes linear [210].

Hence, similarly to (4.77), we obtain the following filter function for the i-th linear segment of $u = f(s)$:

$$\dot{s} + \frac{k_i}{\phi_i} \cdot s = u_i \cdot \mathrm{sgn}(s) + f(\mathbf{x}, t) + d - \ddot{x}_d, \tag{4.86}$$

with

$$u_i = \begin{cases} -\sum\limits_{\nu=1}^{i-1} k_\nu + \dfrac{k_i}{\phi_i} \cdot \sum\limits_{\nu=1}^{i-1} \phi_\nu & \text{if } i \geq 2, \\ 0 & \text{if } i = 1; \end{cases} \tag{4.87}$$

k_ν, $\phi_\nu > 0$, $i = 1, \ldots, n$; n is the number of linear segments.

Equation (4.86) is a state dependent filter with different break frequencies k_i/ϕ_i. In the following we dicuss how to choose the different slopes of the nonlinear operating line which closely corresponds to an appropriate choice of the membership functions for the rule-antecedent.

At a large directional distance s between the state vector $\mathbf{e}$ and the switching line $s = 0$ any unmodelled frequencies are not able to cause a change of sign of the control output variable. Therefore, for a large distance s between the current state and the switching line we may choose a larger control output value than in the case of a small distance.

The result is that the condition

$$\frac{k_i}{\phi_i} \leq \lambda \tag{4.88}$$

has to be fulfilled only for $i = 1$, i.e., in the vicinity of the origin of the phase space.

The tracking quality is guaranteed by the maximum values

$$K_{Fuzz}|_{\max} = \sum_{\nu=1}^{i-1} k_\nu \quad \text{and} \quad \Phi_{\max} = \sum_{\nu=1}^{i-1} \phi_\nu \tag{4.89}$$

as long as

$$\frac{K_{Fuzz}|_{\max}}{\Phi_{\max}} \leq \lambda. \tag{4.90}$$

However, if we have a special plant P_k with upper bounds

$$\widetilde{F}_k + D_k + v_k \leq \widetilde{F}_{\max} + D_{\max} + v_{\max}, \tag{4.91}$$

we are in a position to define a $K_{Fuzz}|_k \leq K_{Fuzz}|_{\max}$ so that

$$K_{Fuzz}|_k \geq \widetilde{F}_k + D_k + v_k + \eta. \tag{4.92}$$

From this we obtain the corresponding Φ_k for plant P_k. Furthermore, using the FKBC for different plants P_i with

$$K_{Fuzz}|_{\max} \geq K_{Fuzz}|_i \geq \widetilde{F}_i + D_i + v_i + \eta, \tag{4.93}$$

one obtains for each plant P_i a particular Φ_i .

In the following, for the purpose of the design of a SMFC we deal with *qualitative aspects of rule design, input-normalization,* and *output-denormalization*

Qualitative aspects of rule design

We design the rules of the FKBC with respect to the normalized phase plane as follows:

R1. u_N should be negative above the switching line and positive below it.

R2. $|u_N|$ should increase as the distance *dist* between the actual state and the switching line $s = 0$ grows.

R3. $|u_N|$ should increase as the distance d between the actual state and the line perpendicular to the switching line grows, for the following reasons:

- discontinuities at the boundaries of the phase plane can be avoided,
- the central domain of the phase plane can be arrived at very quickly.

R4. Normalized states $e_N, \dot{e}_N$ that fall out of the phase plane should be covered by the maximum values $|u_N|_{\max}$ with the respective sign of u_N.

Input-normalization

Normalization is a linear state-transformation. The actual processing of the fuzzy rules takes place within the normalized phase plane. The switching line $s = 0$ has to be transformed as follows. Within the non-normalized phase plane we have $\lambda \cdot e + \dot{e} = 0$. In the normalized plane we obtain $\lambda_N \cdot e_N + \dot{e}_N = 0$. By means of the relationship

$$e_N = e \cdot N_e, \qquad \dot{e}_N = \dot{e} \cdot N_{\dot{e}}; \tag{4.94}$$

where N_e and $N_{\dot{e}}$ are normalization factors, one obtains directly

$$\lambda = \lambda_N \cdot \frac{N_e}{N_{\dot{e}}}. \tag{4.95}$$

From design rule (4.81) for λ one obtains

$$\lambda_N \cdot \frac{N_e}{N_{\dot{e}}} \ll \nu_{su}. \tag{4.96}$$

Hence, from inequality (4.96) we have obtained the break frequency, above which all frequencies of the unmodelled dynamics and disturbances are located.

Output-scaling

The difference between the FKBC of (4.84) and the sliding mode control of (4.72) is in the compensation term. If there is no sufficient model of the nonlinear part $f(\mathbf{x}, t)$, the estimate of the upper bound of $f(\mathbf{x}, t)$ has to be modified:

$$\widetilde{F} = |f(\mathbf{x}, t)_{\max}|. \tag{4.97}$$

If we now return to (4.74) for the required maximum of K_{Fuzz} we obtain

$$K_{Fuzz}|_{\max} \geq \widetilde{F} + D + v + \eta \tag{4.98}$$

where

$$K_{Fuzz}|_{\max} = \max(K_{Fuzz}(e, \dot{e}, \lambda)). \tag{4.99}$$

From this, the denormalization factor N_u can be calculated as follows. Let $\mu_{U_{\max}}(u)$ be the membership function for the largest possible control value in the normalized phase plane, e.g., $\mu_{PB}(u)$, let $u_{N_{\max}} = \text{Defuzz}(\mu_{U_{\max}}(u))$ be the corresponding defuzzified and normalized control value, and finally let $K_{Fuzz_N}|_{\max} = |u_{N_{\max}}|$. The defuzzification operation (Defuzz) can use any defuzzification method described in Section 3.6. From this and $K_{Fuzz}|_{\max} = N_u^{-1} \cdot K_{Fuzz_N}|_{\max}$, the denormalization factor N_u follows directly:

$$N_u = \frac{K_{Fuzz_N}|_{\max}}{K_{Fuzz}|_{\max}}. \tag{4.100}$$

In conclusion, a SMFC is faster and more robust with respect to changes of system parameters than a sliding mode control with boundary layer for the following reasons: the frequency λ is obligatory for all plants P_i considered. From equation (4.88) one obtains a guaranteed $\Phi_{\max}$ for all plants P_i. In addition to this, for each plant P_i, a $\Phi_i \leq \Phi_{\max}$ can be ensured if the upper bounds $\widetilde{F}_i, D_i$ and v_i are predefined.

Furthermore, the control output variable u varies with the distance of the state vector from the line perpendicular to the switching line. This achieves a smooth transition behavior of the control output at the boundaries of the phase plane. Finally, the variation of the slope of the operating line $u = f(s)$ permits a fast approach of the state vector to the switching line combined with smooth behavior close to the line.

On the other hand, a sliding mode control with boundary layer is designed for only one system and its upper bounds, and within the boundary layer, the approach velocity of the state vector to the switching line is constant. Therefore, disturbances in the control loop and fast changes of the desired values require a longer settling time than under fuzzy control.

If condition (4.88) is not satisfied, the system becomes more sensitive to unmodelled frequencies. For sufficiently large Φ (Φ being half the width of boundary layer), this fact does not significantly affect the system's behavior since no sign change of the control output variable u can be caused. However, in the vicinity of the switching line undesired fast sign changes of u can occur especially close to the boundary of the normalized phase plane. It is therefore useful to build a boundary layer for the FKBC. According to equations (4.76) and (4.85), the modified FKBC with boundary layer has the property:

$$u = -\lambda \cdot \dot{e} - K_{Fuzz}(e, \dot{e}, \lambda) \cdot \text{sat}(s/\Phi). \tag{4.101}$$

This gives the following advantages:

1. At the boundary of the $(e, \dot{e})$-plane, drastic changes of u can be avoided
2. Inside the $(e, \dot{e})$-plane and close to the switching line, condition (4.88) is fulfilled if Φ is chosen as follows. Let $|u|_{\text{max}}$ be the maximum value of the crisp (defuzzified) value of u. Then Φ has to be $\Phi = |u|_{\text{max}}/\lambda$.
3. With this design one is able to reduce the number of rules, i.e., one can choose a minimum set of rules generating. However, these generate a relatively rough gradation of u. This can be remedied by a boundary layer, serving as a linear element near the switching line.

It must be mentioned, however, that a boundary layer leads to a steady-state error if the controller close to zero has linear transfer characteristic. In this case the boundary layer should therefore be switched off in the area near to zero. The FKBC with boundary layer provides adaptive tracking quality even under changing process parameters. It achieves this both inside and outside the layer, together with a filtering of unmodelled frequencies. Finally, a well-designed FKBC with boundary layer gives a smoother control than the sliding mode control with boundary layer.

FKBC with Boundary Layer of Higher Order

A FKBC of higher order can be derived from a higher order sliding mode control with boundary layer in analogy to the method just shown for FKBC and sliding mode control. From the i-th rule of an n-th order FKBC,

$$\begin{aligned} R^{(i)}: &\ \textit{if}\ (e \text{ is } LE^{(i)})\ \textit{and}\ (\dot{e} \text{ is } L\dot{E}^{(i)})\ \textit{and}\ \ldots \textit{and}\ (e^{(n-1)} \text{ is } LE^{(i)}_{n-1}) \\ &\ \textit{then}\ u \text{ is } LU^{(i)}, \end{aligned}$$

it is evident that also in the high-order case one has to pay attention to the right choice of normalization factors. We obtain the normalization factors $N_e, N_{\dot{e}}, \ldots, N_{e^{(n-1)}}$ through the following consideration: with respect to the non-normalized phase plane we have

$$(\frac{\mathrm{d}}{\mathrm{d}t} + \lambda)^{(n-1)} e =$$

$$e^{(n-1)} + \binom{n-1}{1} \lambda e^{(n-2)} + \binom{n-1}{2} \lambda^2 e^{(n-3)} + \ldots + \lambda^{n-1} e = 0, \qquad (4.102)$$

and for the normalized phase plane,

$$(\frac{\mathrm{d}}{\mathrm{d}t} + \lambda_N)^{(n-1)} e_N =$$

$$e_N{}^{(n-1)} + \binom{n-1}{1} \lambda_N e_N{}^{(n-2)} + \binom{n-1}{2} \lambda_N{}^2 e_N{}^{(n-3)} + \ \ldots + \lambda_N{}^{n-1} e_N = 0. \qquad (4.103)$$

With normalization (scaling) factors, N_e, $N_{\dot{e}}$, ..., (4.103) can be rewritten as

$$e^{(n-1)} + \binom{n-1}{1} \lambda_N \cdot \frac{N_{e^{(n-2)}}}{N_{e^{(n-1)}}} e^{(n-2)} + \ldots + \lambda_N{}^{n-1} \frac{N_e}{N_{e^{(n-1)}}} e = 0. \qquad (4.104)$$

The comparison of the coefficients of equations (4.103) and (4.104) leads to

$$\lambda = \lambda_N \cdot \frac{N_{e^{(n-2)}}}{N_{e^{(n-1)}}}; \; \lambda^2 = \lambda_N{}^2 \cdot \frac{N_{e^{(n-3)}}}{N_{e^{(n-1)}}}; \; \ldots; \; \lambda^{n-1} = \lambda_N{}^{n-1} \cdot \frac{N_e}{N_{e^{(n-1)}}}. \qquad (4.105)$$

From this we obtain the design rule

$$\frac{N_{e^{(n-2)}}}{N_{e^{(n-1)}}} = \ldots = \frac{N_e}{N_{\dot{e}}} = \frac{\lambda}{\lambda_N}. \qquad (4.106)$$

This means that after the selection of the parameters λ, λ_N and N_e all normalization factors from $N_{\dot{e}}$ to $N_{e^{(n-1)}}$ are already determined. A simple method for building a FKBC of higher order is the following: According to rule $R^{(3)}$ for the two-dimensional case let

$$\mathbf{n}_N = \frac{\left(1, \binom{n-1}{1} \lambda_N, \binom{n-1}{2} \lambda_N^2, \ldots, \lambda_N^{n-1}\right)^T}{\left|\left(1, \binom{n-1}{1} \lambda_N, \binom{n-1}{2} \lambda_N^2, \ldots, \lambda_N^{n-1}\right)^T\right|} \qquad (4.107)$$

be the normal vector of the sliding surface within the normalized $(n-1)$-dimensional phase plane. Furthermore, assume that $s_p = \mathbf{e}_N{}^T \cdot \mathbf{n}_N$ is the projection of the state vector $\mathbf{e}_N$ on the direction of the normal vector $\mathbf{n}_N$. Hence, one obtains the shortest Euclidean distance d (the perpendicular) between $\mathbf{e}_N$ and the direction of $\mathbf{n}_N$,

$$d = \sqrt{|\mathbf{e}_N|^2 - s_p{}^2}. \qquad (4.108)$$

d \ s_p	NB	NM	NS	NZ	PZ	PS	PM	PB
B	PB	PB	PB	PB	NB	NB	NB	NB
M	PB	PB	PB	PM	NM	NB	NB	NB
S	PB	PB	PM	PS	NS	NM	NB	NB
Z	PB	PM	PS	PZ	NZ	NS	NM	NB

Fig. 4.25. (s_p-d)-table.

Now, s_p and the distance d between the direction of the normal vector and the sliding surface are evaluated by if-then rules. We attach to each of the variables s_p, d, and u_N a fuzzy set and obtain from this the (s_p-d)-table in Fig. 4.25. The attributes in the cells of this table describe the magnitude of the control value. They are filled out according to the design rules in the last section. In the two-dimensional case this table corresponds uniquely to the phase plane of the system described in terms of error and change-of-error. For higher dimensions the table is an aggregation of the phase space so that the correspondence between s_p and d on the one hand and the vector $\mathbf{e_N}$ on the other hand is not unique any more.

The i-th rule of a set of rules can be stated as

$$R^{(i)}: \textit{if } s_p \text{ is } L_{s_p}^{(i)} \textit{ and } d \text{ is } L_d^{(i)} \textit{ then } u \text{ is } L_u^{(i)}$$

with linguistic values like *Positive Small, Negative Big,* etc., for s_{p_i}, d_i, and u_i. Although the computational effort for s_p and d seems to be rather large, this controller reduces the number of rules considerably in contrast to a controller in which there are more than two state variables. According to equation (4.101) the structure of a FKBC with boundary layer and partial compensation can be defined as

$$u = -\sum_{k=1}^{n-1} \binom{n-1}{k} \lambda^k \cdot e^{(n-k)} - K_{Fuzz}(\mathbf{e}, \lambda) \cdot \text{sat}(s/\Phi). \tag{4.109}$$

The result is that *all design rules postulated in the previous section are also applicable to FKBC of higher order.*

The following example shows the Sliding Mode FKBC (SMFC) for a second order system.

Example 4.8 The task is to follow a contour by means of a force adaptive robot manipulator which is used for robot guided grinding of castings, welding tasks and other industrial applications. Suppose that the a-priori knowledge of the contour is incomplete. The robot effector (tool, gripper) has to follow the surface keeping a constant desired force. The actual force between effector and surface is measured with the help of a force sensor. The force error between desired value and actual value is given to the FKBC whose output (control variable) is a position correction of the robot arm. Assume for simplicity that the whole dynamics of the system is determined by the the robot effector and the surface. A further simplification is the restriction to one degree of freedom which means that we only have one coordinate in our model. Furthermore, let the effector be modelled as a mass-spring-damper system and the surface as a spring. The corresponding differential equation is

$$m \cdot \ddot{y} + c \cdot (\dot{y} - \dot{y}_O) - m \cdot g + K_1 \cdot (y - y_O - \widetilde{D}) + K_2 \cdot (y - y_c + r + A) = 0. \quad (4.110)$$

Equation (4.110) is a force balance equation (see Fig. 4.26):

$$F_T + F_D - F_G + F_F + F_O = 0 \quad (4.111)$$

with

F_T – inertial force

F_D – damping force

F_G – weighting force

F_F – spring force

F_O – reaction force within the surface.

Let the sensor force be represented by the spring force F_F. Then having that

$$F_F = K_1(y - y_O - \widetilde{D}); \quad \dot{F}_F = K_1(\dot{y} - \dot{y}_O); \quad \ddot{F}_F = K_1(\ddot{y} - \ddot{y}_O), \quad (4.112)$$

we obtain the equation for the spring (sensor) force F_F:

$$\ddot{F}_F = -\frac{c}{m} \cdot \dot{F}_F + K_1 \cdot g - \frac{K_1}{m} \cdot F_F - \frac{K_1}{m} \cdot F_O - K_1 \cdot \ddot{y}_O. \quad (4.113)$$

In this equation,

$$u = -K_1 \cdot \ddot{y}_O \quad (4.114)$$

is the control output variable which has to be produced by the controller. The specific parameters are:

$$C = 50 \text{ kg s}^{-1}$$
$$m = 0.05 \text{ kg}$$
$$K_1 = 5000 \text{ kg s}^{-2}$$
$$T_{sample} = 0.01 \text{ s}.$$

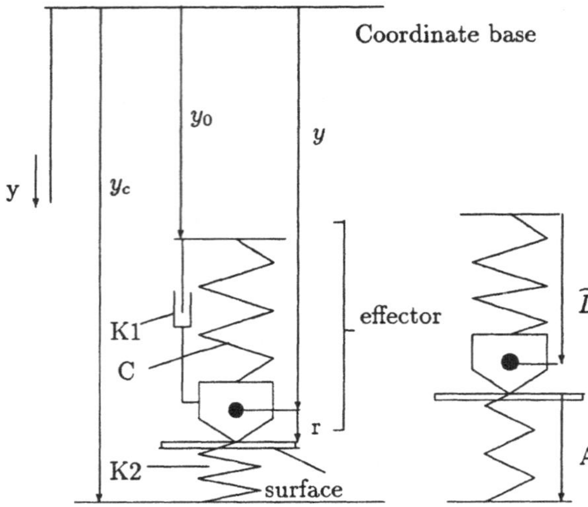

Fig. 4.26. Simplified model of the robot effector acting upon a surface.

The comparison between equations (4.60) and (4.113) implies that

$$x(t) \quad \text{becomes} \quad F_F(t), \tag{4.115}$$

$$x^{(n)}(t) \quad \text{becomes} \quad \ddot{F}_F, \tag{4.116}$$

$$f(\mathbf{x}, t) \quad \text{becomes} \quad -\frac{c}{m} \cdot \dot{F}_F + K_1 \cdot g - \frac{K_1}{m} \cdot F_F - \frac{K_1}{m} \cdot F_O. \tag{4.117}$$

Furthermore, we have

$$s = \lambda e + \dot{e} \tag{4.118}$$

$$e = F_F - F_{Fd}. \tag{4.119}$$

The controller corresponding to (4.101) has the form

$$u = -\lambda \cdot \dot{e} - K_{Fuzz}(e, \dot{e}, \lambda) \cdot \text{sat}(s/\Phi). \tag{4.120}$$

The $\widetilde{F}$ in equation (4.62) is equal to

$$\widetilde{F} = -\frac{c}{m} \cdot \dot{F}_F + K_1 \cdot g - \frac{K_1}{m} \cdot F_F - \frac{K_1}{m} \cdot F_O. \tag{4.121}$$

Within the normalized phase plane, s_{P_N} and the distance d_N described in the last section are

$$s_{P_N} = (e_N + \dot{e}_N)/\sqrt{2} \tag{4.122}$$

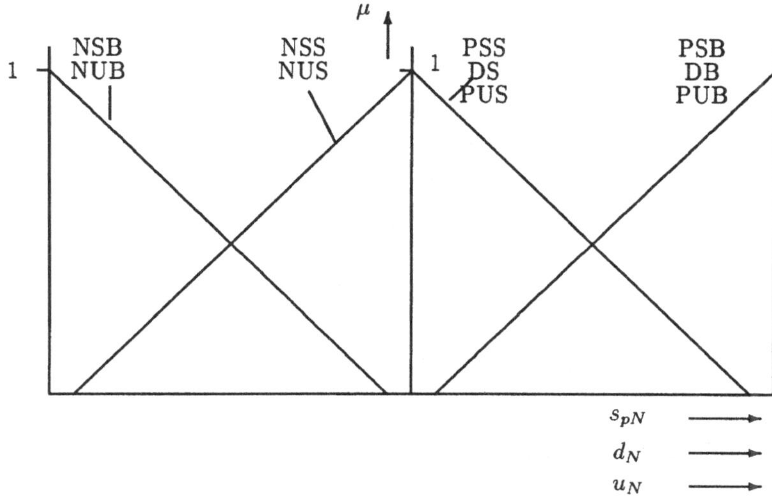

Fig. 4.27. Membership functions for s_{pN}, d_N, and u_N.

and

$$d_N = |(-e_N + \dot{e}_N)/\sqrt{2}|. \tag{4.123}$$

For the normalized values s_{p_N} , d_N and the normalized control variable u_N the following linguistic values are defined:

PSS – s_{p_N} is positive small
PSB – s_{p_N} is positive big
NSS – s_{p_N} is negative small
NSB – s_{p_N} is negative big
DS – d_N is small
DB – d_N is big
PUS – u_N is positive small
PUB – u_N is positive big
NUS – u_N is negative small
NUB – u_N is negative big.

Figure 4.27 shows the corresponding membership functions. The respective (s_{p_N}-d_N)-table is shown in Fig. 4.28. From this table we obtain the rules for the calculation of the control u_N:
when $s_{p_N} > 0$

if PSS *and* DS *then* NUS
if ((PSS *and* DB) *or* PSB) *then* NUB

when $s_{p_N} \leq 0$

d_N \ s_{pN}	NSB	NSS	PSS	PSB
DB	PUB	PUB	NUB	NUB
DS	PUB	PUS	NUS	NUB

Fig. 4.28. (s_{p_N}-d_N)-table for the contour tracking problem.

if NSS *and* DS *then* PUS
if ((NSS *and* DB) *or* NSB) *then* PUB.

The following FKBC parameters have been chosen:

$$\lambda = 50 \text{ s}^{-1} \qquad N_e = 20 \text{ kg}^{-1} \text{ m}^{-1} \text{ s}^2$$

and from λ and N_e

$$\begin{aligned} N_{\dot{e}} &= 0.4 \text{ kg}^{-1} \text{ m}^{-1} \text{ s}^3 \\ N_u &= 0.01 s^4/\text{kg m} \\ \Phi &= 2.5 \text{ m s}^{-1} \\ F_d &= 10 \text{ N} \\ v &= 0.2 \text{ m s}^{-1} \quad \text{(feed rate).} \end{aligned}$$

Figure 4.29 shows the tracking simulation of a sloping sinusoidal contour. The robot arm together with the force sensor has been modelled by the mass-spring-damper system of equation (4.110).

4.4.3 Sugeno FKBC

The Control Structure

One of the FKBC, differing from other FKBC types, has been developed by Sugeno and Takagi [200]. This controller consists of rules of the form

$R^{(1)}$: *if* (x_1 is $LX_1^{(1)}$) *and* ... *and* (x_1 is $LX_n^{(1)}$) *then* $u_1 = f_1(x_1, x_2, \ldots, x_n)$

$\vdots$

$R^{(m)}$: *if* (x_1 is $LX_1^{(m)}$) *and* ... *and* (x_1 is $LX_n^{(m)}$) *then* $u_m = f_1(x_1, x_2, \ldots, x_n)$.

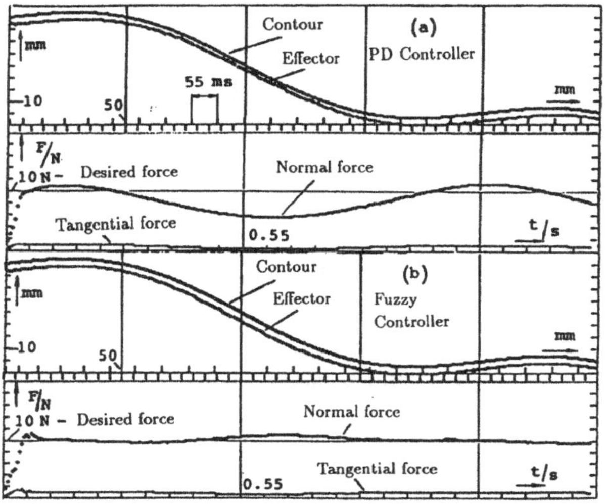

Fig. 4.29. Simulation results of the contour tracking.

Inference with these rule is done as follows. Let us consider the first rule. First, for given input values of the process state variables $x_1^*, \ldots, x_n^*$, their degrees of membership $\mu_{LX_i^{(1)}}(x_i^*)$, $(i = 1, \ldots, n)$, are calculated and are called rule-antecedent weight or rule-weight. Second, the minimum of these n degrees of membership is taken and denoted as μ_{u_1}. Third, the value of u_1 is computed as $u_1^* = f_1(x_1^*, x_2^*, \ldots, x_n^*)$. Thus for each rule $R^{(i)}$ one obtains a tuple (μ_{u_i}, u_i^*) Finally, the combined output of all rules is the weighted normalized sum of all pairs,

$$u = \frac{\sum_{i=1}^{m} \mu_{u_i} \cdot u_i^*}{\sum_{i=1}^{m} \mu_{u_i}}. \tag{4.124}$$

Example 4.9 Consider a Sugeno FKBC consisting of two rules and let the inputs be $x_1^* = 4$ and $x_2^* = 60$.

$R^{(1)}$: *if* x_1 is BIG *and* x_2 is MEDIUM *then* $u_1 = x_1 - 3 \cdot x_2$
$R^{(2)}$: *if* x_1 is SMALL *and* x_2 is BIG *then* $u_2 = 4 + 2 \cdot x_1$.

From Fig. 4.30 we then obtain:

$$\mu_{\text{BIG}}(x_1^*) = 0.3 \qquad \mu_{\text{BIG}}(x_2^*) = 0.35$$

and

$$\mu_{\text{SMALL}}(x_1^*) = 0.7 \qquad \mu_{\text{MED}}(x_2^*) = 0.75.$$

For the degrees of match or rule-weights of $R^{(1)}$ and $R^{(2)}$ we obtain

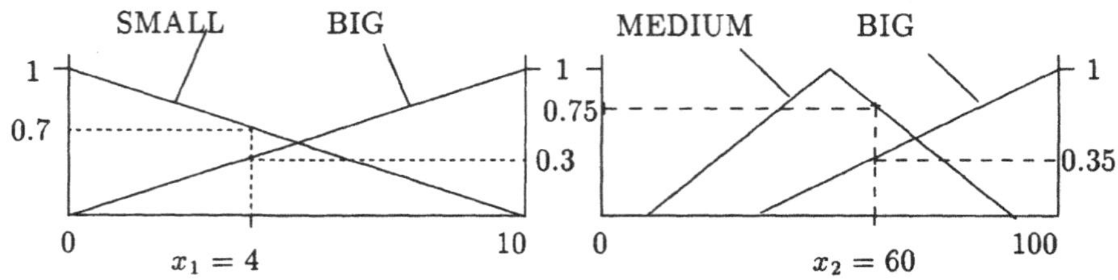

Fig. 4.30. Fuzzification procedure.

$$\min(0.3, 0.75) = 0.3 \qquad \min(0.7, 0.35) = 0.35. \tag{4.125}$$

Furthermore, for the rule-consequents of rule $R^{(1)}$ and $R^{(2)}$ we have

$$u_1 = 4 - 3 \cdot 60 = -176 \qquad u_2 = 4 + 2 \cdot 4 = 12. \tag{4.126}$$

So the two pairs corresponding to each rule are $(0.3, -176)$ and $(0.35, 12)$ Thus by taking the weighted normalized sum we get

$$u = \frac{0.3 \cdot (-176) + 0.35 \cdot 12}{0.3 + 0.35} = -74.77. \tag{4.127}$$

From this we directly realize the correspondence between the Sugeno FKBC and the second kind of *rule oriented representation* of conventional transfer elements in Section 4.2. Namely, with $x_1, \ldots, x_n$ as the states of the system every rule-consequent corresponds to a linear controller with constant parameters. Furthermore, special areas of the state space are dominated by only one rule, whereas in other areas rules are strongly overlapping. This means that the controller is a linear filter with variable coefficients:

Let the output of rule $R^{(i)}$ be

$$u_i = c_{1i} \cdot x + c_{2i} \cdot \dot{x} + \ldots + c_{ni} \cdot x^{(n-1)}. \tag{4.128}$$

Then the output is computed by

$$u = \frac{\sum_i (c_{1i} \cdot x + c_{2i} \cdot \dot{x} + \ldots + c_{ni} \cdot x^{(n-1)}) \cdot \mu_{u_i}}{\sum_i \mu_{u_i}}. \tag{4.129}$$

From this we obtain

$$u = \frac{\sum_i c_{1i} \cdot \mu_{u_i}}{\sum_i \mu_{u_i}} \cdot x + \frac{\sum_i c_{2i} \cdot \mu_{u_i}}{\sum_i \mu_{u_i}} \cdot \dot{x} + \ldots + \frac{\sum_i c_{ni} \cdot \mu_{u_i}}{\sum_i \mu_{u_i}} \cdot x^{(n-1)}. \tag{4.130}$$

This equation describes a filter of n-th order whose coefficients depend on the state vector $\mathbf{x} = (x, \dot{x}, \ldots, x^{(n-1)})^T$. With the help of this method, to every region in the state space a special linear controller (or linear filter) expressed as a single rule can be attached. The coefficients of this controller are nonlinear functions of the state vector $\mathbf{x}$ and the rule-antecedent. So, it is possible to weaken or amplify certain frequencies according to the system's state. Although state adaptive filtering can already be done by conventional switching of the controllers (filters) (see Section 4.2.3), in contrast to the conventional way, the fuzzy method leads to a smooth transition between the individual controllers (filters).

In the linear case the i-th Sugeno FKBC rule of page 186 is represented as

$R^{(i)}$: *if* $\mathbf{x}$ is $\mathbf{L^i}$ *then* $u = \mathbf{c_i^T} \cdot \mathbf{x}$,

where

$$\mathbf{c_i} = (c_1, c_2, \ldots, c_n)_i^T, \tag{4.131}$$

with the notation

$\mathbf{x}$ is $\mathbf{L^i}$ is an abbreviation for (x_1 is $LX_1^{(i)}$) and ... and (x_n is $LX_n^{(i)}$).

Since u is not restricted to linear functions, the function f_i in rule $R^{(i)}$ from page 186 can be either a linear or nonlinear function of $\mathbf{x}$:

$R^{(i)}$: *if* $\mathbf{x}$ is $\mathbf{L^i}$ *then* $u = f_i(\mathbf{x})$,

which expressed linguistically says that

if state vector $\mathbf{x}$ has the property $\mathbf{L^i}$
then apply controller f_i.

Like the Mamdani FKBC, the controller input (process state vector) is normalized, which gives a high flexibility with regard to the membership functions to be used. However, this normalization procedure is done only in the *rule-antecedent part*. The *rule-consequent part* works with non-normalized states so that denormalization is not needed. Furthermore, explicit defuzzification can also be omitted since the result of each rule is already crisp and the total result is determined by the weighted and normalized sum of the result of each rule; see equation (4.124).

Stability analysis

For stability analysis [209] we first consider the system described by the following nonlinear differential equation (see also equation (4.60)):

$$x^{(n)} = f(\mathbf{x}) + u \tag{4.132}$$

where the state vector

$$\mathbf{x} = (x_1, x_2, \ldots, x_n)^T = (x, \dot{x}, \ldots, x^{(n-1)})^T \tag{4.133}$$

is identical with the process output vector and $f(\mathbf{x})$ is a nonlinear function of $\mathbf{x}$. Now, we linearize (4.132) around an arbitrary set-point $\mathbf{x_d}$ so that

$$x^{(n)} = f(\mathbf{x_d}) + \mathbf{a^T} \cdot (\mathbf{x} - \mathbf{x_d}) + u \tag{4.134}$$

with

$$\mathbf{a} = (a_1, a_2, \ldots, a_k, \ldots, a_n)^T \tag{4.135}$$

and

$$a_k = \left. \frac{\partial f(\mathbf{x})}{\partial x_k} \right|_{\mathbf{x}=\mathbf{x_d}} . \tag{4.136}$$

Rewriting equation (4.134) as a state equation we obtain

$$\dot{\mathbf{x}} = \mathbf{b_1} + \mathbf{A} \cdot (\mathbf{x} - \mathbf{x_d}) + \mathbf{b_2} \cdot u \tag{4.137}$$

where

$$\mathbf{A} = \begin{pmatrix} 0 & 1 & 0 & \ldots & 0 \\ 0 & 0 & 1 & \ldots & 0 \\ \vdots & \vdots & \vdots & \ddots & \vdots \\ 0 & 0 & \ldots & 0 & 1 \\ a_1 & a_2 & \ldots & \ldots & a_n \end{pmatrix}; \quad \mathbf{b_1} = (0, \ldots, f(\mathbf{x_d}))^T; \quad \mathbf{b_2} = (0, \ldots, 1)^T. \tag{4.138}$$

Now, we approximate the state space by a fuzzy state space consisting of overlapping subspaces called fuzzy regions REG_i. Region REG_i is represented by the rule-antecedent

$$\mathbf{x_d} \text{ is } \mathbf{L^i} \tag{4.139}$$

or

$$(x_{d1} \text{ is } LX_1^{(i)}) \textit{ and } \ldots \textit{ and } (x_{dn} \text{ is } LX_n^{(i)}). \tag{4.140}$$

It has to be emphasized that, instead of the current state $\mathbf{x}$, the *desired* state $\mathbf{x_d}$ is applied . In this way we approximate the system (4.137) in terms of the Sugeno FKBC approach in which each region corresponds to one particular rule:

$$R^{(i)}: \textit{if } \mathbf{x_d} \text{ is } \mathbf{L^i} \textit{ then } \dot{\mathbf{x}} = \mathbf{b_1} + \mathbf{A_i} \cdot (\mathbf{x} - \mathbf{x_d}) + \mathbf{b_2} \cdot u.$$

To perform a stability analysis we first have to prove the stability of the closed-loop for each linearized region REG_i. Then, we investigate the stability of the total system using *Lyapunov's direct method.*

Connection of System and Controller

We have obtained a state equation and a control law for each region REG_i, ($i = 1, \ldots, m$), according to the system and the controller, respectively. REG_i is described by

System *if* $\mathbf{x_d}$ is $\mathbf{L^i}$ *then* $\dot{\mathbf{x}} = \mathbf{b_1} + \mathbf{A_i} \cdot (\mathbf{x} - \mathbf{x_d}) + \mathbf{b_2} \cdot u$,

Controller *if* $\mathbf{x_d}$ is $\mathbf{L^i}$ *then* $u = \mathbf{c_i^T} \cdot (\mathbf{x} - \mathbf{x_d}) - \hat{f}(\mathbf{x_d})$, where $\hat{f}(\mathbf{x_d})$ denotes the estimate of $f(\mathbf{x_d})$, and $\mathbf{c_i} = (c_1, c_2, \ldots, c_n)^T$ is a $(n \times 1)$ gain vector to be designed.

The resulting equations for system and controller are

$$\textbf{System} \qquad \dot{\mathbf{x}} = \sum_{i=1}^{m} \mu_i \cdot (\mathbf{b_1} + \mathbf{A_i} \cdot (\mathbf{x} - \mathbf{x_d}) + \mathbf{b_2} \cdot u)$$

$$\textbf{Controller} \qquad u = \sum_{i=1}^{m} \mu_i \cdot (\mathbf{c_i^T} \cdot (\mathbf{x} - \mathbf{x_d}) - \hat{f}(\mathbf{x_d}))$$

From this one obtains

$$\dot{\mathbf{x}} = \sum_{i=1}^{m} \mu_i \cdot (\widetilde{\mathbf{A}}_\mathbf{i} \cdot (\mathbf{x} - \mathbf{x_d}) + \Delta \mathbf{b_1}) \tag{4.141}$$

with

$$\begin{aligned}
&\widetilde{\mathbf{A}}_\mathbf{i} = \mathbf{A_i} + \mathbf{b_2} \cdot \mathbf{c_i^T} \\
&\Delta \mathbf{b_1} = \mathbf{b_1} - \hat{\mathbf{b}}_\mathbf{1} \\
&\hat{\mathbf{b}}_\mathbf{1} = (0, \ldots, \hat{f}(\mathbf{x_d}))^T \\
&\mu_i = \mu_i(\mathbf{x_d}); \qquad i = 1 \ldots m.
\end{aligned}$$

Local stability in the centers of the regions

Local stability in a center of a region is reached if the real parts of the eigenvalues of $\widetilde{\mathbf{A}}_\mathbf{i}$ are negative. The design of the still unknown gain vector $\mathbf{c_i^T}$ can be done by solving the equation

$$|\widetilde{\mathbf{A}}_\mathbf{i} - \mathbf{I} \cdot s| = 0 \tag{4.142}$$

where s denotes the Laplace operator and $\mathbf{I}$ denotes the unity matrix.

Then, the gain vector $\mathbf{c_i^T}$ can be obtained by different methods, e.g., pole placement or root locus method. As we will see in the next paragraph we are able to investigate the stability of the system around an arbitrary set-point $\mathbf{x_d}$ without considering the weights of the rules.

Local stability around any set-point $\mathbf{x_d}$

Local stability around any set-point $\mathbf{x_d}$ can be investigated by using *Lyapunov's direct method.* Let

$$\dot{\mathbf{e}} = \widetilde{\mathbf{A}}_\mathbf{i} \cdot \mathbf{e} \tag{4.143}$$

be the local homogeneous state equation around any operating point $\mathbf{x_i}$ where $\mathbf{e} = \mathbf{x} - \mathbf{x_d}$, $\widetilde{\mathbf{A}}_\mathbf{i}$ is a stable matrix, and $\dot{\mathbf{x}}_\mathbf{d} = \mathbf{0}$. Furthermore, let

$$\dot{\mathbf{e}} = \sum_{i=1}^{m} \mu_i \cdot \widetilde{\mathbf{A}}_\mathbf{i} \cdot \mathbf{e} / \sum_{i=1}^{m} \mu_i \tag{4.144}$$

be the global homogeneous state equation. Next, let

$$V(\mathbf{e}) = \mathbf{e^T} \cdot \mathbf{P} \cdot \mathbf{e} \tag{4.145}$$

be a Lyapunov function with

$$\begin{aligned} V(\mathbf{0}) &= 0 \\ V(\mathbf{e}) &> 0 \qquad \forall \mathbf{e} \geq 0 \\ V(\mathbf{e}) &\to \infty \qquad \text{as} \quad |\mathbf{e}| \to \infty. \end{aligned}$$

The equilibrium point $\mathbf{e} = \mathbf{0}$ is *asymptotically stable* if $\dot{V}(\mathbf{e})$ is negative definite with

$$\dot{V}(\mathbf{e}) < 0 \quad \text{for} \quad \mathbf{e} \neq \mathbf{0}. \tag{4.146}$$

For the global system we have

$$\begin{aligned} \dot{V} &= \dot{\mathbf{e}}^{\mathbf{T}} \cdot \mathbf{P} \cdot \mathbf{e} + \mathbf{e}^{\mathbf{T}} \cdot \mathbf{P} \cdot \dot{\mathbf{e}} \\ &= \mathbf{e}^{\mathbf{T}} \cdot (\sum_{i=1}^{m} \mu_i \cdot \widetilde{\mathbf{A}}_{\mathbf{i}}^{\mathbf{T}} / \sum_{i=1}^{m} \mu_i) \cdot \mathbf{P} \cdot \mathbf{e} + \mathbf{e}^{\mathbf{T}} \cdot \mathbf{P} \cdot (\sum_{i=1}^{m} \mu_i \cdot \widetilde{\mathbf{A}}_{\mathbf{i}} / \sum_{i=1}^{m} \mu_i) \cdot \mathbf{e} \\ &= \sum_{i=1}^{m} [\mu_i \cdot \mathbf{e}^{\mathbf{T}} \cdot (\widetilde{\mathbf{A}}_{\mathbf{i}}^{\mathbf{T}} \cdot \mathbf{P} + \mathbf{P} \cdot \widetilde{\mathbf{A}}_{\mathbf{i}}) \cdot \mathbf{e}] / \sum_{i=1}^{m} \mu_i. \end{aligned} \tag{4.147}$$

Now we obtain a sufficient stability condition if there is a common matrix $\mathbf{P}$ for which V is positive definite and if for each matrix $\widetilde{\mathbf{A}}_{\mathbf{i}}$, $(i = 1 \dots m)$,

$$\mathbf{Q} = \widetilde{\mathbf{A}}_{\mathbf{i}}^{\mathbf{T}} \cdot \mathbf{P} + \mathbf{P} \cdot \widetilde{\mathbf{A}}_{\mathbf{i}} \tag{4.148}$$

leads to a negative definite $\dot{V}$. *It is remarkable that the weight of a specific rule does not affect the stability of the closed-loop system.* The following example deals with the check of local stability according to a simple nonlinear system.

Example 4.10 We consider a one-link robot manipulator with the dynamic equation

$$J \cdot \ddot{\phi} + m \cdot g \cdot l \cdot \sin\phi + D \cdot \dot{\phi} = u, \tag{4.149}$$

where

J — moment of inertia
m — mass at the end of the link
l — length of the link
g — gravitation constant
D — damping constant
ϕ - angle of link with respect to its null position.

Equation (4.149) is linearized with respect to two selected regions REG_0 and REG_1 and their corresponding angles ϕ_0 and ϕ_1. Let $\Delta\phi_0 = \phi - \phi_0$ and $\Delta\phi_1 = \phi - \phi_1$. Then we obtain

$$\begin{aligned} \sin(\phi_0 + \Delta\phi_0) &\approx \sin\phi_0 + \cos\phi_0 \cdot \Delta\phi_0, \\ \sin(\phi_1 + \Delta\phi_1) &\approx \sin\phi_1 + \cos\phi_1 \cdot \Delta\phi_1. \end{aligned} \tag{4.150}$$

From (4.149) and (4.150) we get

$$
\begin{aligned}
J \cdot \ddot{\phi} + D \cdot \dot{\phi} + m \cdot g \cdot l \cdot \cos\phi_0 \cdot \phi &= u_0 - m \cdot g \cdot l \cdot (\sin\phi_0 - \phi_0 \cdot \cos\phi_0), \\
J \cdot \ddot{\phi} + D \cdot \dot{\phi} + m \cdot g \cdot l \cdot \cos\phi_1 \cdot \phi &= u_1 - m \cdot g \cdot l \cdot (\sin\phi_1 - \phi_1 \cdot \cos\phi_1). \qquad (4.151)
\end{aligned}
$$

Further, let the control laws in the particular regions be

$$
\begin{aligned}
u_0 &= K_{1_0} \cdot (\phi - \phi_d) + K_{2_0} \cdot (\dot{\phi} - \dot{\phi}_d) + K_{3_0} \cdot \int (\phi - \phi_d) \mathrm{d}t + m \cdot g \cdot l \cdot \sin\phi_0, \\
u_1 &= K_{1_1} \cdot (\phi - \phi_d) + K_{2_1} \cdot (\dot{\phi} - \dot{\phi}_d) + K_{3_1} \cdot \int (\phi - \phi_d) \mathrm{d}t + m \cdot g \cdot l \cdot \sin\phi_1.
\end{aligned}
$$

Substitution of u_0 and u_1 and differentation of (4.151) yields

$$
\begin{aligned}
& J \cdot \phi^{(3)} + (D - K_{2_i}) \cdot \ddot{\phi} + (m \cdot g \cdot l \cdot \cos\phi_i - K_{1_i}) \cdot \dot{\phi} - K_{3_i} \cdot \phi \\
& \quad = -(K_{1_i} \cdot \dot{\phi}_d + K_{2_i} \cdot \ddot{\phi}_d + K_{3_i} \cdot \phi_d), \qquad i = 0, 1. \qquad (4.152)
\end{aligned}
$$

By means of pole placement we design the parameters K_{1_0} ,K_{1_1}, K_{2_0}, K_{2_1}, K_{3_0}, K_{3_1} in such a way that we reach *local stability* around $\phi = \phi_0$ and $\phi = \phi_1$, respectively. From (4.152) we obtain the following state equation:

$$
\begin{aligned}
\dot{x}_1 &= x_2 \\
\dot{x}_2 &= x_3 \\
\dot{x}_3 &= -\frac{1}{J} \cdot (D - K_{2_i}) \cdot x_3 - \frac{1}{J} \cdot (m \cdot g \cdot l \cdot \cos\phi_i - K_{1_i}) \cdot x_2 \\
& \quad + \frac{1}{J} \cdot K_{3_i} \cdot x_1 + C_i \qquad (4.153)
\end{aligned}
$$

with $x_1 = \phi$, $x_2 = \dot{\phi}$ and $x_3 = \ddot{\phi}$; $i = 0, 1$, and

$$
C_i = -\frac{1}{J}(K_{1_i} \cdot \dot{\phi}_d + K_{2_i} \cdot \ddot{\phi}_d + K_{3_i} \cdot \phi_d). \qquad (4.154)
$$

From this we get the state equation in vector form,

$$
\dot{\mathbf{x}} = \widetilde{\mathbf{A}}_{\mathbf{i}} \cdot \mathbf{x} + \mathbf{c}_{\mathbf{i}}, \qquad (4.155)
$$

with $\mathbf{x} = (x_1, x_2, x_3)^T$,

$$
\widetilde{\mathbf{A}}_{\mathbf{i}} = \begin{pmatrix} 0 & 1 & 0 \\ 0 & 0 & 1 \\ \frac{K_{3_i}}{J} & -\frac{1}{J} \cdot (m \cdot g \cdot l \cdot \cos\phi_i - K_{1_i}) & -\frac{1}{J} \cdot (D - K_{2_i}) \end{pmatrix}, \qquad (4.156)
$$

$$
\mathbf{c}_i = (0, 0, C_i)^T. \qquad (4.157)
$$

Stability is reached for a common matrix $\mathbf{P}$ with positive definite V:

$$
V = \mathbf{x}^{\mathbf{T}} \cdot \mathbf{P} \cdot \mathbf{x} > 0 \qquad (4.158)
$$

and a negative definite $\dot{V}$:

$$
\dot{V} = \mathbf{x}^{\mathbf{T}} \cdot (\widetilde{\mathbf{A}}_{\mathbf{i}}^{\mathbf{T}} \cdot \mathbf{P} + \mathbf{P} \cdot \widetilde{\mathbf{A}}_{\mathbf{i}}) \cdot \mathbf{x} = \mathbf{x}^{\mathbf{T}} \cdot \mathbf{Q}_{\mathbf{i}} \cdot \mathbf{x} < 0. \qquad (4.159)
$$

With

$$\mathbf{P} = \begin{pmatrix} p_{11} & p_{12} & p_{13} \\ p_{21} & p_{22} & p_{23} \\ p_{31} & p_{32} & p_{33} \end{pmatrix} \tag{4.160}$$

and

$$a_{31} = \frac{K_{3_i}}{J}; \quad a_{32} = -\frac{1}{J}\cdot(m\cdot g\cdot l\cdot\cos\phi_i - K_{1_i}); \quad a_{33} = -\frac{1}{J}\cdot(D - K_{2_i}) \tag{4.161}$$

we get $\mathbf{Q_i} =$

$$\begin{pmatrix} (p_{13}+p_{31})a_{31} & p_{11}+p_{13}a_{32}+p_{32}a_{31} & p_{12}+p_{13}a_{33}+p_{33}a_{31} \\ p_{11}+p_{31}a_{32}+p_{23}a_{31} & p_{12}+p_{21}+(p_{32}+p_{23})a_{32} & p_{22}+p_{13}+p_{23}a_{33}+p_{33}a_{32} \\ p_{21}+p_{31}a_{33}+p_{33}a_{31} & p_{22}+p_{31}+p_{32}a_{33}+p_{33}a_{32} & p_{23}+p_{32}+2p_{33}a_{33} \end{pmatrix}. \tag{4.162}$$

According to Sylvester's theorem, we have a negative definite $\dot{V}$ with

$$\mathbf{Q} = \begin{pmatrix} q_{11} & q_{12} & q_{13} \\ q_{21} & q_{22} & q_{23} \\ q_{31} & q_{32} & q_{33} \end{pmatrix} \tag{4.163}$$

if the following determinants fulfill the conditions:

$$q_{11} < 0; \qquad \begin{vmatrix} q_{11} & q_{12} \\ q_{21} & q_{22} \end{vmatrix} < 0; \tag{4.164}$$

$$\begin{vmatrix} q_{11} & q_{12} & q_{13} \\ q_{21} & q_{22} & q_{23} \\ q_{31} & q_{32} & q_{33} \end{vmatrix} < 0; \tag{4.165}$$

Example 4.11 Let the system's parameters be

$$\begin{aligned} J &= 10 \text{ kg m}^2 \\ m &= 10 \text{ kg} \\ l &= 1 \text{ m} \\ D &= 5 \text{ kg m}^2 \text{ s}^{-1} \\ \phi_0 &= 30^\circ \\ \phi_1 &= 60^\circ. \end{aligned}$$

For the regions REG_0 and REG_1 we get the following homogeneous differential equations:

$$REG_0: \quad \phi^{(3)}+(0.5-0.1\cdot K_{2_0})\cdot\ddot{\phi}+(8.43-0.1\cdot K_{1_0})\cdot\dot{\phi}-0.1\cdot K_{3_0}\cdot\phi = 0, \tag{4.166}$$

$$REG_1: \quad \phi^{(3)}+(0.5-0.1\cdot K_{2_1})\cdot\ddot{\phi}+(4.905-0.1\cdot K_{1_1})\cdot\dot{\phi}-0.1\cdot K_{3_1}\cdot\phi = 0. \tag{4.167}$$

For each region we choose the poles at

$$s1 = s2 = -1; \qquad s3 = -10. \tag{4.168}$$

From this we obtain two parameter sets:

$$K_{1_0} = -125.7; \quad K_{2_0} = -115; \quad K_{3_0} = -100; \tag{4.169}$$

$$K_{1_1} = -161; \quad K_{2_1} = -115; \quad K_{3_1} = -100. \tag{4.170}$$

The resulting differential equation is in any case

$$\phi^{(3)} + 12 \cdot \ddot{\phi} + 21 \cdot \dot{\phi} + 10 \cdot \phi = 0. \tag{4.171}$$

First, we prove the *local stability* at the centers of REG_0 and REG_1 by means of *Hurwitz's criterion*. With

$$A_0 \cdot \phi^{(3)} + A_1 \cdot \ddot{\phi} + A_2 \cdot \dot{\phi} + A_3 \cdot \phi = 0 \tag{4.172}$$

we have stability if

$$A_1 > 0; \qquad \begin{vmatrix} A_1 & A_3 \\ A_0 & A_2 \end{vmatrix} > 0. \tag{4.173}$$

With

$$\begin{vmatrix} 12 & 10 \\ 1 & 24.5 \end{vmatrix} > 0 \quad \text{and} \quad \begin{vmatrix} 12 & 10 \\ 1 & 17.47 \end{vmatrix} > 0 \tag{4.174}$$

this criterion is fulfilled. Now, we check the variants

Var 1 : $a_{31} = -10;\ a_{32} = -24.5;\ a_{33} = -12$
Var 2 : $a_{31} = -10;\ a_{32} = -17.47;\ a_{33} = -12;$

on stability around any set-point in state space. The result is that with the matrix

$$\mathbf{P} = \begin{pmatrix} 1 & -0.3 & 0.5 \\ -0.3 & 1 & -0.3 \\ 0.5 & -0.3 & 1 \end{pmatrix} \tag{4.175}$$

the two variants lead to a negative definite $\dot{V}$. Thus, the system is locally stable for any arbitrary set-point.

5. Adaptive Fuzzy Control

5.1 Introduction

Most of the real-world processes that require automatic control are nonlinear in nature. That is, their parameter values alter as the operating point changes, over time, or both. As conventional control schemes are linear, a controller can only be tuned to give good performance at a particular operating point or for a limited period of time. The controller needs to be retuned if the operating point changes, or retuned periodically if the process changes with time. This necessity to retune has driven the need for adaptive controllers that can automatically retune themselves to match the current process characteristics. An excellent introduction to "conventional" adaptive control systems is by Åström [8].

FKBC are nonlinear and so they can be designed to cope with a certain amount of process nonlinearity. However, such design is difficult, especially if the controller must cope with nonlinearity over a significant portion of the operating range of the process. Also, the rules of the FKBC do not, in general, contain a temporal component, so they cannot cope with process changes over time. So there is a need for adaptive FKBC as well.

There is still contention as to what exactly constitutes an adaptive controller [8], and there is no consensus on the terminology to use in describing adaptive controllers. However, to enable the clear classification of the differing attempts at developing adaptive FKBC, we will introduce some definitions and terminology.

Adaptive controllers generally contain two extra components on top of the standard controller itself. The first is a "process monitor" that detects changes in the process characteristics. It is usually in one of two forms:

- a performance measure that assesses how well the controller is controlling, or

- a parameter estimator that constantly updates a model of the process.

The second component is the adaptation mechanism itself. It uses information passed to it by the process monitor to update the controller parameters and so adapts the controller to the changing process characteristics. Adaptive controllers can be classified as performance-adaptive or parameter-adaptive depending on which type of process monitor they employ [184]. Both types of process monitor will be seen in the adaptive FKBC that follow.

FKBC contain a number of sets of parameters that can be altered to modify the controller performance. These are:

- the scaling factors for each variable,
- the fuzzy set representing the meaning of linguistic values,
- the if-then rules.

A non-adaptive FKBC is one in which these parameters do not change once the controller is being used on-line. If any of these parameters are altered on-line, we will call the controller an adaptive FKBC. Each of these sets of parameters has been used as the controller parameters to be adapted in different adaptive FKBC. For the purposes of this chapter, adaptive FKBC that modify the fuzzy set definitions or the scaling factors will be called self-tuning controllers. Altering these parameters essentially fine tunes an already working controller. Adaptive FKBC that modify the rules will be called self-organizing controllers. They can either modify an existing set of rules, in which case they are similar to self-tuning controllers, or they can start with no rules at all and "learn" their control strategy as they go.

Detailed descriptions of example adaptive FKBC are given in Section 5.3. In the following section the design options available when building an adaptive FKBC will be given in terms of choosing a suitable process monitor and adaptation mechanism.

5.2 Design and Performance Evaluation

As described above, the adaptive component of an adaptive controller consists of two parts, namely,

1. the process monitor,
2. the adaptation mechanism.

The process monitor looks for changes in process characteristics and the adaptation mechanism alters the controller parameters on the basis of any detected changes. The nature of the performance monitor is not dictated by whether the underlying controller is fuzzy or not. So identical performance monitors can be used in both adaptive FKBC and non-fuzzy adaptive controllers. The adaptation mechanism, on the other hand, is specifically designed for altering the parameters of a FKBC. These two components will now be described in more detail.

5.2.1 The Performance Monitor

Parameter Estimators

Changes in process characteristics can either be detected through on-line identification of process model, or by assessment of the controlled response of the process. A process model is a mathematical description of the process that gives values of the process-outputs, given the current process-state and the inputs to the process. The process-inputs come from the controller. An on-line identifier requires the assumption of a particular form of process model.

The most commonly used model in controller design, fuzzy or otherwise, is the single-input, single-output, linear, first-order plus dead-time model described by the transfer function

$$\frac{\bar{y}(s)}{\bar{u}(s)} = \frac{K_P e^{-t_d s}}{\tau s + 1}, \tag{5.1}$$

where $\bar{y}(s)$ is the Laplace transform of the process-output, $\bar{u}(s)$ is the Laplace transform of the process-input, K_P is the gain, t_d is the dead-time, and τ is the time constant. This type of model is used in the self-tuning regulator [8], one of the most popular non-fuzzy adaptive controllers. The process monitor must continually estimate values for the process parameters, K_P, τ and t_d. This model has also been used in adaptive FKBC [208, 89].

A discrete-time, single-input, single-output (SISO), linear model is the time-series model of the form

$$y_{t+1} = \sum_{i=0}^{n} a_i(t) y_{t-i} + \sum_{j=0}^{m} b_j(t) u_{t-j}, \tag{5.2}$$

where y_t and u_t are the process-output and input at time t, respectively. In this case the process monitor must estimate values for the parameters $a_i(t)$ and $b_j(t)$ from collections of input-output data, $\{u_k, y_k\}_{k=1}^{K}$. This type of model has also been used in an adaptive FKBC [14].

The model can also be a fuzzy model of the process. Such a model consists of a set of rules of the same form as FKBC rules, but which describe the linguistic values of the process-output for given linguistic values of the process-input and process-state variable. For a multi-input, multi-output system (MIMO), with process-state vector $\mathbf{x} = (x_1, x_2, \ldots, x_n)^T$, $(x_j \in \mathcal{X}_j)$, process-input vector $\mathbf{u} = (u_1, u_2, \ldots, u_m)^T$, and $(u_k \in \mathcal{U}_k)$, process-output vector $\mathbf{y} = (y_1, y_2, \ldots, y_\ell)^T$, $(y_r \in \mathcal{Y}_r)$, a fuzzy model consists of rules of the form:

if $(x_1$ is $LX_1^{(i)}$ and $\ldots$ and x_n is $LX_n^{(i)})$
and $(u_1$ is $LU_1^{(i)}$ and $\ldots$ and u_m is $LU_m^{(i)})$
then $(y_1$ is $LY_1^{(i)}$ and $\ldots$ and y_ℓ is $LY_\ell^{(i)})$,

where i designates the i-th rule, and $LX_j^{(i)}$ $(j = 1, \ldots, n)$, $LU_k^{(i)}$ $(k = 1, \ldots, m)$, and $LY_r^{(i)}$ $(r = 1, \ldots, \ell)$ are linguistic values of the process-state, process-input and process-output variables, respectively. In the fuzzy model the meaning of these linguistic values is represented by fuzzy sets $\widetilde{LX}_j^{(i)}$, $\widetilde{LU}_k^{(i)}$ and $\widetilde{LY}_r^{(i)}$. These fuzzy sets are defined on the respective universes of discourse $\mathcal{X}_j$, $\mathcal{U}_k$ and $\mathcal{Y}_r$. The meaning of such a rule is given as a fuzzy relation $\widetilde{R}^{(i)}$ on $\mathcal{X} \times \mathcal{U} \times \mathcal{Y}$, where

$$\begin{aligned} \mathcal{X} &= \mathcal{X}_1 \times \ldots \times \mathcal{X}_n \\ \mathcal{U} &= \mathcal{U}_1 \times \ldots \times \mathcal{U}_m \\ \mathcal{Y} &= \mathcal{Y}_1 \times \ldots \times \mathcal{Y}_\ell . \end{aligned} \tag{5.3}$$

In the particular case of a Mamdani-type implication we will have that the fuzzy relation $\widetilde{R}^{(i)}$ is given as

$$\widetilde{R}^{(i)} = \left(\widetilde{LX}_1^{(i)} \times \ldots \times \widetilde{LX}_n^{(i)}\right) \times \left(\widetilde{LU}_1^{(i)} \times \ldots \times \widetilde{LU}_m^{(i)}\right) \times \left(\widetilde{LY}_1^{(i)} \times \ldots \times \widetilde{LY}_\ell^{(i)}\right). \tag{5.4}$$

For the whole set of rules we have

$$\widetilde{R} = \bigcup_i \widetilde{R}^{(i)}. \tag{5.5}$$

If we denote

$$\begin{aligned} \widetilde{L\mathbf{x}} &= \left(\widetilde{LX}_1 \times \ldots \times \widetilde{LX}_n\right) \\ \widetilde{L\mathbf{u}} &= \left(\widetilde{LU}_1 \times \ldots \times \widetilde{LU}_m\right) \\ \widetilde{L\mathbf{y}} &= \left(\widetilde{LY}_1 \times \ldots \times \widetilde{LY}_\ell\right), \end{aligned} \tag{5.6}$$

then $\widetilde{R}^{(i)} = \widetilde{L\mathbf{x}} \times \widetilde{L\mathbf{u}} \times \widetilde{L\mathbf{y}}$. The *fuzzy model* of the process is given as

$$\widetilde{L\mathbf{y}} = \left(\widetilde{L\mathbf{x}} \times \widetilde{L\mathbf{u}}\right) \circ \widetilde{R} = \widetilde{L\mathbf{x}} \circ \widetilde{L\mathbf{u}} \circ \widetilde{R}. \tag{5.7}$$

Example 5.1 A *fuzzy model* of a gas furnace, for example, where the input is the methane feed rate, and the output is the CO_2 concentration in the outlet gases, may consist of rules such as:

> *if* the current CO_2 concentration is MEDIUM
> *and* the previous methane feedrate was LOW
> *then* the next CO_2 concentration will be JUST HIGH.

This should be contrasted with a *FKBC if-then rule* for the same process. Such a rule would be of the form:

> *if* the previous CO_2 concentration was MEDIUM
> *and* the current CO_2 concentration is JUST HIGH
> *then* change the methane feed rate with a SMALL INCREASE.

Identification of a fuzzy process model involves estimation of the fuzzy relation, $\widetilde{R}$, from process input-output data. $\widetilde{R}$ is also called fuzzy relation matrix or relation matrix for short. There is an extensive literature covering techniques for doing this [43, 159, 160, 161, 98, 10, 204]. A particular method is given in detail in Section 5.3.4 where an adaptive FKBC that uses on-line identification of a fuzzy process model is described [81]. Other useful references that address the wider issues of fuzzy modelling, as well as fuzzy identification, include [163, 202, 203, 216, 217].

An interesting comparison has been made between time-series models and fuzzy models. A popular basis for the comparison of model identification techniques is the gas furnace data of Box and Jenkins [28]. This data consists of 296 pairs of input-output data that relate the flow rate of methane gas into the furnace to the concentration of carbon dioxide in the outlet gases. Box and Jenkins used this data to demonstrate the development of time-series models of the form shown in (5.2). Their model, obtained by a least-squares fit of the model parameters to the data, is

$$y_t = 0.55 \cdot y_{t-1} - 0.52 \cdot u_{t-3} - 0.4 \cdot u_{t-4} - 0.51 \cdot u_{t-5} + N_t, \tag{5.8}$$

$$N_t - 1.53 \cdot N_{t-1} + 0.63 \cdot N_{t-2} = a_t, \tag{5.9}$$

where y_t is the concentration of carbon dioxide in the outlet gases and u_t is the methane flow rate, at time t. The N_t is the noise in the system, and a_t is white noise. This is a rather complicated six-parameter, third-order model. It gives a fitting factor of 0.202, where the fitting factor is calculated as the sum of the squared error between the model output and the data, divided by the number of data points. For further details on how this model is derived, the reader should consult [28].

This same data set has been used to test fuzzy identification methods [81, 159, 217, 216]. Using a simpler model structure that gives

$$y_t = f(y_{t-1}, u_{t-4}), \tag{5.10}$$

fuzzy models that are nearly as accurate as the time-series model have been developed using fuzzy identification on the gas furnace data. The models have the form

$$\widetilde{LY}_t = \widetilde{LY}_{t-1} \circ \widetilde{LU}_{t-4} \circ \tilde{R}, \tag{5.11}$$

where $\tilde{R}$ is the fuzzy relation. Models using different numbers and shapes of fuzzy sets for each linguistic variable have been tried. A particular model, that gives a fitting factor of 0.47 to the data, in which five fuzzy sets were defined to cover each universe of discourse, is shown as a rule table in Table 5.1. Reading from the table, one rule is, for example:

if y_{t-1} is JUST_LOW *and* u_{t-4} is LOW *then* y_t is MEDIUM.

Adaptive controllers that use on-line identification of a process model as their performance monitor are known as parameter-adaptive controllers, since their adaptation mechanisms rely on an estimate for values of the process model parameters.

Performance Measures

The alternative type of process monitor forms an assessment of controller performance based on readily measured variables. For the regulatory control problem, where the aim is to keep a process-state variable at its specified set-point, a number of performance-related variables are of potential use. These include [198]:

Table 5.1. Rule table for fuzzy model of the gas furnace data. The table entries are y_t. (Notation: L=low, M=medium, H=high, J=just) (from Graham [81])

		y_{t-1}				
		LOW	JUST_LOW	MEDIUM	JUST_HIGH	HIGH
	LOW	JL	M	M	H	H
	JUST_LOW	L	JL	M	JH	H
u_{t-4}	MEDIUM	L	JL	M	JH	H
	JUST_HIGH	L	JL	M	JH	H
	HIGH	L	L	M	M	JH

- Overshoot
- Rise time
- Settling time
- Decay ratio
- Frequency of oscillations of the transient
- Integral of the square error
- Integral of the absolute value of the error
- Integral of the time-weighted absolute error
- Gain and phase margins.

The choice of one or more performance measures depends on the type of response the control system designer wishes to achieve. Many of these measures have been used in adaptive FKBC [13, 105, 14, 11, 230, 150, 128, 169]. The final output of the performance monitor, as seen by the adaptation mechanism, is either the actual values of the performance measures (e.g., [13]), or a performance index derived from the performance measures (e.g., [169]).

For more complex supervisory control problems, the controller performance is measured by the level of satisfaction of any number of different goals and constraints. The overall performance is derived as a suitable confluence of the goals and constraints that takes into account the relative priorities involved. A fuzzy performance measure where goals and constraints are described by performance fuzzy sets is used in the adaptive FKBC of [81]. This is based on the theory of fuzzy decision making [18]. A more detailed description of this performance measure is given in Section 5.3.4.

Adaptive controllers that use a measure of controller performance as their performance monitor are known as performance-adaptive controllers.

5.2.2 The Adaptation Mechanism

The adaptation mechanism must modify the controller parameters to improve the controller performance on the basis of the output from the process monitor. Adaptation mechanisms for FKBC can be classified according to which parameters are adjusted. Parameters that can be adjusted include the scaling factors with which controller input and output values are mapped onto the universe of discourse of the fuzzy set definitions. The set definitions will often be defined on a normalized universe from, say, -1.0 to $+1.0$, as illustrated in Fig. 5.1.

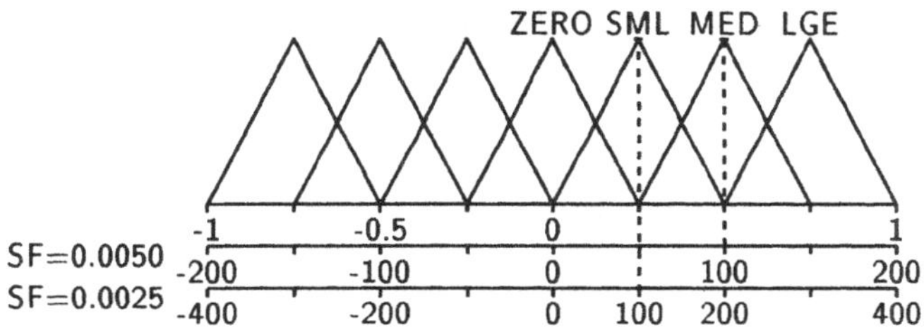

Fig. 5.1. The effect of altering a scaling factor.

The real values of the input variables may actually range from, say, -200 to $+200$, and so need to be scaled. If, for example, the input value is multiplied by a scaling factor of 0.005, the input is mapped to the universe of discourse as shown by the middle scale in Fig. 5.1. In this case an input value of 100 is classified as MEDIUM. Altering the scaling factor changes the classification of an input value. For example, with a scaling factor of 0.0025 a value of 100 is now classified as SMALL, as shown by the bottom scale of Fig. 5.1. This reduces the sensitivity of the controller to the input, and so reduces the controller gain. In this way, altering scaling factors is similar to gain tuning in standard PID-controllers.

Another gain tuning mechanism is to alter the shapes of the fuzzy sets. An example where the sets are altered to increase the sensitivity of the controller to small values of the input is shown in Fig. 5.2.

Whereas altering the scaling factors alters the gain uniformly across the entire input universe, changing the shapes of specific fuzzy sets allows the gain to be modified within specific regions of the universe. Adaptive controllers that adjust the fuzzy set definitions or scaling factors are called self-tuning controllers.

A third set of parameters that can be altered are the if-then rules themselves. For example, a rule may be changed from

if temperature is HIGH *then* SMALL INCREASE in cooling

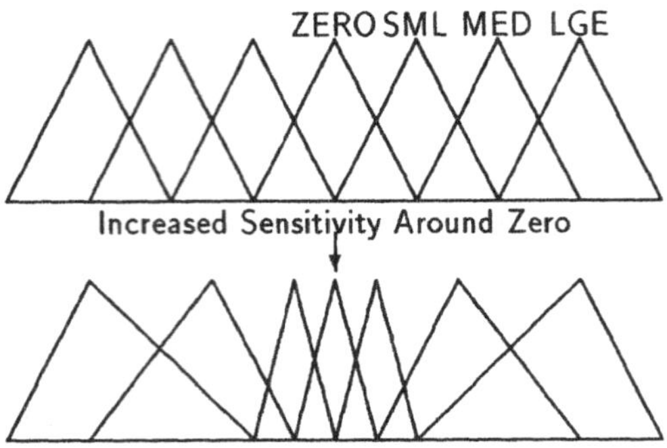

Fig. 5.2. Adapting fuzzy set definitions.

to

if temperature is HIGH *then* LARGE INCREASE in cooling

to adapt to changing operating conditions. Adaptive controllers that alter the rules are called self-organizing controllers.

Altering Scaling Factors

Quite simple schemes for altering the scaling factors to meet various performance criteria can be devised. An example is the work of Yamashita et al. [230]. Their control problem was temperature control during start-up of a packed-bed catalytic reactor. The temperature is controlled by altering the flow rate of hydrogen gas to the reactor. This process is highly exothermic and nonlinear, making it a difficult control problem. During startup the catalyst temperature must be increased to its operating point by increasing the hydrogen gas flow rate. The temperature increases slowly at first, and then rises abruptly with the onset of ignition. In order to prevent overshoot and oscillations in the temperature, the controller gain must be kept low during this period, so that only small changes in gas flow rate are made for large changes in temperature. However, if this low gain is retained once the operating temperature is reached, small, low frequency oscillations in the temperature result, due to the low sensitivity of the controller to fluctuations in the temperature. The controller gain consequently needs to be increased. Yamashita et al. designed a Mamdani FKBC with the error and change-of-error of the temperature as the inputs, and the change in hydrogen gas flow rate to the reactor as the output. They used the following scheme to automatically increase the controller gain once the operating temperature was reached by altering the scaling factors for the error and change-of-error.

The performance measure is the average of the squared error over the previous three sampling times. At sample time, k, a scaling factor modifier, C_k, is calculated as a function of the performance measure, P_k, according to the set of linguistic rules:

if P_k is VERY LARGE *then* C_k is VERY SMALL
if P_k is LARGE *then* C_k is SMALL
if P_k is MEDIUM *then* C_k is MEDIUM
if P_k is SMALL *then* C_k is LARGE.

The scaling factors for the error (GE) and change-of-error (GΔE) are then calculated via

$$\mathrm{GE}_k = C_k\,\mathrm{GE}_0 \tag{5.12}$$

$$\mathrm{G\Delta E}_k = C_k\,\mathrm{G\Delta E}_0, \tag{5.13}$$

where GE_0 and $\mathrm{G\Delta E}_0$ are fixed initial values.

These rules for C_k can be implemented in a fuzzy way, but Yamashita et al. used a set of crisp relations that specified a specific crisp value of C_k for different ranges of P_k. The rules have the effect of increasing the controller gain by increasing the scaling factors, as the average squared error decreases as the process is maintained around its set-point. Results with an experimental reactor showed the expected improvement in control with this simple adaptive scheme.

Various schemes for altering the scaling factors on the basis of process models have also been devised [208, 89]. For example, Hayashi [89] has derived a set of equations for calculating the input and output scaling factors for a PI-like FKBC from the parameters of the first-order plus dead-time model of the process given by equation (5.1).

Modifying Fuzzy Set Definitions

A similar tuning mechanism is to alter the shapes of the fuzzy sets defining the meaning of linguistic values. There has been some argument [131, 215] that changing the fuzzy set definitions should not be used to tune the controller. The fuzzy set definitions are not arbitrary but are chosen to reflect the meaning of the linguistic values taken by the variables. While this is certainly true for the broad shapes of the sets, small modifications can still be made without endangering the underlying linguistic meaning. Work has shown [150, 128, 77, 13] that this can be an effective means of tuning a FKBC.

Recent work has centred on the use of mathematical optimization techniques to alter the set definitions so that the output from the FKBC matches a suitable set of reference data as closely as possible [150, 128]. This procedure is carried out off-line and so tunes the controller before it is used. No subsequent on-line adaptation is performed, so the controllers are not strictly adaptive FKBC. However, the technique is closely allied to the adaptive methods discussed in this chapter, and it has been demonstrated that it can be used on-line [77]. The basic tuning method is detailed in Section 5.3.1.

A truly adaptive FKBC that modifies the set definitions on-line has been developed by Bartolini et al. [13]. The controller is similar to [230] in that a simple algorithm has been devised to alter the set definitions in response to a number of different performance measures, such as the average squared error. For example, if the average error is consistently below set-point, the degree to which values of the error are recognized as "negative" is incremented, and the corresponding degrees for "OK" and "positive" are decremented. This has the effect of increasing the controller's ability to detect "negative" values of the error. Bartolini et al. applied this adaptive controller to the control of a simulation of a continuous casting plant. A detailed description is given in Section 5.3.2.

Self-Organizing Controllers

A number of schemes for self-organizing FKBC have been developed. The original attempt is that of Mamdani and his coworkers [130, 132, 169, 231]. Their idea is to try to identify which rule is responsible for the current poor control performance, and then to replace it with a better rule. This requires both a mechanism for determining which rule is incorrect, and one for determining an improved rule to take its place. The performance monitor assesses the controller performance on the basis of the error and change-of-error of the process-output variables compared with what is desirable from a regulatory control perspective. The final performance monitor output is an indication of the change required in the process-output variable to achieve good control. For example, if a variable is well above set-point and moving further away from set-point, then its performance is very poor and a large change is required. The adaptation mechanism must translate the required changes in the process-output variables into required changes in the process-input variables (control-outputs). This can be done using a simple incremental model of the process that relates changes in process-inputs to changes in process-outputs (how such a model can be obtained is described in Section 5.3.3). By inverting such a model, so that it gives the changes in process-inputs required to achieve given changes in process-outputs, a new control rule can be formulated. Some knowledge of the process order and dead times is used to identify which past control-outputs are responsible for the current poor performance, and so should be corrected. This controller can modify a predefined set of rules, or it can start with no rules at all and "learn" its control policy as it goes.

A model-based self-organizing FKBC has been developed by Graham and Newell [81, 82, 83]. It is a parameter-adaptive controller whose performance monitor performs on-line identification of a fuzzy process model. There is no explicit adaptation mechanism as the model-based controller uses the fuzzy process model directly in calculating the appropriate control-output. The performance of the controller is adapted to the process as the on-line identification modifies the model to make it a more accurate representation of the process.

The controller can start with no model at all and "learn" its control policy on-line in a similar fashion to the self-organizing controller (SOC) of Mamdani.

Both of these self-organizing controllers are described in more detail in Section 5.3.

5.2.3 Performance Evaluation

Little serious work has been done on performance evaluation of adaptive FKBC. All application of the controllers has been to simulations of processes, or to laboratory-scale equipment. The results of these applications, however, are encouraging. Every adaptive controller showed improvement in performance over a non-adaptive version of the same controller, or over a conventional PID-controller, when comparisons were made. The adaptive FKBC also performed well compared with "conventional" adaptive controllers. The adaptive fuzzy controller for a catalytic reactor [230] outperformed a model-reference adaptive controller (MRAC). The FKBC of [77] learned a satisfactory control scheme faster than a comparable neural network-based controller. Finally, the performance-adaptive controller of [13] did as well as a self-tuning regulator (STR) in the control of a continuous casting plant.

Since few results have been obtained for the stability of non-adaptive FKBC, especially when they are nonlinear, it is not surprising that the stability of adaptive FKBC is still basically an open question. Procyk and Mamdani [169] carried out an empirical study of the effect of various design parameters on the performance of their self-organizing controller. These results are discussed in Section 5.3.3.

Batur and Kasparian [14] demonstrate that the circle stability criterion [173] can be used to establish the stability of their adaptive FKBC, provided that the process being controlled is linear and bounds are placed on the maximum and minimum change in control-output that can be made at each sampling time.

Further results, in the near future, are likely to be of an empirical nature. Application of these controllers to pilot-scale plants is needed to give a good evaluation of their potential worth.

5.3 The Main Approaches to Design

In the following paragraphs some of the main approaches to the design of adaptive FKBC are described in detail. The examples chosen cover the full range of different types of adaptive controllers, as described by the classifications given earlier.

5.3.1 Membership Function Tuning using Gradient Descent

This method relies on having a set of training data against which the controller is tuned. If a reliable set of controller input-output data is available, it is possible to tune the membership functions used in the FKBC rules using a numerical optimization procedure. The basic example of this is given by Nomura et al. [150] where they use gradient descent to tune simple membership functions.

The FKBC from Nomura et al.

Their FKBC consists of a set of n if-then rules of the form

$$\text{Rule } i\text{: } \textit{if } x_1 \text{ is } LX_1^{(i)} \text{ and } \ldots \text{ and } X_m \text{ is } LX_m^{(i)} \textit{ then } u \text{ is } LU^{(i)}, \tag{5.14}$$

where $x_1, x_2, \ldots, x_m$ are the controller inputs (process-state variables), u is the control-output variable, i is the rule number, $LX_1^{(i)}, \ldots, LX_m^{(i)}$ are the linguistic values of the rule-antecedent, and $LU^{(i)}$ is the linguistic value of the rule-consequent.

The membership functions, $\widetilde{LX}_j^{(i)}$, of the antecedent part are simply triangles described by a peak value, a_{ij}, and a support, b_{ij}, in the universe of discourse, as shown in Fig. 5.3. The membership function is thus given by

$$\widetilde{LX}_j^{(i)}(x) = 1 - \frac{2|x - a_{ij}|}{b_{ij}} \tag{5.15}$$

where x is a crisp controller input in the domain $\mathcal{X}_j$ of the input variable x_j.

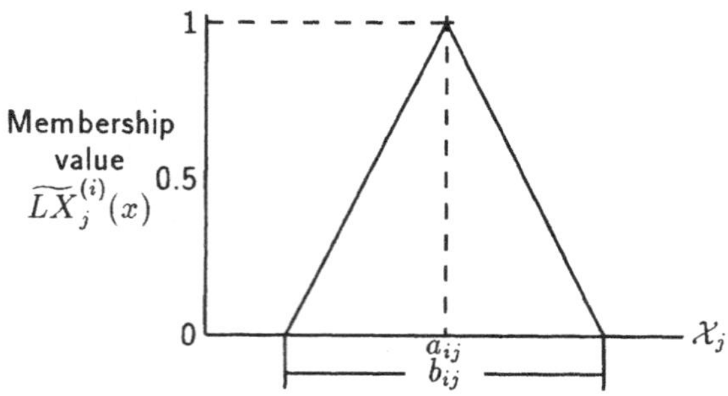

Fig. 5.3. Triangular fuzzy set (redrawn from Nomura *et al.* [150]).

The control-output membership function, $\widetilde{LU}^{(i)}$, is a fuzzy singleton set defined on the real number, u_i, i.e., $\widetilde{LU}^{(i)} = 1/u_i$.

Using the max-product composition (see Section 2.2.4) and the Center-of-Area defuzzification method, the non-fuzzy output u^* from the rule set is given by

$$u^* = \frac{\sum_{i=1}^{n} \mu_i \cdot u_i}{\sum_{i=1}^{n} \mu_i}, \tag{5.16}$$

where μ_i is given by the product of the membership degrees of the input variables in the i-th rule

$$\mu_i = \widetilde{LX}_1^{(i)}(x_1) \cdot \widetilde{LX}_2^{(i)}(x_2) \cdot \ldots \cdot \widetilde{LX}_m^{(i)}(x_m), \tag{5.17}$$

where x_j is a crisp controller input value from the domain $\mathcal{X}_j$ of the input variable X_j.

The Gradient Descent Algorithm

If a set of operating data is available that describes the desirable control-output, u^r, for various values of the process-state, $x_1^r, \ldots, x_m^r$, the fuzzy controller can be optimized by minimizing some criterion on the error between the FKBC output given by (5.16) and the desired output given by the reference data. By substitution of (5.15) and (5.17) into (5.16), we have an equation for the control-output, u, in terms of the membership function parameters a_{ij}, b_{ij} and u_i, $i = 1, \ldots, n$; $j = 1, \ldots, m$. These are the parameters to be tuned by the optimization procedure.

Nomura et al. have chosen to minimize the objective function, E, given by

$$E = \frac{1}{2}(u - u^r)^2, \tag{5.18}$$

where u^r is the desired real-valued control-output as given by the reference data, and u is the FKBC output, for a particular process-state.

One of the simplest methods for solving this optimization problem is the steepest descent algorithm (see, for example [180]). This is an iterative algorithm that seeks to decrease the value of the objective function with each iteration. It relies on the fact that from any point the objective function decreases most rapidly in the direction of the negative gradient vector of its parameters at that point. If we have $E(Z)$, where $Z = (z_1, z_2, \ldots, z_p)$, then this vector is

$$\left(-\frac{\partial E}{\partial z_1}, -\frac{\partial E}{\partial z_2}, \ldots, -\frac{\partial E}{\partial z_p} \right).$$

If $z_i(t)$ is the value of the i-th parameter at iteration t, the steepest descent algorithm seeks to decrease the value of the objective function by modifying the parameter values via

$$z_i(t+1) = z_i(t) - K\frac{\partial E(Z)}{\partial z_i}, \qquad i = 1, \ldots, p, \tag{5.19}$$

where K is a constant which controls how much the parameters are altered at each iteration (choosing a suitable K can be difficult – consult a textbook on minimization techniques, such as [180], for details). As the iterations proceed, the objective function converges to a local minimum.

In this case, the objective function parameters we wish to alter are the membership function parameters a_{ij}, b_{ij} and u_i, i.e.,

$$(z_1, \ldots, z_p) = (a_{11}, \ldots, a_{nm}, b_{11}, \ldots, b_{nm}, u_1, \ldots, u_n), \qquad p = 2nm + n. \tag{5.20}$$

Substituting (5.17) and (5.16) into (5.18) gives the objective function in terms of the membership functions,

$$E = \frac{1}{2} \cdot \left(\frac{\sum_{i=1}^{n} (\prod_{j=1}^{m} \widetilde{LX}_j^{(i)}(x_j^r)) \cdot u_i}{\sum_{i=1}^{n} (\prod_{j=1}^{m} \widetilde{LX}_j^{(i)}(x_j^r))} - u^r \right)^2 . \tag{5.21}$$

The steepest descent algorithm gives the following iterative equations for the parameter values,

$$a_{ij}(t+1) = a_{ij}(t) - K_a \cdot \frac{\partial E}{\partial a_{ij}}, \qquad i = 1, \ldots, n;\ j = 1, \ldots, m, \tag{5.22}$$

$$b_{ij}(t+1) = b_{ij}(t) - K_b \cdot \frac{\partial E}{\partial b_{ij}}, \qquad i = 1, \ldots, n;\ j = 1, \ldots, m, \tag{5.23}$$

$$u_i(t+1) = u_i(t) - K_u \cdot \frac{\partial E}{\partial u_i}, \qquad i = 1, \ldots, n. \tag{5.24}$$

Taking the partial derivatives of E yields the following equations,

$$a_{ij}(t+1) = a_{ij}(t) - \frac{K_a \cdot \mu_i}{\sum_{k=1}^{n} \mu_k} \cdot (u - u^r) \cdot (u_i(t) - y) \cdot \mathrm{sgn}(x_j^r - a_{ij}(t)) \cdot \frac{2}{b_{ij}(t) \cdot \widetilde{LX}_j^{(i)}(x_j^r)} \tag{5.25}$$

$$b_{ij}(t+1) = b_{ij}(t) - \frac{K_b \cdot \mu_i}{\sum_{k=1}^{n} \mu_k} \cdot (u - u^r) \cdot (u_i(t) - y) \cdot \frac{(1 - \widetilde{LX}_j^{(i)}(x_j^r))}{\widetilde{LX}_j^{(i)}(x_j^r)} \cdot \frac{1}{b_{ij}(t)} \tag{5.26}$$

$$u_i(t+1) = u_i(t) - \frac{K_u \cdot \mu_i}{\sum_{k=1}^{n} \mu_k} \cdot (u - u^r). \tag{5.27}$$

The Tuning Procedure

The FKBC rules are extracted from the process operators. The membership functions are defined initially such that the input domains are divided equally, by a suitable choice of the a_{ij}, and the sets overlap, by a suitable choice of the b_{ij} (for an example see the top half of Fig. 5.2). The u_i are chosen to give a suitable range of control-outputs covering, for example, a large decrease, small decrease, no change, small increase, and a large increase.

Once a set of reliable controller input/output data $(x_1^r, \ldots, x_m^r, u^r)$ has been collected, a possible optimization procedure is as follows:

1. the rules are fired on the input data $(x_1^r, \ldots, x_m^r)$ to obtain the antecedent value, μ_i, for each rule, and the real-valued control-output, u,
2. parameters u_i are updated using (5.27),
3. rule firing is repeated using the new values of u_i,

4. parameters a_{ij} and b_{ij} are updated by (5.25) and (5.26), using the new values of u_i, μ_i and u,
5. inference error $D(t) = 1/2 \cdot (u(t) - u^r)^2$ is calculated,
6. if the change-of-error $|D(t) - D(t-1)|$ is suitably small, the optimization is complete; otherwise it is repeated from step 1.

This procedure will modify the actual values, u_i, used for the controller outputs, and will change the centre, a_{ij}, and width, b_{ij}, of the antecedent fuzzy sets. The new fuzzy sets may look something like those seen in the bottom half of Fig. 5.2.

Example 5.2 Nomura et al. have applied this to several simulated systems, including the problem of a mobile robot avoiding a moving obstacle, with some success. The robot problem was for the robot to move from its starting point to a fixed target point while avoiding a single obstacle moving across its path. Only the steering angle of the robot was controlled, while its driving speed was constant. The dynamics of the robot were not considered. The robot received four inputs:

1. distance to the obstacle,
2. angle between the robot and the obstacle, from a fixed baseline,
3. distance to the target,
4. angle between the robot and the target.

Five membership functions were defined for each of the four inputs, and an initial set of 625 rules was provided.

Training data was obtained by an operator manually controlling the robot steering angle during trials involving two different obstacle movement patterns. Typical training runs are shown in Fig. 5.4. Training yielded 66 sets of input-output data. This data was then used to tune the membership functions, using the algorithm given above. Once trained, the robot was trialled on automatically avoiding the obstacle. Results are shown in Fig. 5.5. In panels (a) and (b), the robot is seen avoiding the obstacle for the same obstacle trajectories as used in the training runs. The automatic robot tracks are smoother than the operator controlled tracks. In the remaining panels, the robot is shown avoiding the obstacle moving along trajectories for which the robot was not trained. The robot still manages to avoid the obstacle while travelling along smooth trajectories.

Alternative Methods

The algorithm described is among the simplest for this type of self-tuning of membership functions, but it serves to illustrate the ideas. Alternative algorithms could include more sophisticated optimization methods and more general membership functions. Work along these lines is already being done, for

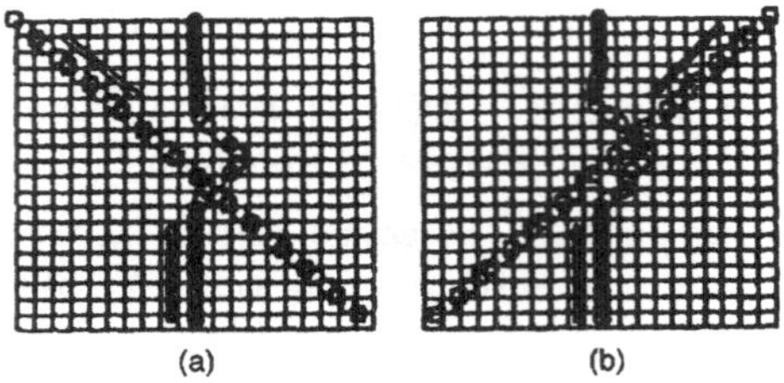

Fig. 5.4. Robot tracks taught by operator (from Nomura et al. [150]).

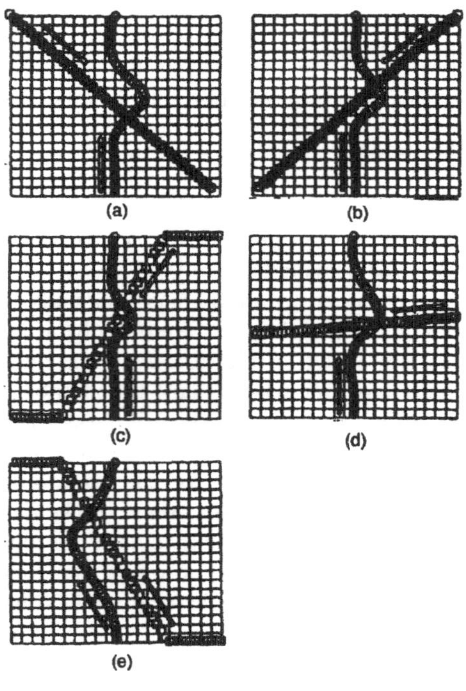

Fig. 5.5. Robot tracks controlled by tuned rules (from Nomura et al. [150]).

example by Maeda et al. [128]. Their Supervised Learning Algorithm is, they claim, some 40 times faster than the gradient descent method.

On-line Gradient Descent

The gradient descent tuning algorithm can be used on-line to form an adaptive FKBC, if suitable reference data can be generated. An architecture that achieves this has been developed by Glorennec [77]. It is illustrated in Fig. 5.6.

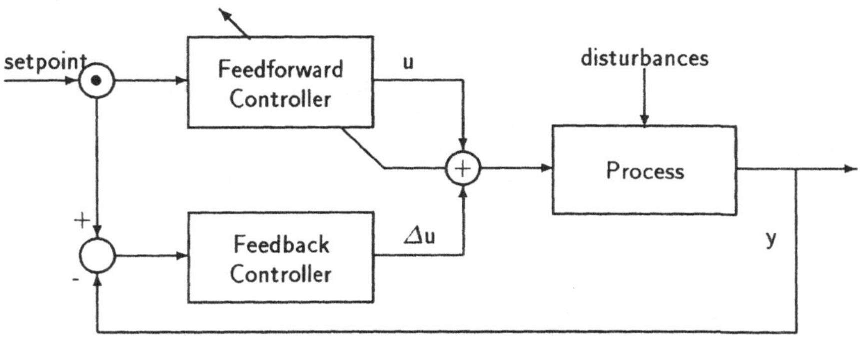

Fig. 5.6. Adaptive feedforward/feedback fuzzy controller (redrawn from Glorennec [77]).

The FKBC to be tuned is a feedforward controller. Given the required set-point, it calculates the required process input using an inverse fuzzy process model. Instead of specifying the desired control-output for a given process state, the inverse fuzzy process model predicts the process-output to be expected in time due to the current and previous process-states and control-outputs. In the absence of unmeasured disturbances, the quality of its control is dependent on the accuracy of the inverse fuzzy process model it employs. If the process-input it calculates is wrong, errors will result, and the PI-like FKBC will come into play to drive the process to set-point. As it does so, the corrections it makes to the process input, Δu, can be used as the error data, $u - u^r$, for the steepest descent tuning algorithm, to tune the feedforward controller's membership functions. In this way the feedforward controller will learn the correct process-input for a particular set-point. Glorennec chooses to only tune the rule-consequent membership functions, via equation (5.27). In principal, the full algorithm could be used to tune all the membership functions.

Glorennec has applied this scheme to a simulation of a mixer tap. The problem is to control the temperature and flow rate of the output water stream by manipulating the cold and hot water openings. He initially generated a set of process data by measuring the temperature and flow rate that resulted from various hot and cold water openings. This data was then used to tune the feedforward controller off-line. The resultant controller performed well. The on-line architecture was then used on the same problem, with all the consequent membership functions of the feedforward controller set to zero. This meant that initially the feedforward controller had no control knowledge. The final tuned feedforward controller performed identically to the controller tuned off-line.

5.3.2 Membership Function Tuning Using Performance Criteria

One of the earliest examples of a self-tuning FKBC is that of Bartolini et al. [13]. It is an example of a performance adaptive FKBC. Unlike the gradient descent method described above, which is essentially for use off-line, this controller modifies its membership functions on-line according to various controller performance criteria. So it is truly an adaptive FKBC. Bartolini et al. applied this controller to a simulation of a continuous casting plant. Comparison with standard PID-controller and a conventional adaptive control scheme showed the performance adaptive FKBC worked very well.

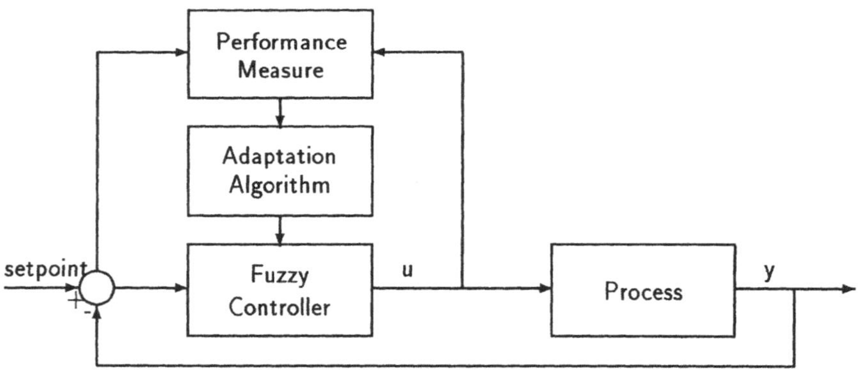

Fig. 5.7. Performance adaptive fuzzy controller.

The Controller

The structure of the controller is shown in Fig. 5.7. Only the single-input, single-output (SISO) control problem is considered. The aim is to maintain a single process-state variable at set-point. The controller is a PD-like FKBC, with the inputs being the error and change-of-error, and its output being the required change in the manipulated (control) variable. The performance monitor uses a set of six performance criteria to assess the FKBC while the controller is operating on-line. The indices are:

1. average square error, $\bar{e}^2$,

2. average error, $\bar{e}$,
3. average absolute error, $|\bar{e}|$,
4. maximum absolute error, $|e|_{max}$,
5. number, n_1, of consecutive variations in control-output,
6. number, n_2, of variations in control-output during a certain time interval.

The indices are evaluated over a fixed observation period, the length of which is a tuning parameter for the controller. As the primary aim of the controller is to maintain the process at set-point, the first four performance indices are of major importance. The last two indices express the secondary control objective of reducing the number of command variations.

The adaptation algorithm concerned with improving the set-point control is shown in Fig. 5.8. Four different adaptation actions are taken depending on the values of the first four performance indices. If the average error is too large, then adaptation action (a) or (b) is carried out depending on whether the average error is below or above set-point, respectively. Adaptation action (c) is taken if the average squared error is too large, indicating imprecise control. Finally, adaptation action (d) is taken if the error becomes too large, even for a single sampling instant. All the adaptation actions result in changes to the fuzzy set membership function definitions for the error, change-of-error, and change in control-output.

The adaptation action (a) is concerned with improving the controller performance when the process is consistently below set-point. This can be done, for instance, by increasing the degree to which values of the error are recognized as negative, i.e., below the set-point. Modification of the membership functions for the error (E) to achieve this is illustrated in Fig. 5.9. The modifications due to adaptation action (b) are exactly symmetrical with those of (a).

Adaptation actions (c) and (d) can be taken to be identical as they have the basic aim of improving the sensitivity of the controller. This can be done by sharpening the membership function definitions for the error, as illustrated in Fig. 5.10.

The adaptation algorithm for minimizing command variations is shown in Fig. 5.11. Adaptation action (e) has the aim of decreasing the sensitivity of the controller, and so it is exactly symmetrial with adaptation actions (c) and (d).

Controller Design

These adaptation algorithms contain a large number of parameters whose values must be chosen by the designer. Currently there are no specific selection criteria for these parameters. However, the following qualitative observations can be made. The aim of the adaptation process is to provide quick controller adaptation without causing instability or oscillations. In general, only small changes should be made to the fuzzy set definitions at any one time. In addition, the adaptation procedure need not be carried out at every sampling time. Bartolini

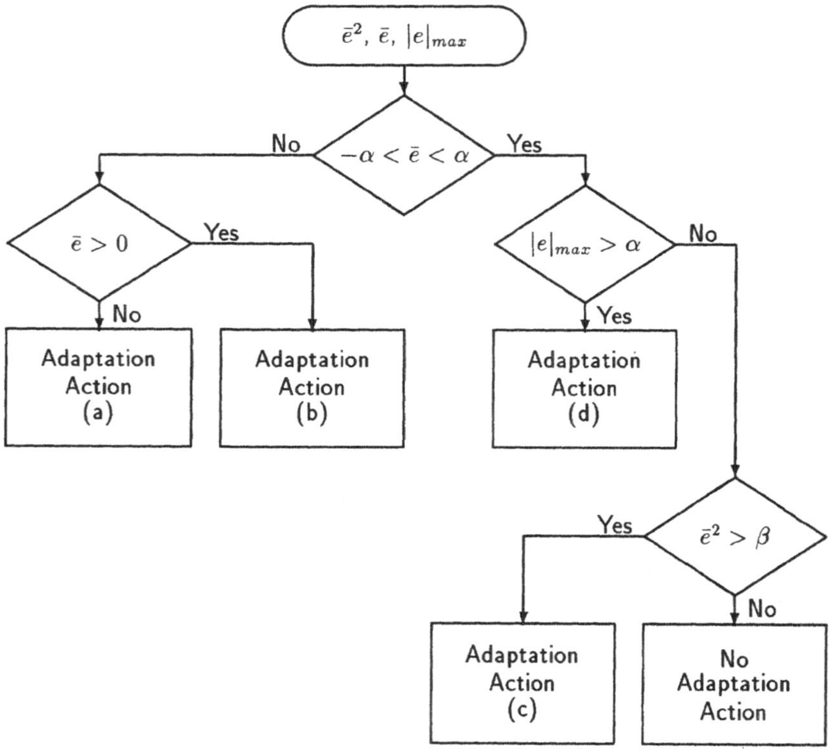

Fig. 5.8. Adaptation algorithm to improve setpoint control (redrawn from Bartolini et al. [13]).

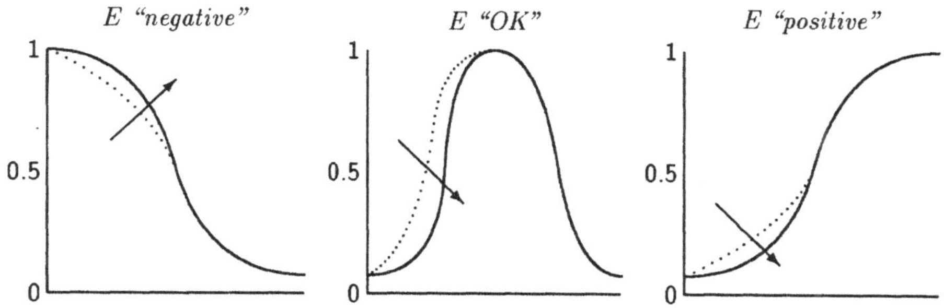

Fig. 5.9. Adaptation action (a).

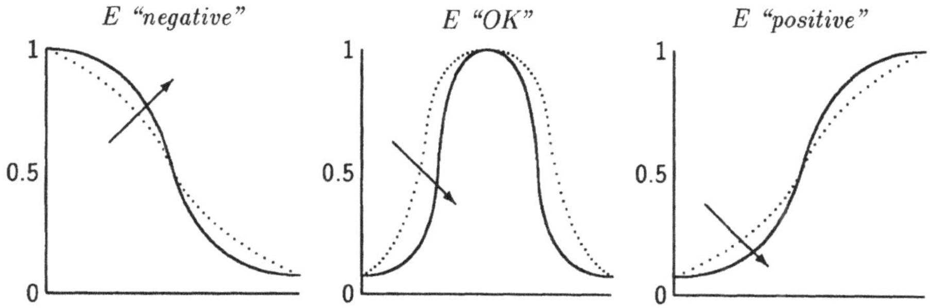

Fig. 5.10. Adaptation actions (c) and (d).

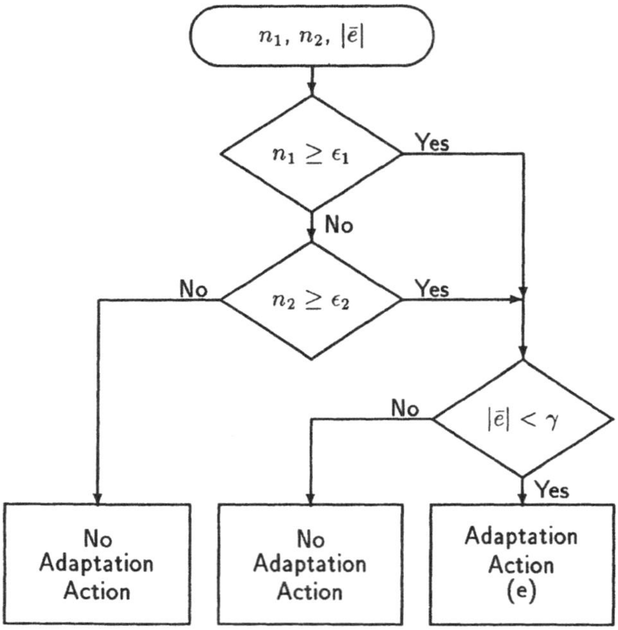

Fig. 5.11. Adaptation algorithm to reduce the number of control variations (redrawn from Bartolini et al. [13]).

et al. chose to monitor for excessive command variations are every sampling time so that this could be quickly corrected, but monitored set-point control performance at a slower rate. Adapting for set-point errors too aggressively could lead to instability.

The adaptation is done by modifying the shapes of the membership functions in proportion to the undesired effects that are being corrected. The threshold values α, β, ϵ_1, ϵ_2, and γ must be carefully chosen as they can have a large effect on the overall control system performance. The parameters α and β are concerned with specifying what levels of the set-point error criteria are acceptable before adaptation is required. If these parameters are set too low, then the controller will be constantly trying to adapt, and may never be able to satisfy the required performance. On the other hand, if the parameters are set too high, the controller may give unnecessarily bad performance as little adaptation will be carried out. Bartolini et al. demonstrated with a computer simulation of the control of a continuous casting plant that increasing β leads to an increase in the average square error. However, it also leads to a decrease in the number of command variations. Thus, choosing β involves a trade-off between competing performance criteria. The same considerations apply to α, and to the parameters controlling the adaptation to minimize command variations. The parameters, n_1 and n_2, specifying the level of command variation that is intolerable are chosen on the grounds of energy conservation requirements. Adaptation to minimize command variation only occurs when the set-point error is less than a level specified by γ, i.e., only when the error is sufficiently low that excessive control-output is not justified. If γ is too high, then too much adaptation to minimize control-output may take place, leading to an insensitive controller that provides poor set-point control. If γ is too low, then no adaptation to minimize control-output may ever take place.

The time window over which the performance indices are calculated is also important. The longer the interval, the more significant are the values of $\bar{e}$, $\bar{e}^2$, and $|e|$, but the more sluggish is the adaptation system's response. Consideration must be taken of the noisiness of the signals from which the error measurements are being made when deciding on an appropriate time window. The noisier the signals, the longer the time window should be to filter out the effects of the noise.

In summary, the choice of parameter values is process specific and requires careful matching with the controller performance objectives.

5.3.3 The Self-Organizing Controller

The major work on self-organizing controllers is that of Mamdani and his colleagues [130, 132, 169]. A detailed description of their self-organizing controller (SOC) is given in [169].

Like the controller of Bartolini et al., the self-organizing controller is also a performance-adaptive FKBC. However, in this case it is the rules themselves that are adapted, not the fuzzy set definitions, or the scaling factors. The structure of the controller is similar. but this time it also includes a model of the process. A block diagram is given in Fig. 5.12. The controller is of the standard double-input, single-output type, with the error (e) and change in error (Δe) as inputs, and process-input or control-output (u) as the output. The universes of discourse are discrete and normalized, with the appropriate scaling factors (GE, GΔE and GU, respectively) being chosen manually, i.e., they are not adapted as part of the adaptation mechanism.

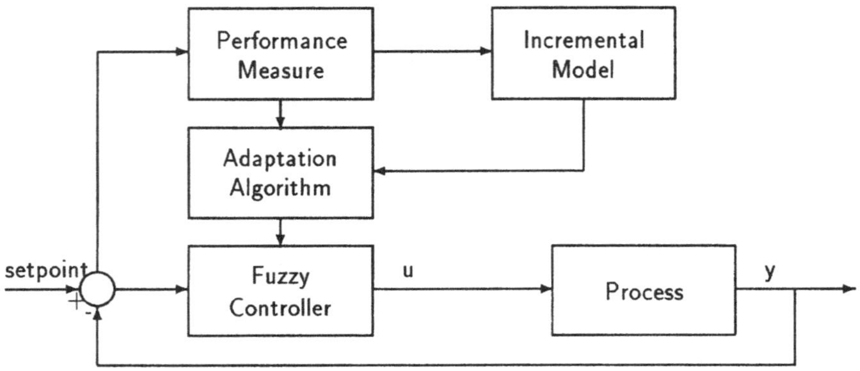

Fig. 5.12. Self-organising fuzzy controller.

Performance Monitor

The performance monitor consists of a performance measure that specifies the basic regulatory control requirements, namely a sufficiently fast approach to set-point, good damping when close to the set-point and a measure of tolerance around the set-point. The requirements with respect to this performance measure can be expressed as a set of rules that define the performance of the system at each sampling time, given the values of the error and change-of-error. For example:

if e is VERY LARGE *and* Δe is NEGATIVE VERY LARGE
then P is SATISFACTORY
if e is VERY LARGE *and* Δe is VERY LARGE
then P is VERY POOR,

where P is the performance measure.

For use in the adaptive controller, the output of the performance monitor is not given as a value of the performance, but as a value for the correction required at the process-output to obtain good performance. For example, the rules given above can be rewritten as rules describing the changes required to achieve good performance:

if e is VERY LARGE *and* Δe is NEGATIVE VERY LARGE
then C is ZERO
if e is VERY LARGE *and* Δe is VERY LARGE
then C is NEGATIVE VERY LARGE,

where C is the required correction. In other words, if the process output is a long way from set-point, but is moving rapidly towards set-point, then no change is required as the process is already moving as fast as possible towards its set-point. If, on the other hand, the error is large and is growing rapidly, then a large change is required to get the process at least moving towards its set-point. The entire performance monitor is described by the decision table shown in Table 5.2.

Table 5.2. Performance monitor decision table. The table entries are the required changes in the process-output. (Notation: Z=zero, S=small, M=medium, L=large, J=just, V=very, P=positive, N=negative) (from Procyk and Mamdani [169])

		Change in error (Δe)												
		Towards set-point								Away from set-point				
		NVL	NL	NJL	NM	NJM	NS	Z	S	JM	M	JL	L	VL
	NVL	Z	Z	Z	Z	Z	Z	VL	VL	VL	VL	VL	VL	VL
Below	NL	Z	Z	Z	JM	JM	M	VL	VL	VL	VL	VL	VL	VL
set-	NJL	Z	Z	Z	JM	JL	L	VL	VL	VL	VL	VL	VL	VL
point	NM	Z	Z	Z	JM	JM	M	JL	JL	JL	JL	L	L	VL
	NJM	Z	Z	Z	Z	Z	Z	JM	JM	JM	M	JL	L	VL
	NS	Z	Z	Z	Z	Z	Z	S	S	S	JM	M	JL	L
Error	NZ	Z	Z	Z	Z	Z	Z	Z	Z	Z	S	JM	M	JL
(e)	PZ	Z	Z	Z	Z	Z	Z	Z	Z	Z	NS	NJM	NM	NJL
	S	Z	Z	Z	Z	Z	Z	NS	NS	NS	NJM	NM	NJL	NL
	JM	Z	Z	Z	Z	Z	Z	NJM	NJM	NJM	NM	NJL	NL	NVL
Above	M	Z	Z	Z	NJM	NJM	NM	NJL	NJL	NJL	NJL	NL	NL	NVL
set-	JL	Z	Z	Z	NJM	NJL	NL	NVL	NVL	NVL	NVL	NVL	NVL	NVL
point	L	Z	Z	Z	NJM	NJM	NM	NVL	NVL	NVL	NVL	NVL	NVL	NVL
	VL	Z	Z	Z	Z	Z	Z	NVL	NVL	NVL	NVL	NVL	NVL	NVL

To update the controller, what is needed is the appropriate correction for the control-output. Thus the required correction in the process-output must be related to changes to the process-input, and hence the control-output. This can be done using a simple incremental model of the process that relates changes in process-inputs to changes in process-outputs. This model can simply be a matrix of coefficients related to the system Jacobian matrix. For example, suppose the process is a simple dynamical system in which changes in the process-outputs are related to the process-inputs via the equation

$$\dot{\mathbf{y}} = f(\mathbf{u}), \tag{5.28}$$

where $f(\cdot)$ is a nonlinear function. This system can be linearized around a particular operating point, $(\mathbf{y}_0, \mathbf{u}_0)$, by taking the first term of the Taylor series expansion of the above equation

$$f(\mathbf{u}_0 + \delta\mathbf{u}) \approx f(\mathbf{u}_0) + \nabla_{\mathbf{u}} f|_{\mathbf{u}_0} \delta\mathbf{u}, \tag{5.29}$$

giving

$$\delta\dot{\mathbf{y}} = f(\delta\mathbf{u}) = \nabla_{\mathbf{u}} f|_{\mathbf{u}_0}\delta\mathbf{u} = J\delta\mathbf{u}, \tag{5.30}$$

where $\delta\mathbf{y}$ and $\delta\mathbf{u}$ are small changes in the process-outputs and inputs, respectively. The matrix J is the system's Jacobian matrix. For a two-input, two-output system

$$\begin{pmatrix} \delta\dot{y}_1 \\ \delta\dot{y}_2 \end{pmatrix} = \begin{pmatrix} \frac{\partial f_1}{\partial u_1} & \frac{\partial f_1}{\partial u_2} \\ \frac{\partial f_2}{\partial u_1} & \frac{\partial f_2}{\partial u_2} \end{pmatrix} \begin{pmatrix} \delta u_1 \\ \delta u_2 \end{pmatrix}, \tag{5.31}$$

where $\mathbf{y} = (y_1, y_2)^T$, $\mathbf{u} = (u_1, u_2)^T$, and $f = (f_1, f_2)^T$.

The change in outputs $\Delta\mathbf{y}$ over a single sampling time, T, due to a change in inputs $\Delta\mathbf{u}$ is approximated by

$$\Delta\mathbf{y} \approx T\delta\dot{\mathbf{y}} = TJ\Delta\mathbf{u} = M\Delta\mathbf{u}. \tag{5.32}$$

The incremental model is given by the matrix of coefficients, M. For a single-input, single-output (SISO) system, the matrix M collapses to a single coefficient which, after normalization, is unity. There is no necessity for the model to be an accurate representation of the process. If it is not accurate, the adaptor will need to make repeated corrections for the same poor performance, which will slow down the learning process. However, the controller should still adapt towards improved performance.

If the performance monitor specifies, at sampling time nT, the required output corrections, $\Delta\mathbf{y}(nT) = \mathbf{c}(nT)$, the required changes in the process-inputs, or input reinforcements, are given by

$$\mathbf{r}(nT) = M^{-1}\mathbf{c}(nT). \tag{5.33}$$

For the single-input, single-output case this reduces to

$$r(nT) = c(nT), \tag{5.34}$$

i.e., the required input reinforcements are simply taken to be equal to the required output corrections.

Adaptation Algorithm

These input reinforcements are the amount that must be added to the process-inputs, i.e., control-outputs, to compensate for the current poor performance. The question remains, however, as to which control-outputs are responsible for the current poor performance. This depends on the dynamics of the process. High-order processes with large time lags will require control-outputs from a long time in the past to be adjusted. Low-order processes with short time lags will require control-outputs much nearer the present to be adjusted. Which control-outputs are corrected must be chosen in advance by the control system designer. Correcting the control-outputs in the FKBC means altering the rule-consequents of the appropriate rules.

Suppose that the current sampling time is n, and the control-output taken m sampling times in the past is the major influence on the current control performance. The numerical values of the error and change-of-error at that time were $e(nT-mT)$ and $\Delta e(nT-mT)$, respectively. The control-output was $u(nT-mT)$. These values can be described by the fuzzy sets $\widetilde{LE}_{n-m}$, $\widetilde{L\Delta E}_{n-m}$, and $\widetilde{LU}_{n-m}$, defined as

$$\widetilde{LE}_{n-m}(e) = \begin{cases} 1 & \text{if } e = e(nT-mT) \\ 0, & \text{otherwise,} \end{cases} \tag{5.35}$$

$$\widetilde{L\Delta E}_{n-m}(\Delta e) = \begin{cases} 1 & \text{if } \Delta e = \Delta e(nT-mT) \\ 0, & \text{otherwise,} \end{cases} \tag{5.36}$$

$$\widetilde{LU}_{n-m}(u) = \begin{cases} 1, & \text{if } u = u(nT-mT) \\ 0 & \text{otherwise.} \end{cases} \tag{5.37}$$

The controller response m sampling times in the past can be expressed as the set of n rules

$$\textit{if } e \text{ is } LE^{(i)}_{n-m} \textit{ and } \Delta e \text{ is } L\Delta E^{(i)}_{n-m} \textit{ then } u \text{ is } LU^{(i)}_{n-m}. \tag{5.38}$$

Also, suppose the performance monitor classifies the current performance as poor, and specifies that the control-output taken m sampling times ago should be modified by an input reinforcement of $r(nT)$. That is, the control-output for that time should have been $v(nT-mT) = u(nT-mT)+r(nT)$, not $u(nT-mT)$. If this new control-output is described by the fuzzy set

$$\widetilde{LV}_{n-m}(u) = \begin{cases} 1 & \text{if } u = u(nT-mT) + r(nT) \\ 0, & \text{otherwise,} \end{cases} \tag{5.39}$$

the correct controller response is described by the set of n rules

$$\textit{if } e \text{ is } LE^{(i)}_{n-m} \textit{ and } \Delta e \text{ is } L\Delta E^{(i)}_{n-m} \textit{ then } u \text{ is } LV^{(i)}_{n-m}. \tag{5.40}$$

To adapt the FKBC to reflect this new response, the i-th rule (a bad rule) that specifies the control-output, $LU^{(i)}_{n-m}$, should be deleted, and the new rule (good rule) specifying the control-output, $LV^{(i)}_{n-m}$, should be inserted in its place.

Suppose at sampling time n the complete rule set is described by the relation matrix $\widetilde{R}_n$. The meaning of the individual bad and good rules (equations (5.38) and (5.40)) can be expressed by the fuzzy relation matrices given by

$$\widetilde{R}_{\text{bad}} = \widetilde{LE}_{n-m} \times \widetilde{L\Delta E}_{n-m} \times \widetilde{LU}_{n-m} \tag{5.41}$$

$$\widetilde{R}_{\text{good}} = \widetilde{LE}_{n-m} \times \widetilde{L\Delta E}_{n-m} \times \widetilde{LV}_{n-m} \tag{5.42}$$

where $\times$ is the Cartesian product.

A new relation matrix, $\widetilde{R}_{n+1}$, needs to be generated that no longer includes the bad relation, $\widetilde{R}_{\text{bad}}$, but does include the new relation, $\widetilde{R}_{\text{good}}$. This requirement can be expressed linguistically (symbolically) as

$$R_{n+1} = (R_n \text{ but not } R_{\text{bad}}) \text{ else } R_{\text{good}}. \tag{5.43}$$

This expression is not unique as a way the relation matrix can be updated. For example,

$$R_{n+1} = (R_n \text{ else } R_{\text{good}}) \text{ but not } R_{\text{bad}}, \tag{5.44}$$

would do just as well. It can be argued that the first method is to be preferred as it removes the bad relation first before including the new relation. The meaning of this symbolically expressed modification procedure is

$$\widetilde{R}_{n+1} = (\widetilde{R}_n \cap \widetilde{R}'_{\text{bad}}) \cup \widetilde{R}_{\text{good}}, \tag{5.45}$$

in other words, in $\widetilde{R}_{n+1}$ all elements corresponding to $\widetilde{R}'_{\text{bad}}$ have been replaced by elements of $\widetilde{R}_{\text{good}}$.

Example 5.3 Consider a simple numerical example. To aid in presentation we consider a proportional-only controller in which the control-output is only a function of the error, and not the change-of-error. In this case the FKBC is described by the equation

$$\widetilde{U} = \widetilde{LE}^* \circ \widetilde{R}, \tag{5.46}$$

where $\widetilde{R}$ is the fuzzy relation describing the meaning of the whole set of rules. $\widetilde{LE}^*$ is the fuzzified crisp input for error, which was denoted (Section 2.4.1) as μ^*_{LE}. $\widetilde{U}$ is the overall control-output (the union of the clipped output fuzzy sets in Section 2.4.1). Suppose both the error and control-output are defined on discrete universes consisting of the elements: $\{-2,-1,0,1,2\}$. The relation matrix, $\widetilde{R}$, is a 5×5-matrix, and suppose at sample time n, it is

$$\widetilde{R}_n = \begin{bmatrix} 1.0 & 0.8 & 0.9 & 0.6 & 0.5 \\ 0.3 & 0.3 & 0.3 & 0.3 & 0.3 \\ 0.2 & 0.2 & 0.2 & 0.2 & 0.2 \\ 0.1 & 0.1 & 0.1 & 0.1 & 0.1 \\ 0.0 & 0.0 & 0.0 & 0.0 & 0.0 \end{bmatrix}.$$

Suppose at sampling time $(n-m)$, the error was -1. This can be described by the fuzzy set

$$\widetilde{LE}^*_{n-m} = \begin{bmatrix} 0.0 & 1.0 & 0.0 & 0.0 & 0.0 \end{bmatrix},$$

i.e., the membership function for the fuzzy set $0.0/-2 + 1.01/-1 + 0.0/0 + 0.0/1 + 0.0/2$. If the controller relation, $\widetilde{R}_{n-m}$, m sampling times in the past was still the same as $\widetilde{R}_n$, then the control-output

$$\widetilde{U}_{n-m} = \widetilde{LE}^*_{n-m} \circ \widetilde{R}_{n-m} = \begin{bmatrix} 0.3 & 0.3 & 0.5 & 1.0 & 0.3 \end{bmatrix}.$$

This gives an actual control-output of 1, if the Middle-of-Maxima defuzzification method (Section 3.6) is used to extract a numerical value from the fuzzy set, $\widetilde{U}_{n-m}$. But suppose that at sampling time n, the performance monitor says that the control-output m sampling times in the past should be reinforced by 1, i.e., the control-output should have been $v = u + 1 = 2$. By fuzzifying the bad control-output, $u = 1$, and the good control action, $v = 2$, we can generate the bad and good relations of equations (5.41) and (5.42):

$$\widetilde{LE}^*_{n-m} = \begin{bmatrix} 0.0 & 1.0 & 0.0 & 0.0 & 0.0 \end{bmatrix}$$

$$\widetilde{U}_{n-m} = \begin{bmatrix} 0.0 & 0.0 & 0.0 & 1.0 & 0.0 \end{bmatrix}$$

$$\widetilde{R}_{\text{bad}} = \widetilde{LE}^*_{n-m} \times \widetilde{U}_{n-m} = \begin{bmatrix} 0.0 & 0.0 & 0.0 & 0.0 & 0.0 \\ 0.0 & 0.0 & 0.0 & 1.0 & 0.0 \\ 0.0 & 0.0 & 0.0 & 0.0 & 0.0 \\ 0.0 & 0.0 & 0.0 & 0.0 & 0.0 \\ 0.0 & 0.0 & 0.0 & 0.0 & 0.0 \end{bmatrix}$$

$$\widetilde{LE}^*_{n-m} = \begin{bmatrix} 0.0 & 1.0 & 0.0 & 0.0 & 0.0 \end{bmatrix}$$

$$\widetilde{V}_{n-m} = \begin{bmatrix} 0.0 & 0.0 & 0.0 & 0.0 & 1.0 \end{bmatrix}$$

$$\widetilde{R}_{\text{good}} = \widetilde{LE}^*_{n-m} \times \widetilde{V}_{n-m} = \begin{bmatrix} 0.0 & 0.0 & 0.0 & 0.0 & 0.0 \\ 0.0 & 0.0 & 0.0 & 0.0 & 1.0 \\ 0.0 & 0.0 & 0.0 & 0.0 & 0.0 \\ 0.0 & 0.0 & 0.0 & 0.0 & 0.0 \\ 0.0 & 0.0 & 0.0 & 0.0 & 0.0 \end{bmatrix}$$

The controller relation matrix can now be updated according to the adaptation procedure specified in (5.45):

$$\begin{aligned} \widetilde{R}_{n+1} &= (\widetilde{R}_n \cap \widetilde{R}'_{\text{bad}}) \cup \widetilde{R}_{\text{good}} \\ &= \left(\begin{bmatrix} 1.0 & 0.8 & 0.9 & 0.6 & 0.5 \\ 0.3 & 0.3 & 0.3 & 0.3 & 0.3 \\ 0.2 & 0.2 & 0.2 & 0.2 & 0.2 \\ 0.1 & 0.1 & 0.1 & 0.1 & 0.1 \\ 0.0 & 0.0 & 0.0 & 0.0 & 0.0 \end{bmatrix} \cap \begin{bmatrix} 1.0 & 1.0 & 1.0 & 1.0 & 1.0 \\ 1.0 & 1.0 & 1.0 & 0.0 & 1.0 \\ 1.0 & 1.0 & 1.0 & 1.0 & 1.0 \\ 1.0 & 1.0 & 1.0 & 1.0 & 1.0 \\ 1.0 & 1.0 & 1.0 & 1.0 & 1.0 \end{bmatrix} \right) \\ &\quad \cup \begin{bmatrix} 0.0 & 0.0 & 0.0 & 0.0 & 0.0 \\ 0.0 & 0.0 & 0.0 & 0.0 & 1.0 \\ 0.0 & 0.0 & 0.0 & 0.0 & 0.0 \\ 0.0 & 0.0 & 0.0 & 0.0 & 0.0 \\ 0.0 & 0.0 & 0.0 & 0.0 & 0.0 \end{bmatrix} \\ &= \begin{bmatrix} 1.0 & 0.8 & 0.9 & 0.6 & 0.5 \\ 0.3 & 0.3 & 0.3 & 0.0 & 1.0 \\ 0.2 & 0.2 & 0.2 & 0.2 & 0.2 \\ 0.1 & 0.1 & 0.1 & 0.1 & 0.1 \\ 0.0 & 0.0 & 0.0 & 0.0 & 0.0 \end{bmatrix}. \end{aligned}$$

So if an error of -1 occurs again, the controller will now produce the control-output

$$\widetilde{U} = \begin{bmatrix} 0.3 & 0.3 & 0.3 & 0.0 & 1.0 \end{bmatrix},$$

which defuzzifies to the actual control-output of 2, as required.

Though this performance adaptive FKBC has been described for a single-input, single-output (SISO) system, it can easily be extended to the multivariable case. The number of variables increases and the control action of more than one manipulated variable may need to be corrected. In this case the above rule modification procedure is simply repeated for each of the control-outputs that need correction.

Controller Design

Procyk and Mamdani have applied the controller to a wide range of single and multivariable simulated processes, with encouraging results. Starting with no rules at all, they trained their controllers with repeated control runs involving the same set-point change. In general, the controllers provided fast convergence to a well performing set of rules and proved to be relatively insensitive to parameter values, such as the scaling factors, the delay term, and the incremental model of the process. Convergence was defined as no further rule changes taking place, and was generally achieved after five to seven runs. Convergence may not take place due to noise, or a poor choice of controller parameters, but the controller could still give good control performance. The controller could be started with no rules at all, but initial performance was significantly improved if at least a crude set of rules was provided at the start.

Their adaptation mechanism requires a "delay in reward" parameter that specifies which past control-outputs should be corrected. This parameter introduces phase advance into the system which results in oscillations around the set-point that are too large to allow convergence to take place, if the "delay in reward" parameter is too small, i.e., a control action too close to the present is corrected. The parameter must approximately match the time delays in the process for good performance.

The inputs to the controller, the error and change-of-error, are multiplied by scaling factors GE and GΔE, respectively. The output from the controller is multiplied by factor GU. The values of GE and GΔE have a strong effect on the performance measure, and hence on the adaptive response. Increasing GE and GΔE makes the performance measure more sensitive around the set-point, but less sensitive to the rise time. Decreasing these parameters has the opposite effect. Hence there are low limits on GE and GΔE that are defined by the amount of tolerance on rise time, steady-state error and oscillations around the set-point the control system designer is willing to accept. Similarly, there are high limits above which the performance measure becomes too sensitive to allow convergence. The output scaling factor GU affects the gain of the controller and so modifies the overall control response.

The work of Yamazaki and Mamdani [231] has established some guidelines for choosing the "delay in reward" term, and the scaling factors of the controller, to help overcome problems with limit cycling, poor settling time and occasional instability. On the basis of a first-order plus dead-time model of the process to be controlled, the following rules should be followed:

- the 'delay in reward' parameter should be chosen to be somewhere between the process dead-time and the delay time to the peak value of the impulse response of the process;

- GU should be chosen to give a reasonable rise time, but also fine control around the set-point, e.g., a minimum change in the control-output of 1%, and a maximum change of 6% of the possible full range of control-output in one sampling time;

- GE should be between 4 and 6 to give a tolerance band of error around the set-point of approximately 1% full scale, for an error range of 0–100% to be mapped to 13 quantized levels;

- the GΔE to GE ratio should be between 4 and 6, so that a small change-of-error, relative to the error, is mapped to the same quantized level.

The accuracy of the incremental model turns out to be not very critical at all, as the adaptation mechanism can compensate for quite large model errors. However, the model should be developed using any available knowledge of the process. For example, it may be possible to carry out step testing of the process-inputs to measure the magnitude of their effects on the outputs. In this way an incremental model of the process can be developed. However, it will not account for multivariable interactions, so will not be completely accurate.

Applications

The self-organizing controller has been applied to a number of interesting experimental systems. Shao [190] applied the SOC to the control of the temperature of a heater, and to DC motor speed control. For the heater, good control was achieved after only three runs. Good DC motor speed control was achieved by the ninth run, but rule modifications continued. These results are illustrated in Fig. 5.13 and Fig. 5.14, respectively.

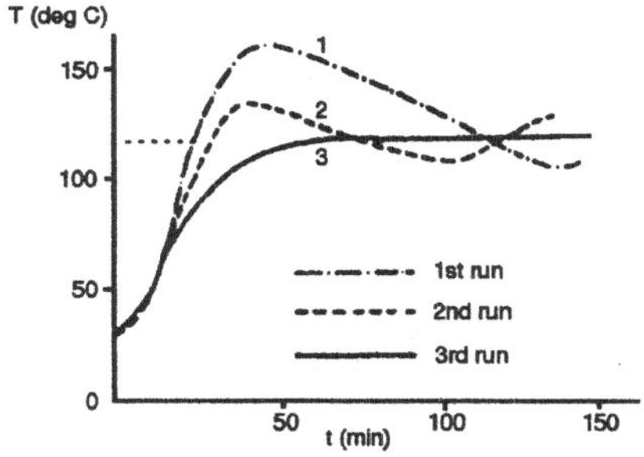

Fig. 5.13. The response of the heater (from Shao [190]).

The application of the SOC to a robot arm [211] required the use of several SOC working in parallel, with each SOC controlling a process with highly non-linear dynamics. Sampling times were only a few milliseconds. The revolving

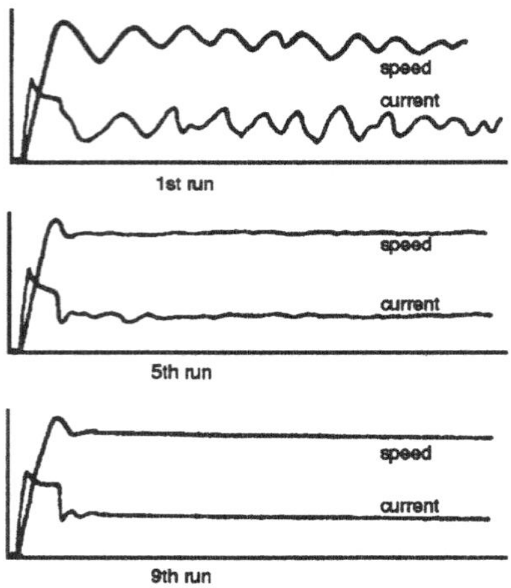

Fig. 5.14. The speed and current responses of the DC motor (from Shao [190]).

joint robot arm had a maximum lifting capacity of 1.5 kg, a maximum slew speed of 35 degrees/sec, and an estimated dominant time constant of 90 milliseconds. The shoulder link was 0.57 metres long, and the arm link was 0.43 metres long. The two-joint SOC algorithm was run on an LSI 11/23 computer. The SOC were tested on step responses and on tracking a two-dimensional square.

The step response test involved moving the shoulder joint clockwise from 82 degrees down to 22 degrees, with the arm staying at a constant angle of -128 degrees. By the second run the rise time and overshoot had improved, and fluctuations in the controller output, evident on the first run, had disappeared. These first two runs are shown in Fig. 5.15.

Tracking the square involved the cooperation of two SOC algorithms, one controlling the shoulder movement, and the other the arm. Each motor-joint system was a single-input, single-output process. The SOC learnt their own rules independently in the face of cross-coupling effects. The square of side 0.4 metres was centred at $x = z = 0.55$ metres from the origin of the vertical $z - x$ plane at the support end of the shoulder link. It took nine seconds to traverse the square. The SOC achieved maximum deviations of 1.97 and 0.87 centimetres in the x and z directions, respectively. This compares with 3.5 and 2.3 centimetres, respectively, for a PID-controller using the same quantization levels for the input error and change-of-error as the SOC. Two responses for the SOC and PID-controllers are shown in Fig. 5.16.

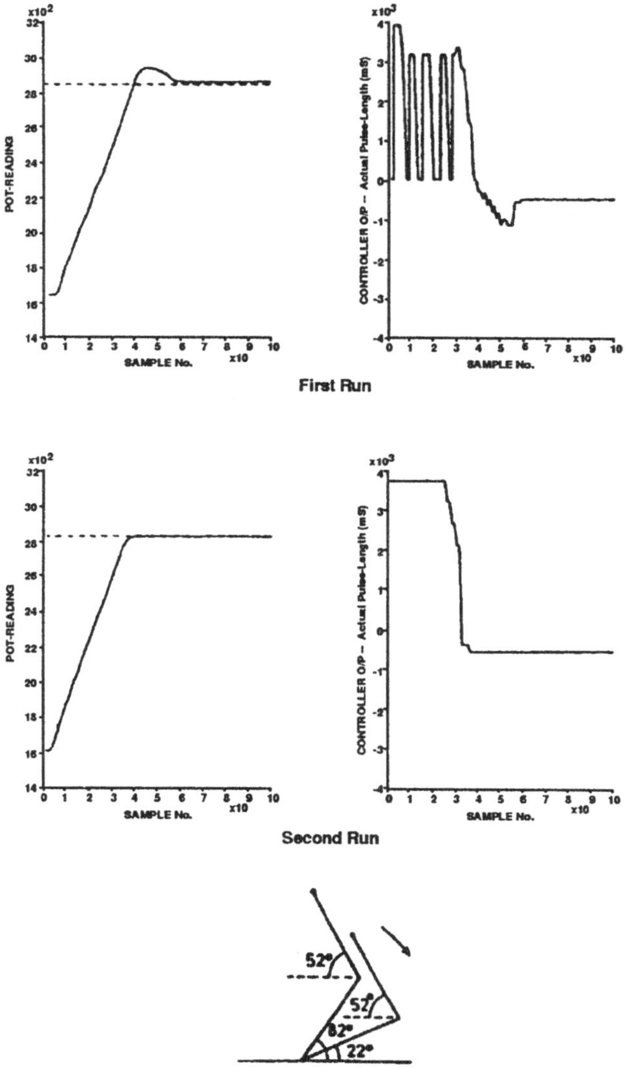

Fig. 5.15. Robot step response test (from Tanscheit and Scharf [211]).

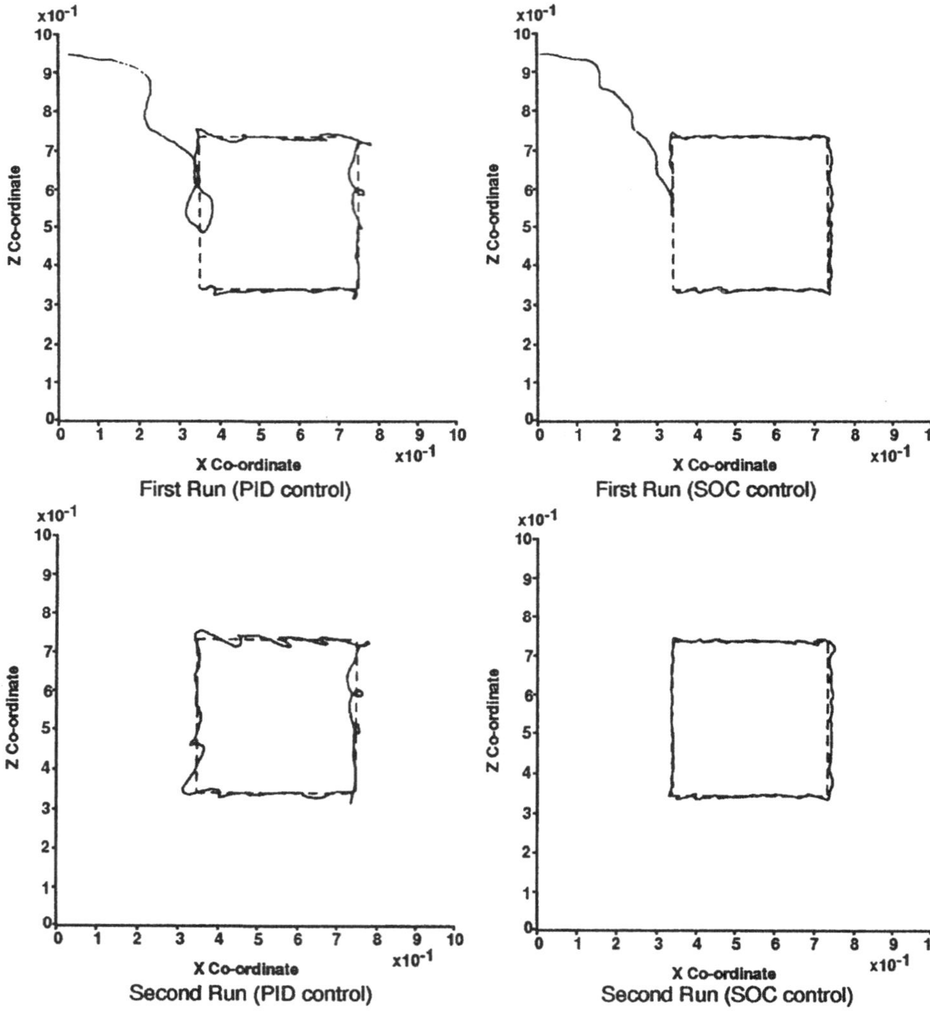

Fig. 5.16. Robot square tracking test (from Tanscheit and Scharf [211]).

5.3.4 A Model Based Controller

A self-organizing system that uses on-line identification of a fuzzy process model for adaptation has been developed by Graham and Newell [81, 82, 83]. The structure of the controller is shown in Fig. 5.17.

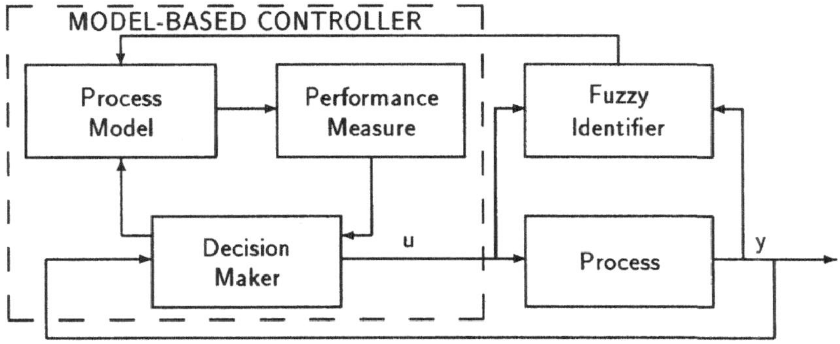

Fig. 5.17. Model-based adaptive fuzzy controller (redrawn from Graham and Newell, [82]).

The FKBC

The FKBC itself is not of the normal type, but is a model based controller (MBC). The controller was originally developed by Hendy [96] and consists of three parts:

1. a fuzzy process model,
2. a controller performance measure,
3. a decision maker.

The fuzzy process model is as described in Section 5.2.1 and consists of a set of if-then rules that are the inverse of the rules found in a Mamdani FKBC. Instead of specifying the desired control-output for a given process state, they predict the process-output to be expected in time due to the current and previous process-states and control-outputs. For example, a fuzzy model of a gas furnace may contain a rule such as:

if the current CO_2 concentration is MEDIUM
and the previous methane feedrate was LOW
then the next CO_2 concentration will be JUST HIGH.

The controller performance measure consists of a group of fuzzy sets. Each fuzzy set describes the required performance of a process variable. The performance of a particular variable is given by the degree of membership of its current value in its performance fuzzy set. For a cement kiln, for example, the performance objectives might be to maximize the amount of material processed by the kiln, i.e., the throughput, while at the same time maintaining the kiln temperature within a specified operating range. Performance fuzzy sets that describe these control objectives are shown in Fig. 5.18.

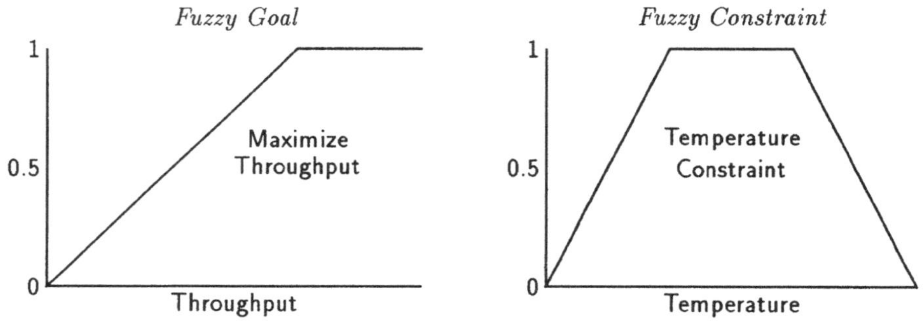

Fig. 5.18. Fuzzy performance measure goals and contraints (redrawn from Graham [81]).

The overall controller performance is given by a confluence rule that combines the individual performance values, for example:

Overall performance = Throughput performance *and*
Temperature performance.

The *and* connective might be implemented as the minimum (MIN decision maker) or product (MULT decision maker) of the individual values. The actual implementation of *and* that is used can influence controller performance. Taking the minimum of the individual performance values results in a controller that will try to improve the worst performing variable, even at the cost of degrading the performance of other variables. The multiplication of the individual values gives a controller that will try to improve the performance of all the variables at the same time. While this ensures that individual performances are generally not degraded, a badly performing variable may only be improved very slowly.

The control-output is calculated by the decision maker, using the fuzzy process model and the performance measure. The aim of the decision maker is to choose, from a predefined, finite set of control-outputs, which action will maximize the performance of the process if no future control-outputs are taken. This is achieved by applying each possible control-output to the fuzzy process model to obtain predictions of the process-state to be expected due to each control-output. The control-output that produces the predicted process state of highest performance, as specified by the performance measure, is chosen as the current control-output.

A set of possible control-outputs for a single control-variable might be {LARGE DECREASE, SMALL DECREASE, NO CHANGE, SMALL INCREASE, LARGE INCREASE}. The number of possible control-outputs grows rapidly with the number of control-variables. For example, for a multivariable controller with two outputs, each of which can be altered by the above set of five control-outputs, there are $5 \times 5 = 25$ possible combinations of control-outputs to be tested by the decision maker. This rises to $5 \times 5 \times 5 = 125$ combinations for three variables.

It is possible that two or more control-outputs might give an equal predicted increase in performance. In this case the decision maker could simply choose one of these control actions at random. Graham [81] suggests employing both the MIN and MULT decision makers to minimize the need for a random choice. The scheme is to first apply the MIN decision maker, with the aim of improving the worst performing variable. If multiple "best" control actions are found, the MULT decision maker is applied to this "best" set to find the control-output that will improve the overall performance the most. If multiple "best" control-outputs are still found, then one of these is chosen at random.

The sequence of operations that comprise the model based controller are summarized by the flowchart shown in Fig. 5.19.

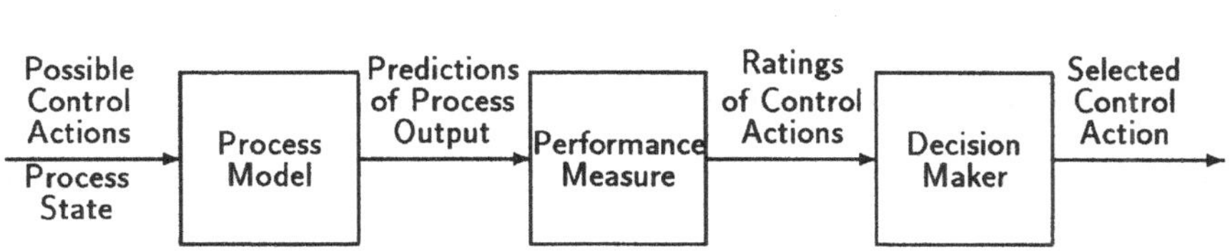

Fig. 5.19. Flowchart of model based controller.

Adaptation Mechanism

Adaptation of the controller is achieved by the use of on-line fuzzy identification of the process model. The controller may start with no model at all, or a predefined model. The on-line identification will build a model, or modify the predefined model to more closely match the process. As the identification proceeds, the performance of the controller improves, because the accuracy of the model predictions of the effect of different control-outputs improves.

The fuzzy model of the process is of the form

$$\widetilde{L\mathbf{y}} = \left(\widetilde{L\mathbf{x}} \times \widetilde{L\mathbf{u}}\right) \circ \tilde{R} = \widetilde{L\mathbf{x}} \circ \widetilde{L\mathbf{u}} \circ \tilde{R}, \tag{5.47}$$

where process-state $\mathbf{x} = (x_1, x_2, \ldots, x_n)^T$, process-input $\mathbf{u} = (u_1, u_2, \ldots, u_m)^T$, predicted process-output $\mathbf{y} = (y_1, y_2, \ldots, y_\ell)^T$, and $\tilde{R}$ is the fuzzy relation. The aim of the fuzzy identification is to identify the relation matrix $\tilde{R}$ from a given collection of input-output data

$$[\widetilde{L\mathbf{y}}_k; \widetilde{L\mathbf{x}}_k, \widetilde{L\mathbf{u}}_k]_{k=1}^K = [\widetilde{LY}_{1k}, \ldots, \widetilde{LY}_{\ell k}; \widetilde{LX}_{1k}, \ldots, \widetilde{LX}_{nk}, \widetilde{LU}_{1k}, \ldots, \widetilde{LU}_{mk}]_{k=1}^K. \tag{5.48}$$

Each set of data can be written as the if-then rule

if (x_1 is LX_{1k} and ... and x_n is LX_{nk})
and (u_1 is LU_{1k} and ... and u_m is LU_{mk})
then (y_1 is LY_{1k} and ... and y_ℓ is $LY_{\ell k}$).

The meaning of such a rule is defined by the fuzzy relation

$$\tilde{R}_k = \widetilde{LX}_{1k} \times \ldots \times \widetilde{LX}_{nk} \times \widetilde{LU}_{1k} \times \ldots \times \widetilde{LU}_{mk} \times \widetilde{LY}_{1k} \times \ldots \times \widetilde{LY}_{lk}. \tag{5.49}$$

If there is a fuzzy relation $\tilde{R}_k^*$ that satisfies

$$\widetilde{L\mathbf{y}}_k = \widetilde{L\mathbf{x}}_k \circ \widetilde{L\mathbf{u}}_k \circ \tilde{R}_k^*, \tag{5.50}$$

then the relation $\tilde{R}_k$ given by (5.49) is a solution to (5.50).

The entire collection of input-output data can be written as the set of if-then rules:

if $\mathbf{x}$ is $L\mathbf{x}_1$ *and* $\mathbf{u}$ is $L\mathbf{u}_1$ *then* $\mathbf{y}$ is $L\mathbf{y}_1$
if $\mathbf{x}$ is $L\mathbf{x}_2$ *and* $\mathbf{u}$ is $L\mathbf{u}_2$ *then* $\mathbf{y}$ is $L\mathbf{y}_2$
⋮ ⋮ ⋮
if $\mathbf{x}$ is $L\mathbf{x}_K$ *and* $\mathbf{u}$ is $L\mathbf{u}_K$ *then* $\mathbf{y}$ is $L\mathbf{y}_K$.

The meaning of this set of rules is defined by the overall fuzzy relation $\tilde{R}$ given by

$$\tilde{R} = \tilde{R}_1 \cup \tilde{R}_2 \cup \ldots \cup \tilde{R}_K, \tag{5.51}$$

where each $\tilde{R}_k$ is given by (5.49). In general, a set of data collected from a real process will be noisy and incomplete, and the fuzzy relation $\tilde{R}$ will not satisfy

$$\widetilde{L\mathbf{y}}_k = \widetilde{L\mathbf{x}}_k \circ \widetilde{L\mathbf{u}}_k \circ \tilde{R}, \qquad k = 1, \ldots, K. \tag{5.52}$$

Hence the relation $\tilde{R}$ is only an approximate, and not necessarily optimal, fuzzy model of the process.

If a suitable set of process data has already been collected, this identification scheme can be used to identify a fuzzy model of the process off-line for use by the fuzzy model based controller. It can also form the initial model for the adaptive version of the controller.

The adaptive fuzzy model-based controller uses on-line fuzzy identification in the following algorithm to obtain controller adaptation:
At time 0:

1. An intial relation matrix $\tilde{R}_0$ is defined.
2. The initial process-state $\widetilde{L\mathbf{x}_0}$, process-output $\widetilde{L\mathbf{y}_0}$, and control-output $\widetilde{L\mathbf{u}_0}$ are established.

At time $k+1$:

1. The relation matrix is updated using fuzzy identification via:

$$\tilde{R}_{k+1} = \tilde{R}_k \cup \left(\widetilde{LX}_{1k} \times \ldots \times \widetilde{LX}_{nk} \times \widetilde{LU}_{1k} \times \ldots \times \widetilde{LU}_{mk} \times \widetilde{LY}_{1k} \times \ldots \times \widetilde{LY}_{\ell k}\right). \tag{5.53}$$

2. The model-based controller calculates a new control-output using predictions from the fuzzy model:

$$\widetilde{L\mathbf{y}} = \widetilde{L\mathbf{x}} \circ \widetilde{L\mathbf{u}} \circ \tilde{R}_{k+1}. \tag{5.54}$$

The initial relation matrix may be completely empty, i.e., $\tilde{R}_0 = 0$, or it may be predefined using, say, identification from process data, as suggested above.

Example 5.4 A numerical example of the identification procedure, for a single-input, single-output (SISO) process where the fuzzy model is of the form

$$\widetilde{LY} = \widetilde{LU} \circ \tilde{R}, \tag{5.55}$$

and starting with an initially empty relation matrix, is as follows:
Step 0:

$$\widetilde{LU}_0 = \begin{bmatrix} 1.0 & 0.3 & 0.2 & 0.1 & 0.0 & 0.0 \end{bmatrix}$$
$$\widetilde{LY}_0 = \begin{bmatrix} 1.0 & 0.8 & 0.9 & 0.6 & 0.5 & 0.7 \end{bmatrix}$$
$$\tilde{R}_0 = \begin{bmatrix} 0.0 & 0.0 & 0.0 & 0.0 & 0.0 & 0.0 \\ 0.0 & 0.0 & 0.0 & 0.0 & 0.0 & 0.0 \\ 0.0 & 0.0 & 0.0 & 0.0 & 0.0 & 0.0 \\ 0.0 & 0.0 & 0.0 & 0.0 & 0.0 & 0.0 \\ 0.0 & 0.0 & 0.0 & 0.0 & 0.0 & 0.0 \\ 0.0 & 0.0 & 0.0 & 0.0 & 0.0 & 0.0 \end{bmatrix}$$

Step 1:

$$\widetilde{LU}_0 = \begin{bmatrix} 1.0 & 0.3 & 0.2 & 0.1 & 0.0 & 0.0 \end{bmatrix}$$
$$\widetilde{LY}_0 = \begin{bmatrix} 1.0 & 0.8 & 0.9 & 0.6 & 0.5 & 0.7 \end{bmatrix}$$
$$\tilde{R}_1 = \tilde{R}_0 \cup (\widetilde{LU}_0 \times \widetilde{LY}_0) = \begin{bmatrix} 1.0 & 0.8 & 0.9 & 0.6 & 0.5 & 0.7 \\ 0.3 & 0.3 & 0.3 & 0.3 & 0.3 & 0.3 \\ 0.2 & 0.2 & 0.2 & 0.2 & 0.2 & 0.2 \\ 0.1 & 0.1 & 0.1 & 0.1 & 0.1 & 0.1 \\ 0.0 & 0.0 & 0.0 & 0.0 & 0.0 & 0.0 \\ 0.0 & 0.0 & 0.0 & 0.0 & 0.0 & 0.0 \end{bmatrix}$$

Step 2:

$$\widetilde{LU}_1 = \begin{bmatrix} 0.3 & 1.0 & 0.2 & 0.1 & 0.0 & 0.0 \end{bmatrix}$$

$$\widetilde{LY}_1 = \begin{bmatrix} 0.3 & 0.3 & 0.5 & 1.0 & 0.3 & 0.3 \end{bmatrix}$$

$$\tilde{R}_2 = \tilde{R}_1 \cup (\widetilde{LU}_1 \times \widetilde{LY}_1) = \begin{bmatrix} 1.0 & 0.8 & 0.9 & 0.6 & 0.5 & 0.7 \\ 0.3 & 0.3 & 0.5 & 1.0 & 0.3 & 0.3 \\ 0.2 & 0.2 & 0.2 & 0.2 & 0.2 & 0.2 \\ 0.1 & 0.1 & 0.1 & 0.1 & 0.1 & 0.1 \\ 0.0 & 0.0 & 0.0 & 0.0 & 0.0 & 0.0 \\ 0.0 & 0.0 & 0.0 & 0.0 & 0.0 & 0.0 \end{bmatrix}$$

Step 3:

$$\widetilde{LU}_2 = \begin{bmatrix} 0.2 & 0.3 & 1.0 & 0.2 & 0.1 & 0.0 \end{bmatrix}$$

$$\widetilde{LY}_2 = \begin{bmatrix} 0.7 & 0.9 & 1.0 & 0.8 & 0.6 & 0.2 \end{bmatrix}$$

$$\tilde{R}_3 = \tilde{R}_2 \cup (\widetilde{LU}_2 \times \widetilde{LY}_2) = \begin{bmatrix} 1.0 & 0.8 & 0.9 & 0.6 & 0.5 & 0.7 \\ 0.3 & 0.3 & 0.5 & 1.0 & 0.3 & 0.3 \\ 0.7 & 0.9 & 1.0 & 0.8 & 0.6 & 0.2 \\ 0.2 & 0.2 & 0.2 & 0.2 & 0.2 & 0.2 \\ 0.1 & 0.1 & 0.1 & 0.1 & 0.1 & 0.1 \\ 0.0 & 0.0 & 0.0 & 0.0 & 0.0 & 0.0 \end{bmatrix}$$

A problem with the fuzzy model, especially if it is initially empty, is that it may be incomplete during the initial stages of on-line identification. This means that it cannot give a prediction for the expected process-output for all possible process-states and control-outputs. For example, in the numerical example given above, the model prediction, after Step 3, from an input

$$\widetilde{LU} = \begin{bmatrix} 0.0 & 0.0 & 0.0 & 0.1 & 0.3 & 1.0 \end{bmatrix},$$

is the output

$$\widetilde{LY} = \begin{bmatrix} 0.1 & 0.1 & 0.1 & 0.1 & 0.1 & 0.1 \end{bmatrix},$$

which provides no information about $\widetilde{LY}$. In this situation the MBC cannot calculate a control-output and so an auxiliary algorithm must be used to decide on an appropriate control-output. This algorithm may be extremely simple, such as to leave the control-output fixed until such time as the MBC is able to make a decision. It could take the form of another controller entirely, such as a conventional PI-controller. To avoid this problem the controller should be started with a complete initial model. This model may be inaccurate but it will be constantly improved by the on-line identification.

Applications

The identification scheme has been used off-line to design a non-adaptive model based controller for the control of silica levels in alumina from the Bayer process.

The adaptive model based controller has been applied with some success to computer simulations of

- a mass storage tank,
- a cement kiln,
- a single effect forced circulation evaporator.

It has also been trialled on a laboratory-scale liquid level rig [82]. The liquid level rig consisted of a main water holding tank which was fed with water from the top, and drained of water through a valve at the bottom. Water was circulated continuously by an electric pump. The height of water in the tank was measured by a pressure transducer attached to the bottom of a standpipe connected to the main tank by a small tube. There was a significant time delay involved in the measurement due to the constriction of the tube. The inlet flow rate to the main tank could be adjusted by a valve on the inlet flow side of the pump. This flow rate was measured by a flow meter. A schematic of the rig is shown in Fig. 5.20.

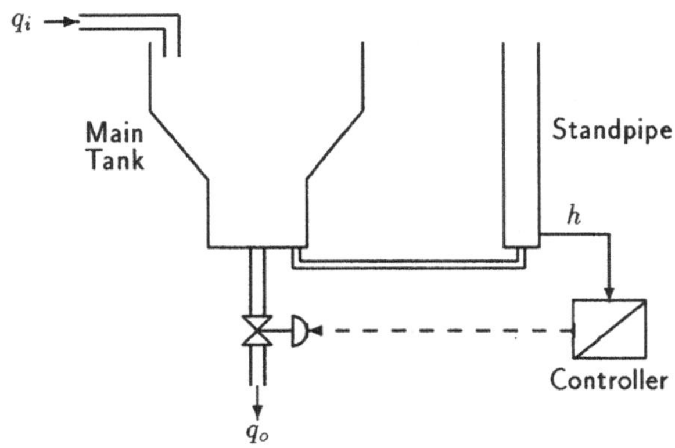

Fig. 5.20. Liquid level rig (redrawn from Graham and Newell [82]).

The control problem was to control the height of water in the tank by adjusting the outlet flow rate. Variations in the inlet flow rate formed a measurable disturbance. The controller was implemented on an NCR Decision Mate personal computer interfaced to the rig.

In mathematical terms, the process consisted of a pure integrator (the main tank) with a first-order measurement lag. Hence the deterministic discrete-time model of the process was

$$h_t = (1+\delta)h_{t-1} - \delta h_{t-2} + \omega_0 q_{t-1} - \omega_1 q_{t-2}, \qquad (5.56)$$

where

$$\delta = exp(-\frac{T}{\tau_m}), \tag{5.57}$$

$$\omega_0 = (T + \tau_m \delta - \tau_m)/A, \tag{5.58}$$

$$\omega_1 = (T\delta + \tau_m \delta - \tau_m)/A, \tag{5.59}$$

and h_t = measure height at sampling time t, q_t = difference between the inlet and outlet flow rates at time t, T = sample period, τ_m = first-order time constant, and A = cross-sectional area of the main tank.

This model could be approximated by a fuzzy model consisting of rules of the form

if h_{t-1} is LH_{t-1} *and* h_{t-2} is LH_{t-2} *and* q_{t-1} is LQ_{t-1} *and* q_{t-2} is LQ_{t-2}
then h_t is LH_t,

where LH and LQ are linguistic values of the height and flow rate difference at different timepoints, respectively. The meaning of each rule is a fuzzy relation $\tilde{R}_H$ given as

$$\tilde{R}_H = \widetilde{LH}_{t-1} \times \widetilde{LH}_{t-2} \times \widetilde{LQ}_{t-1} \times \widetilde{LQ}_{t-2} \times \widetilde{LH}_t. \tag{5.60}$$

To give reasonable resolution, 21 fuzzy sets needed to be defined for the height, and 11 fuzzy sets for the flow rate difference. If this model was represented by a relation matrix $\tilde{R}_H$, so that it was of the form

$$\widetilde{LH}_t = \widetilde{LH}_{t-1} \circ \widetilde{LH}_{t-2} \circ \widetilde{LQ}_{t-1} \circ \widetilde{LQ}_{t-2} \circ \tilde{R}_H, \tag{5.61}$$

the relation matrix would contain $21 \times 21 \times 11 \times 11 \times 21 = 1\,120\,581$ elements. This could not be implemented on the NCR Decision Mate computer. To reduce the size of the relation matrix, a combined deterministic and fuzzy model was defined. This was described by the equation

$$h_t = (1 + \delta)h_{t-1} - \delta h_{t-2} + g(q_{t-1}, q_{t-2}), \tag{5.62}$$

where $g(\cdot)$ is the defuzzified output from a fuzzy model consisting of rules of the form

$$\textit{if } q_{t-1} \text{ is } LQ_{t-1} \text{ and } q_{t-2} \text{ is } LQ_{t-2} \textit{ then } g \text{ is } LG(LQ_{t-1}, LQ_{t-2}). \tag{5.63}$$

The meaning of this rule is given by

$$\tilde{R}_G = \widetilde{LQ}_{t-1} \times \widetilde{LQ}_{t-2} \times \widetilde{LG}(\widetilde{LQ}_{t-1}, \widetilde{LQ}_{t-2}). \tag{5.64}$$

In relation form, the fuzzy model is

$$\widetilde{LG}(\widetilde{LQ}_{t-1}, \widetilde{LQ}_{t-2}) = \widetilde{LQ}_{t-1} \circ \widetilde{LQ}_{t-2} \circ \tilde{R}_G. \tag{5.65}$$

With 21 fuzzy sets defined for the linguistic values of the function $g(\cdot)$, this fuzzy model required a relation matrix with only $11 \times 11 \times 21 = 2541$ elements, and so was easily implemented.

The fuzzy sets for the flow rate differences, $\widetilde{LQ}_{t-1}$ and $\widetilde{LQ}_{t-2}$, are shown in Fig. 5.21. The domain $\mathcal{G}$ of the function $g(\cdot)$ was covered by 21 triangular fuzzy sets arranged in a similar fashion to the flow rate difference sets. The crisp values for these variables, as read from the process, were fuzzified by forming a vector that consisted of their membership function values in each of the defined fuzzy sets. This is illustrated in Fig. 5.21.

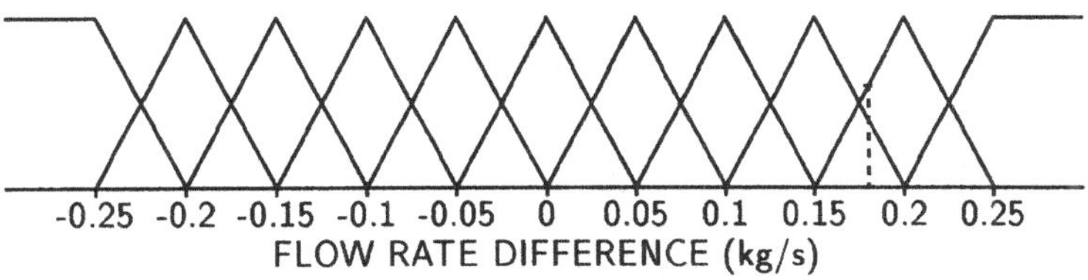

$$q_t = 0.18 \quad \text{(fuzzified)} \Rightarrow \quad Q_t = [0, 0, 0, 0, 0, 0, 0, 0, 0.4, 0.6, 0]$$

Fig. 5.21. Fuzzy sets for the flow rate difference. Example of "fuzzification" of a crisp value shown by dashed line and vector given below figure (redrawn from Graham and Newell [82]).

At each time step, the fuzzy MBC used this combined deterministic and fuzzy model to get predictions for the height to be expected from any one of 9 different outlet flow rates. The performance measure for the height was described by a triangular fuzzy set whose maximum membership function value was at the desired set-point. This is illustrated in Fig. 5.22. The outlet flow rate that was predicted to give the maximal performance was chosen as the current control action. The sequence of calculations made by the MBC at each time step is as follows:

1. read values of the height and inlet flow rates from the process;
2. combine control-outputs (outlet flow rates) with the current inlet flow rate to give 9 possible values for the flow rate difference q_{t-1};
3. fuzzify each of these values, and the stored value of the previous actual flow rate difference q_{t-2};
4. present the fuzzy values of the flow rate difference to the fuzzy model (5.65) to get 9 predictions for $\widetilde{LG}(\widetilde{LQ}_{t-1}, \widetilde{LQ}_{t-2})$;

5. defuzzify each of these predictions, using the centre-of-gravity method, to give 9 values for $g(q_{t-1}, q_{t-2})$;

6. use these values, together with the current value of the height, h_{t-1}, and the previous height, h_{t-2}, in the deterministic model (5.62) to get 9 predictions for the height, h_t, at the next sampling time;

7. test each of these predicted heights against the performance measure;

8. choose the outlet flow rate that gives the predicted height of highest performance.

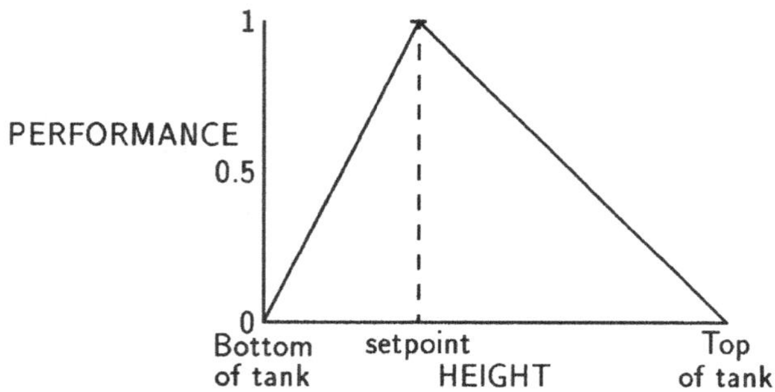

Fig. 5.22. Performance measure for height (redrawn from Graham and Newell, [82]).

To form the adaptive FKBC, the fuzzy model was identified on-line, at each time step, by first calculating the deterministic value $g(\cdot)$ from

$$g(q_{t-1}, q_{t-2}) = h_t - (1 + \delta)h_{t-1} + \delta h_{t-2}. \tag{5.66}$$

The value $g(\cdot)$ was then fuzzified as described above and the fuzzy model was updated via

$$\tilde{G}_t = \tilde{G}_{t-1} \cup \left(\widetilde{LQ}_{t-1} \times \widetilde{LQ}_{t-2} \times \widetilde{LG}\left(\widetilde{LQ}_{t-1}, \widetilde{LQ}_{t-2}\right)\right), \tag{5.67}$$

where $\tilde{G}_t$ is the current model relation, $\tilde{R}_G$.

The adaptive FKBC was started with no model at all. It was trained using a noise sequence on the inlet flow rate that consisted of a normally distributed sequence of random numbers that gave up to a 15% change in the inlet valve position. The noise was applied at each sampling time, immediately after a control action had been taken. Each training run lasted for five minutes, with a sampling time of five seconds. The process was initially at steady-state, with the height in the middle of the lower straight-sided section of the tank. The model identified by the end of each run was used as the initial model on the next run.

During the initial stages of training, the MBC often could not make a control decision, because the fuzzy model relation matrix was too incomplete to allow predictions of the effects of all the possible control-outputs. Formally, the MBC was deemed to be unable to make a control decision if every control-output produced a prediction with a maximum membership function value of less than 0.1. In this situation, the initial steady-state outlet valve position was used as the control-output. This is an example of an auxiliary algorithm.

To speed up the identification of the model, the function $\widetilde{G}(\cdot)$ was assumed to be symmetric around the origin, and linear. For this two-input system, this allowed four updates to be made to the relation matrix at each time step, instead of only one. Consider an example, in which each universe of discourse is covered by 11 fuzzy sets: {*NL, NJL, NM, NJM, NS, Z, S, JM, M, JL, L*}, where *Z*=ZERO, *S*=SMALL, *M*=MEDIUM, *L*=LARGE, *N*=NEGATIVE and *J*=JUST. Suppose that, at sampling time t, we have q_{t-1} is *M*, q_{t-2} is *L* and $g(q_{t-1}, q_{t-2})$ is *JM*. This data is described by the rule:

if q_{t-1} is *M* *and* q_{t-2} is *L* *then* $g(q_{t-1}, q_{t-2})$ is *JM*.

If the relation matrix is updated according to (5.67), we have:

$$\widetilde{G}_t = \widetilde{G}_{t-1} \cup (\widetilde{M} \times \widetilde{L} \times \widetilde{JM}). \tag{5.68}$$

If symmetry is assumed, then three other rules are implied by the above data:

if q_{t-1} is *L* *and* q_{t-2} is *M* *then* $g(q_{t-1}, q_{t-2})$ is *NJM*
if q_{t-1} is *NM* *and* q_{t-2} is *NL* *then* $g(q_{t-1}, q_{t-2})$ is *NJM*
if q_{t-1} is *NL* *and* q_{t-2} is *NM* *then* $g(q_{t-1}, q_{t-2})$ is *JM*,

and the relation matrix can be updated via

$$\widetilde{G}_t = \widetilde{G}_{t-1} \cup (\widetilde{M} \times \widetilde{L} \times \widetilde{JM}) \cup (\widetilde{L} \times \widetilde{M} \times \widetilde{NJM}) \cup (\widetilde{NM} \times \widetilde{NL} \times \widetilde{NJM}) \cup (\widetilde{NL} \times \widetilde{NM} \times \widetilde{JM}). \tag{5.69}$$

It was important to recognize that any modifications made to the relation matrix due these assumptions about the process were not necessarily correct. Hence such modifications were not allowed to change entries in the relation matrix that were due to real process data. Equally, any modifications due to real process data were allowed to overwrite entries due to process assumptions. For example, suppose $\widetilde{G}_t$ has been generated as given above, but the next set of process data is [q_{t-1} is L, q_{t-2} is M, $g(q_{t-1}, q_{t-2})$ is NL]. The assumptions about the process suggest the relation matrix should be updated via

$$\widetilde{G}_{t+1} = \widetilde{G}_t \cup (\widetilde{L} \times \widetilde{M} \times \widetilde{NL}) \cup (\widetilde{M} \times \widetilde{L} \times \widetilde{L}) \cup (\widetilde{NL} \times \widetilde{NM} \times \widetilde{L}) \cup (\widetilde{NM} \times \widetilde{NL} \times \widetilde{NL}). \tag{5.70}$$

However, to avoid modifying the entries due to the previous data set [q_{t-1} is M, q_{t-2} is L, $g(q_{t-1}, q_{t-2})$ is JM], the assumed data set [q_{t-1} is M, q_{t-2} is L, $g(q_{t-1}, q_{t-2})$ is L] is dropped, and the relation matrix is updated via

Table 5.3. Comparison between controllers

Controller	Noise variance	Load variance	Setpoint change IAE
PI	4126	7922	9905
PI/feedforward	1398	741	–
Predefined MBC	1832	446	7155
Identified MBC	852	704	6010

$$\tilde{G}_{t+1} = \tilde{G}_t \cup (\tilde{L} \times \widetilde{M} \times \widetilde{NL}) \cup (\widetilde{NL} \times \widetilde{NM} \times \tilde{L}) \cup (\widetilde{NM} \times \widetilde{NL} \times \widetilde{NL}). \quad (5.71)$$

Training was essentially complete after five runs, by which time the fuzzy relation matrix contained 349 non-zero elements, corresponding to 49 out of a possible 121 different combinations of $\widetilde{LQ}_{t-1}$ and $\widetilde{LQ}_{t-2}$. The model simply covered those combinations of $\widetilde{LQ}_{t-1}$ and $\widetilde{LQ}_{t-2}$ that occurred due to the noise sequence used during training.

The performance of the adaptive FKBC was compared with a standard PI-controller, a PI/feedforward-controller, and a MBC using a model defined off-line using expert knowledge of the process. For these comparisons the on-line identification was turned off, so that the adaptive controller became a normal MBC using the model identified by the end of the third training run. All four controllers were trialled on:

1. the same noise sequence run as used for training,
2. a load change,
3. a set-point change.

The load change was a change of 15% in the inlet flow valve position applied at the tenth sampling time. The set-point change was of 50 mm, also applied at the tenth sampling time. The control results are summarized in Table 5.3.

The actual control runs are shown in Figs. 5.23, 5.24, and 5.25, respectively.

In all cases the MBC using the identified model outperformed both the PI- and the PI/feedforward-controllers, especially for the noise sequence on which it was trained. It also outperformed the MBC using the predefined model, except for the load change, where the predefined MBC was marginally better on the basis of the performance criterion used.

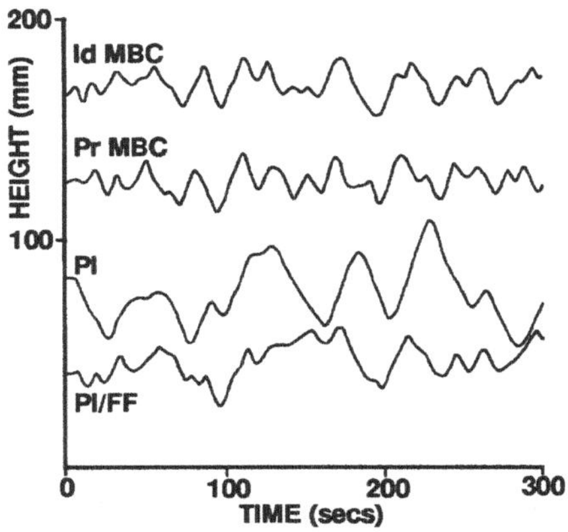

Fig. 5.23. Noise sequence runs with fixed (non-adaptive) controllers (runs shown offset for clarity) (from Graham and Newell [82]).

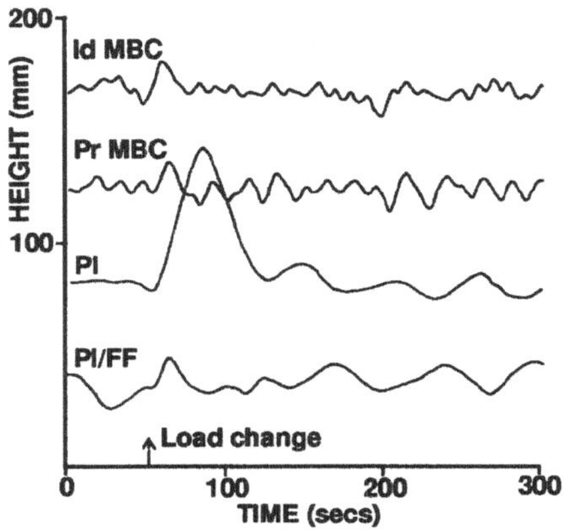

Fig. 5.24. Load changes with fixed (non-adaptive) controllers (runs shown offset for clarity) (from Graham and Newell [82]).

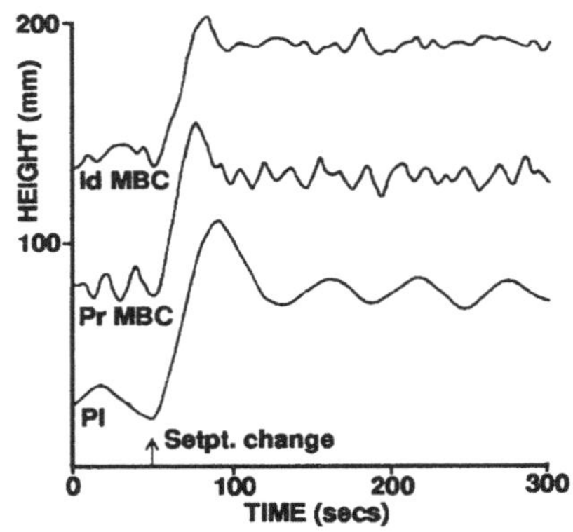

Fig. 5.25. Setpoint changes with fixed (non-adaptive) controllers (runs shown offset for clarity) (from Graham and Newell [82]).

6. Stability of Fuzzy Control Systems

6.1 Introduction

FKBC has been proven to be a powerful tool when applied to the control of processes which are not amenable to conventional, analytic design techniques. The design of most of the existing FKBC has relied mainly on the process operator's or control engineer's experience based heuristic knowledge. Hence, the controller's performance is very much dependent on how good this expertise is. Thus, from the control engineering point of view, the major effort in fuzzy knowledge based control has been devoted to the development of particular FKBC for specific applications rather than to general analysis and design methodologies for coping with the dynamic behavior of control loops. The development of such methodologies is of primary interest for control theory and engineering. In particular, stability analysis is of extreme importance, and the lack of satisfactory formal techniques for studying the stability of process control systems involving FKBC has been considered a major drawback of FKBC.

Fuzzy control systems are essentially nonlinear systems. For this reason it is difficult to obtain general results on the analysis and design of FKBC. Furthermore, the knowledge of the dynamic behavior of the process to be controlled is normally poor. Therefore, the robustness of the fuzzy control system must be studied to guarantee stability in spite of variations in process dynamics.

In this chapter we consider several existing approaches for stability analysis of FKBC. In the fuzzy control literature this type of analysis of FKBC is usually done in the context of the following two views of the system under control:

- Classical nonlinear dynamic systems theory: The system under control is a "non-fuzzy" system, and the FKBC is a particular class of nonlinear controller.

- Dynamic fuzzy systems.

The second view is associated with Zadeh's Extension Principle [237] (see Section 2.2.4) and so far is only of theoretical interest. Research in this area includes stability criteria based on the concept of energy or the controllability of fuzzy systems [76, 117].

The work presented in this chapter corresponds to the first view. We use the control structure shown in Fig. 6.1, where the FKBC is represented by means of

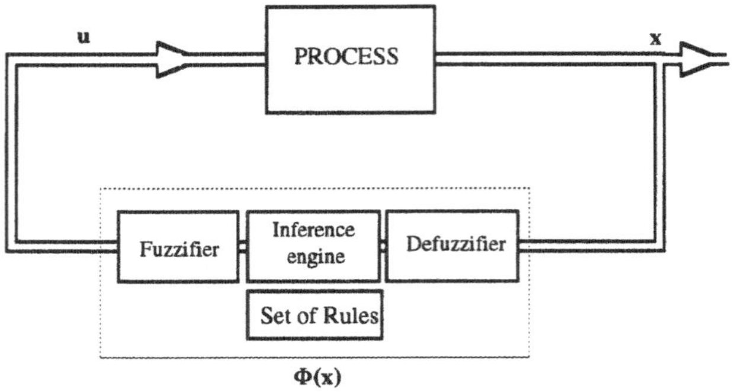

Fig. 6.1. FKBC in a closed-loop.

a nonlinear function $u = \Phi(\mathbf{x})$. As it has been shown in Garcia-Cerezo et al. [73] and in Section 4.2.2, the FKBC is a nonlinear transfer element represented by the function $\Phi(\mathbf{x})$. Then, the structure of Fig. 6.1 can be used to analyze the dynamic behavior of the closed-loop system.

One of the first works dealing with FKBC closed-loop analysis is that of Tong [215]. The analysis is based on the "relation matrix," which is a discrete version of the fuzzy relation, $\tilde{R}$, or μ_R representing the meaning of the rule base. The nonlinear function $\Phi(x)$ can be computed from $\tilde{R}$ by means of fuzzification, composition-based inference, and defuzzification, as we will show in Section 6.2. Tong [216] shows that the relation matrix is dependent only on the set of rules. Thus, the performance of the closed-loop system is improved by modifying these rules.

Braae and Rutherford [29] introduced the concept of linguistic trajectory of closed-loop FKBC systems. They also show the relation between the dynamic behavior of the closed-loop system and the FKBC rules and, at the same time, established a relation between the state space representation of the system to be controlled and the FKBC rules. This relation is related to the notion of control space (state space) partition [71]. The above concepts are examined in Section 6.2.

Analysis of the state space (the state space or phase plane in the two-dimensional case) has also been shown to be a simple but effective tool for simple systems. The "geometric method of stability analysis" [3] is based on a study of the contributions of the vector fields of the plant and controller. The interpretation is very intuitive and can be used to predict the stability under certain conditions. This method is also discussed in Section 6.2.

The above geometric method is formalized with the definition of the so-called stability and robustness indices [4], based on concepts from the "qualitative theory of dynamical systems" [84]. These indices are not only used to determine the stability and robustness of the closed-loop system, but are also

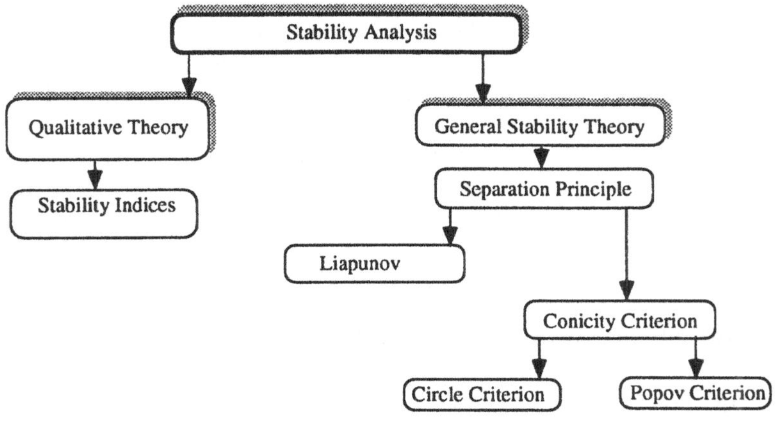

Fig. 6.2. Approaches to stability analysis.

used as a description of its dynamic behavior. Robustness indices are considered in Section 6.3.

The above methods for stability analysis are based on *internal* representation of dynamic systems. In the framework of the general stability theory there are two main directions: (1) stability in the Lyapunov sense which refers to internal representation (the state vector tends toward zero), and (2) input-output stability, which refers to the external representation (relatively small outputs with regard to the input). Section 6.4 is devoted to general concepts in input-output stability.

Vidyasagar [221] shows that, under some restrictive conditions, internal stability implies external stability and vice versa. Safonov [181] established a common conceptual framework for the two families of stability criteria, from which one can derive as particular cases the Lyapunov criteria, and the criteria in the input-output family: conicity, circle and Popov criteria. Figure 6.2 shows the relations among these criteria.

The application of the classical input-output techniques such as the circle and Popov criteria is well known in control theory. The work of Ray and Majumder [173] is one of the first covering the application of input-output stability in FKBC analysis. More precisely, this is done by the application of the circle criterion. Section 6.5 describes this method.

Recently, the conicity criterion has been introduced by Aracil et al. [5] as a method for the stability analysis and design of FKBC. Section 6.6 deals with this method.

It should be noted here that the approaches presented in this chapter assume that a dynamic non-fuzzy model of the process or plant to be controlled is known. In fact, some knowledge about the process dynamics is required for any stability analysis. The important point is that even if the model is only approximately known, some conclusions about the stability of the closed-loop system can still be obtained. Furthermore, if the closed-loop system is robust enough (far from the point of stability loss) it can maintain stability even if the model does not exactly represent the actual dynamic behavior of the process under control.

6.2 The State Space Approach.

One of the first approaches to the stability analysis of FKBC uses the phase plane and was introduced by Braae and Rutherford [30]. The practical applications of this method are restricted to two-dimensional systems due to difficulties in the interpretation of higher order graphical representations of the phase plane.

Stability analysis of a FKBC requires characterization of the relation between the rules and the state space associated with the dynamic system under control. This relationship is based on the relative influence of each rule of the rule base on the control action produced by the FKBC.

Let us consider a rule base composed by rules where the rule-antecedent contains two process state variables (plant variables), x_1 and x_2, representing the controller input variables. The consequent contains a single control output, u. The process state variables, x_1 and x_2, can take n_1 and n_2 linguistic values respectively. Under these conditions, the maximum number of rules contained in the rule base is $n_1 \times n_2$. The region of study for the analysis of the FKBC in the state space is normally bounded by some finite values $x_{\mathrm{min}_i}, x_{\mathrm{max}_i}$ where $i = 1, 2$.

We will now state that the crisp element (x_1, x_2) of the state space belongs to the subspace of the partition associated with rule j, if it holds that

$$\forall j \neq k \; \forall (x_1, x_2) : \mu_{R_j}(x_1, x_2) \geq \mu_{R_k}(x_1, x_2), \tag{6.1}$$

where $\mu_{R_j}(x_1, x_2)$ and $\mu_{R_k}(x_1, x_2)$ are the fuzzy relations representing the meaning of the rule-antecedents of rules j and k respectively. For example, let

Rule j: *if* x_1 is $LX_1^{(j)}$ *and* x_2 is $LX_2^{(j)}$ *then* u is $LU^{(j)}$,
Rule k: *if* x_1 is $LX_1^{(k)}$ *and* x_2 is $LX_2^{(k)}$ *then* u is $LU^{(k)}$.

Then μ_{R_j} and μ_{R_k}, in the case of Mamdani-type implication, are defined as follows

$$\forall x_1, x_2 : \mu_{R_j}(x_1, x_2) = \min(\mu_{LX_1^{(j)}}, \mu_{LX_2^{(j)}}) \tag{6.2}$$

$$\forall x_1, x_2 : \mu_{R_k}(x_1, x_2) = \min(\mu_{LX_1^{(k)}}, \mu_{LX_2^{(k)}}). \tag{6.3}$$

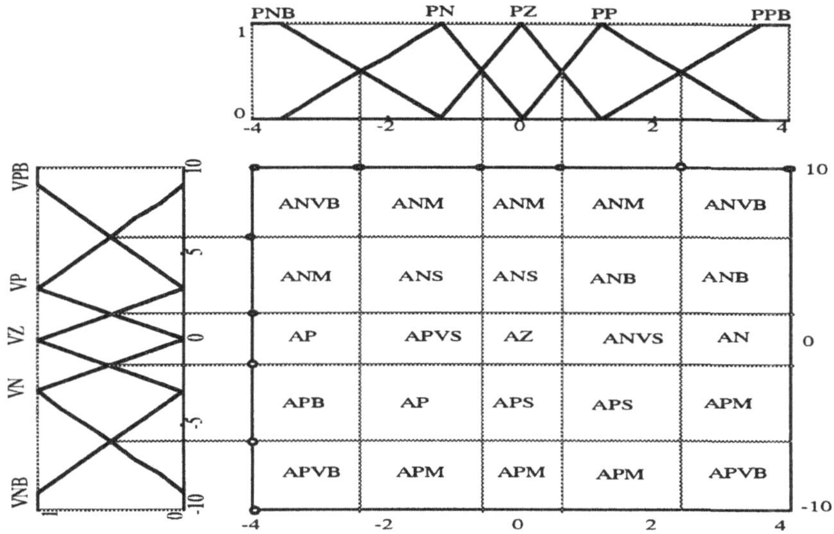

Fig. 6.3. Partition of the input space. In this case $x_{\text{min}_1} = -4$, $x_{\text{max}_1} = 4$, $x_{\text{min}_2} = -10$, $x_{\text{max}_2} = 10$.

The "partition limits" are determined through (6.1). Figure 6.3 shows the determination of the partition for a rule base defined in a table form as shown in Fig. 6.4.

A closed-loop system trajectory can be mapped on the partition space as done in Fig. 6.5. It can be seen there how certain areas of the partition space (the gray partition areas, each area corresponding to a particular FKBC rule) relate to the system trajectory. A sequence of rules, obtained according to the order in which they are fired, forms the so-called *linguistic trajectory* which corresponds to a certain system trajectory. The linguistic trajectory associated with the system trajectory in Fig. 6.5a is given by:

$$\textit{Linguistic trajectory}_1 = (\textit{Rule}_{11}, \textit{Rule}_{16}, \textit{Rule}_{17}, \textit{Rule}_{18}, \textit{Rule}_{19}, \textit{Rule}_{14}, \textit{Rule}_{9}, \textit{Rule}_{8}, \textit{Rule}_{13}).$$

From a design point of view, this method provides interesting guidelines for the analysis of a FKBC. Non-operative rules (non-fired rules in a given mode of operation or working conditions) such as shown in Fig. 6.5b can be easily modified. In Fig. 6.5b, top left corner, only the rules centered around the axis x_1 are fired. A similar situation, this time around the axis x_2, is presented in the top right corner. In the bottom left corner of this figure only the rule covering the origin of the control space is fired. Finally, in the bottom right corner the set of rules does not cover all system trajectories. This dynamic behavior suggests

X_2 \ X_1	PNB	PN	PZ	PP	PPB
VNB	APVB	APM	APM	APM	APVB
VN	APB	AP	APS	APS	APM
VZ	AP	APVS	AZ	ANVS	AN
VP	ANM	ANS	ANS	ANB	ANB
VPB	ANVB	ANM	ANM	ANM	ANVB

Fuzzy set of rules

Fig. 6.4. The variables X_1 and X_2 can for example represent position and velocity. The rule base is in a table form. PNB, PN, PZ, PP and PPB represent linguistic values of X_1, such as negative big, negative, zero, positive, and positive big. VNB, VN, VZ, VP and VPB represent linguistic values of X_2, such as negative big, negative, zero, positive, and positive big. ANVB, ANB, AN, ANM, ANS, AZ, APS, APM, AP, APB and APVB represent linguistic values of the control output such as negative very big, negative big, negative, negative medium, negative small, zero, positive small, positive medium, positive, positive big, and positive very big. The membership functions for these linguistic values are defined in Fig. 6.17.

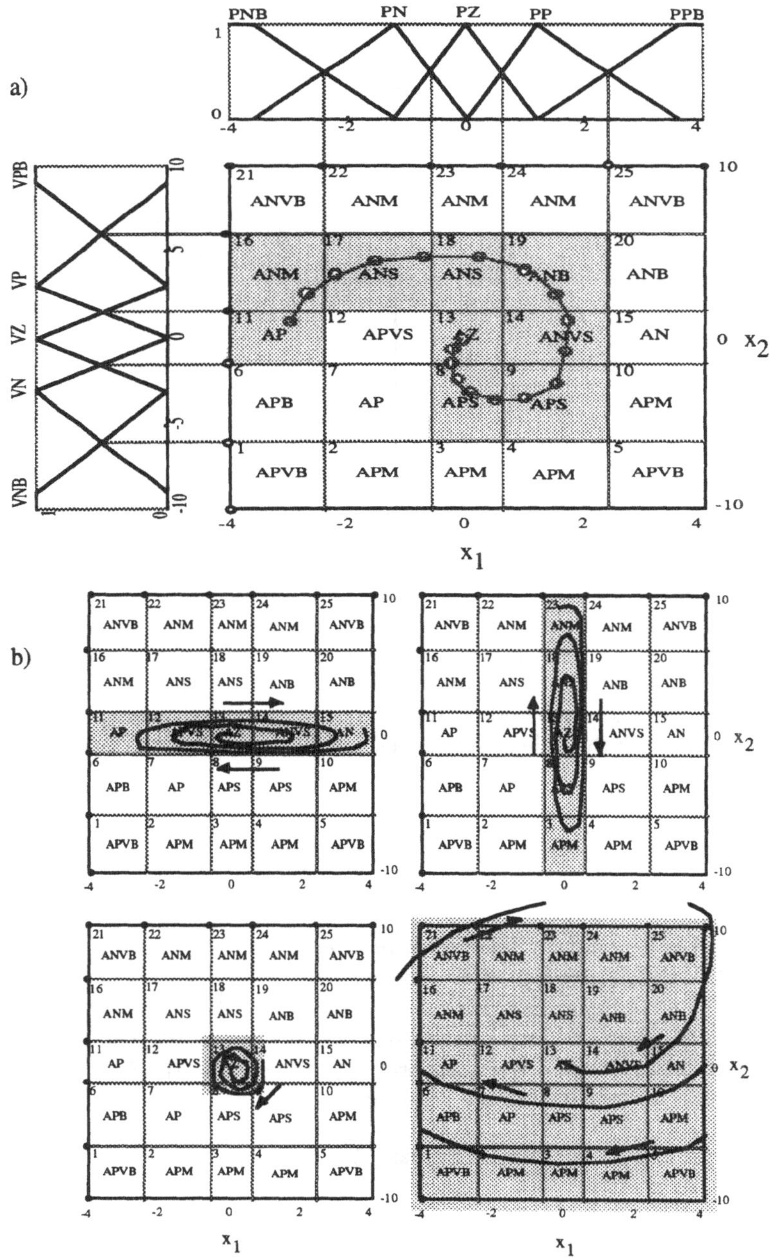

Fig. 6.5. (a) The system trajectory mapped on the partition space. (b) Inadequate coverage of the partition space during the operation of a FKBC. Gray areas correspond to fired rules, white areas correspond to non-fired rules.

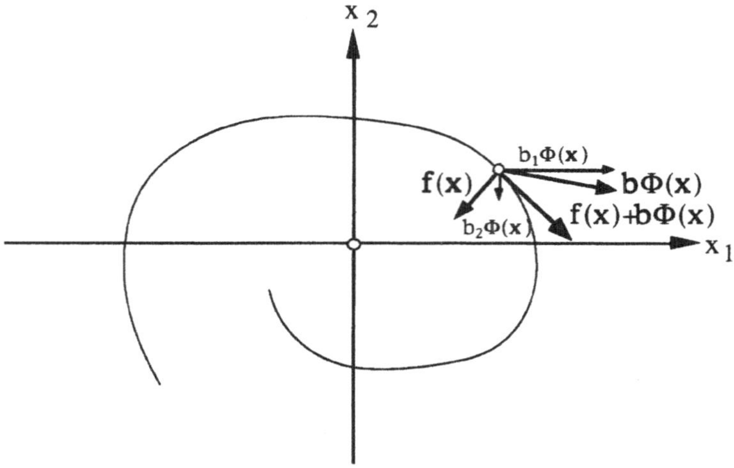

Fig. 6.6. The components of the vector field.

modifications in the fuzzy sets representing the meaning of the linguistic values of x_1 and x_2. Aracil et al. [3] propose a geometric interpretation of the state map. This technique is based on the study of vector fields associated with the plant and the FKBC rule base. Let us consider the closed-loop system from Fig. 6.1 represented as

$$\frac{\mathrm{d}\mathbf{x}}{\mathrm{d}t} = \mathbf{f}(\mathbf{x}) + \mathbf{b}u,$$
$$u = \Phi(\mathbf{x}), \tag{6.4}$$

where $\mathbf{f}(\mathbf{x})$ is a nonlinear function which represents the plant dynamics with $\mathbf{f}(\mathbf{0}) = \mathbf{0}$, $\mathbf{x}$ and $\mathbf{b}$ are vectors of dimension n, u is the scalar control variable, and $\Phi(\mathbf{x})$ is a nonlinear function representing the FKBC with $\Phi(\mathbf{0}) = \mathbf{0}$.

Let $\mu_R(\mathbf{x}, u)$ be the fuzzy relation representing the meaning of the rule base, let *Defuzz* be any defuzzification operator, and let $\mu^*_{\mathbf{x}}$ be the fuzzified input of the controller obtained from the crisp input $\mathbf{x}^*$; then

$$\Phi(\mathbf{x}) = \mathrm{Defuzz}(\mu^*_{\mathbf{x}} \circ \mu_R(\mathbf{x}, u)), \tag{6.5}$$

where $\mathbf{x} \in \mathcal{X}_1 \times \mathcal{X}_2 \times \ldots \times \mathcal{X}_n$ and $u \in \mathcal{U}$.

Closed loop behavior will depend on the nature of $\mathbf{f}(\mathbf{x})$ and $\Phi(\mathbf{x})$. The direction of the vector field associated with the FKBC is determined by coefficients of $\mathbf{b}$, and the magnitude is given by $\mathbf{b} \cdot \Phi(\mathbf{x})$ as shown in Fig. 6.6. One should note the effect which the subspace determined by the condition $\Phi(\mathbf{x}) = 0$ has on the closed loop. This subspace, usually a line (switching line) for second order

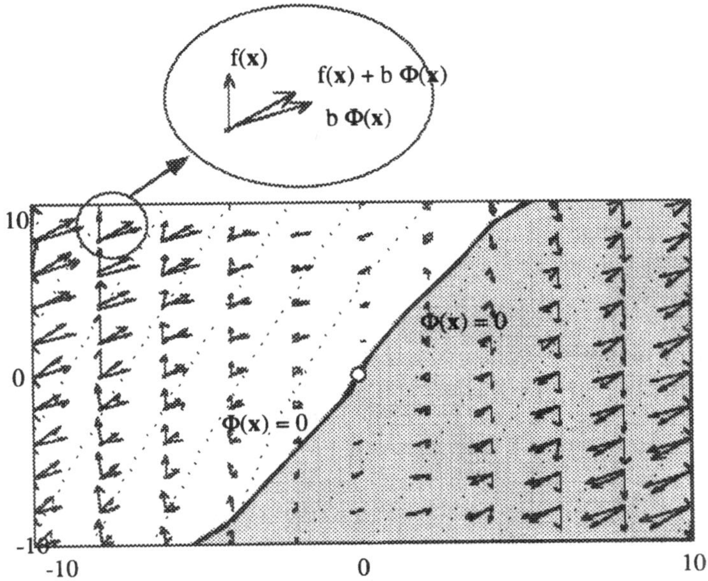

Fig. 6.7. Vector field representation.

systems, determines the separation of the state space in positive and negative control action areas (see Fig. 6.7).

With monotone $\mathbf{f}(\mathbf{x})$ and $\Phi(\mathbf{x})$, stable behavior of the closed-loop system can be predicted under the following conditions

- The open-loop system $\frac{\mathrm{d}\mathbf{x}}{\mathrm{d}t} = f(\mathbf{x})$ is stable.
- The vector field associated with $\mathbf{b} \cdot \Phi(\mathbf{x}) = 0$ tends towards $\Phi(\mathbf{x}) = 0$.

Obviously, for other situations, such as unstable open-loop systems, the characteristics of the vector field must be determined to properly predict the behavior of the system.

However, the above analysis is in line with the previous concepts when considering linguistic trajectories. A simple inspection using this approach suffices to establish stability or instability, in addition to predicting other dynamic phenomena such as limit cycles, isolated areas, oscillations, etc. [71, 3].

To illustrate some of these dynamic phenomena, we can summarize as follows:

a. Stable feedback system: In this case, the vector field obtained as $\Phi(x_1, x_2)$ tries to lead the system trajectories in the direction of the switching curve $\Phi(x_1, x_2) = 0$. When the trajectories approach this curve, $\Phi(x_1, x_2) = 0$, the plant component of the vector field obtains a greater influence which makes the trajectories converge to the equilibrium point (see Fig. 6.8).

b. Critical case: The nonlinear field does not lead the trajectories to the curve $\Phi(x_1, x_2) = 0$, so instability can be avoided if the plant component obtains greater influence than the controller component (see Fig. 6.9).

c. Limit cycles: This case can be considered as a combination of the stable and critical cases (see Fig. 6.10). Limit cycles also depend on the plant component characteristics.

d. Nonlinear vector fields with isolated areas: Isolated areas are regions which behave differently from the dominant area, which is above the switching curve $\Phi(x_1, x_2) = 0$. Isolated areas are normally defined by closed generic curves $\Phi(x_1, x_2) = 0$, that do not contain the plant component equilibrium point, and do not produce field compensation. In this case, the tendency would be to go around these areas. Thus, an output trajectory appears where a large number of trajectories converge (see Fig. 6.11).

The above conclusions about stability are qualitative, but enough to characterize the dynamic behavior of a feedback system, or serve as the basis for adequate selection or modification of rules. For example, the limit cycle in Fig. 6.10 could be avoided by modifying the rule-consequents of the rules $R8$ and $R13$ in order to generate a control output with the opposite sign. Thus "u is PS" in rule $R8$ should become "u is NS" and "u is NS" in rule $R13$ should become "u is PS" The same can be done with the rule $R10$ in order to eliminate the isolated area shown in Fig. 6.11.

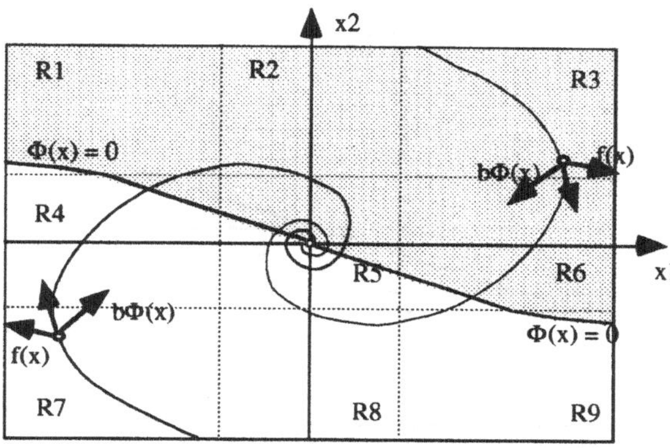

R1: If x1 is N and x2 is P then u is NB
R2: If x1 is Z and x2 is P then u is NM
R3: If x1 is P and x2 is P then u is NS
R4: If x1 is N and x2 is Z then u is PM
R5: If x1 is Z and x2 is Z then u is Z
R6: If x1 is P and x2 is Z then u is NM
R7: If x1 is N and x2 is N then u is PS
R8: If x1 is Z and x2 is N then u is PM
R9: If x1 is P and x2 is N then u is PB

Fig. 6.8. Stable feedback system.

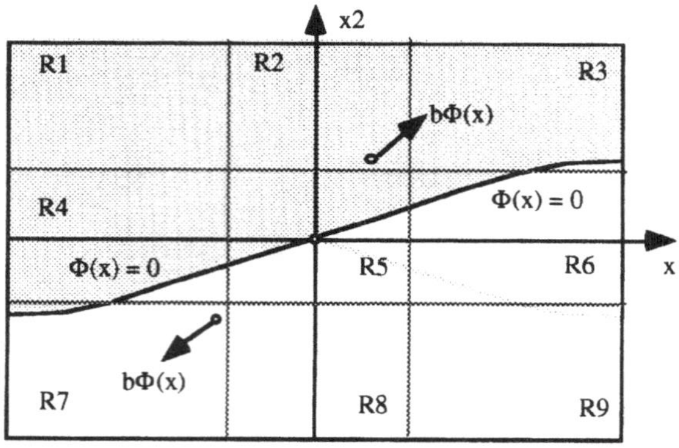

R1: If x1 is N and x2 is P then u is PB
R2: If x1 is Z and x2 is P then u is PM
R3: If x1 is P and x2 is P then u is PS
R4: If x1 is N and x2 is Z then u is PM
R5: If x1 is Z and x2 is Z then u is Z
R6: If x1 is P and x2 is Z then u is NS
R7: If x1 is N and x2 is N then u is NS
R8: If x1 is Z and x2 is N then u is NM
R9: If x1 is P and x2 is N then u is NB

NB, NM, N, NS, Z, PS, P, PM, PB represents negative big, negative medium, negative, negative small, zero, positive small, positive, positive medium, and positive big respectively.

Fig. 6.9. Critical case.

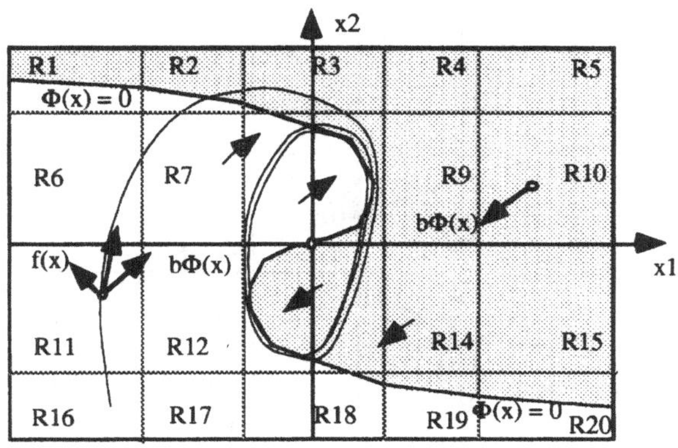

R1: If x1 is NB and x2 is PB then u is PS
R2: If x1 is N and x2 is PB then u is PS
R3: If x1 is Z and x2 is PB then u is PM
R4: If x1 is P and x2 is PB then u is PM
R5: If x1 is PB and x2 is PB then u is PB

R6: If x1 is NB and x2 is P then u is NM
R7: If x1 is N and x2 is P then u is NM
R8: If x1 is Z and x2 is P then u is PS
R9: If x1 is P and x2 is P then u is PM
R10: If x1 is PB and x2 is P then u is PM

R11: If x1 is NB and x2 is N then u is NM
R12: If x1 is N and x2 is N then u is NM
R13: If x1 is Z and x2 is N then u is NS
R14: If x1 is P and x2 is N then u is PM
R15: If x1 is PB and x2 is N then u is PM

R16: If x1 is NB and x2 is NB then u is NB
R17: If x1 is N and x2 is NB then u is NM
R18: If x1 is Z and x2 is NB then u is NM
R19: If x1 is P and x2 is NB then u is NS
R20: If x1 is PB and x2 is NB then u is NS

Fig. 6.10. Presence of a limit cycle.

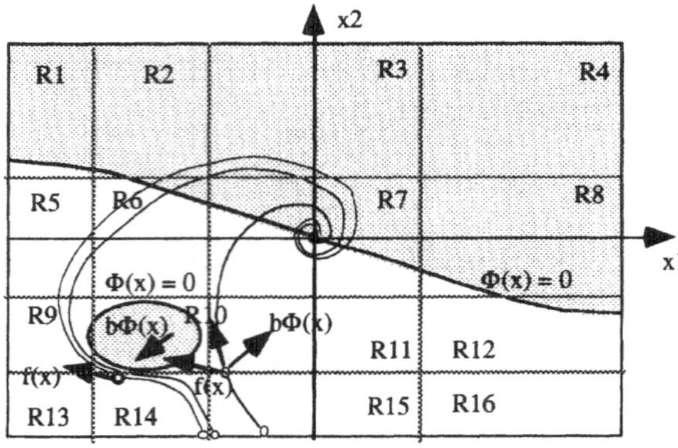

R1: If x1 is NB and x2 is P then u is PB
R2: If x1 is N and x2 is P then u is PB
R3: If x1 is Z and x2 is P then u is PM
R4: If x1 is P and x2 is P then u is PS
R5: If x1 is NB and x2 is Z then u is NM
R6: If x1 is N and x2 is Z then u is NM
R7: If x1 is Z and x2 is Z then u is Z
R8: If x1 is P and x2 is Z then u is PM
R7: If x1 is NB and x2 is N then u is NM
R7: If x1 is N and x2 is N then u is PM
R8: If x1 is Z and x2 is N then u is NM
R9: If x1 is P and x2 is N then u is NS
R7: If x1 is NB and x2 is NB then u is NB
R7: If x1 is N and x2 is NB then u is NB
R8: If x1 is Z and x2 is NB then u is NM
R9: If x1 is P and x2 is N Bthen u is NM

NB, NM, N, NS, Z, PS, P, PM, PB represents negative big, negative medium, negative, negative small, zero, positive small, positive, positive medium, and positive big respectively.

Fig. 6.11. Nonlinear dynamics with isolated areas.

6.3 Stability and Robustness Indices

In the previous section we have presented an approximation of the problem of FKBC stability, and its geometrical interpretation in the state space.

In this section we formalize the problem of stability analysis by using concepts extracted from the qualitative theory of nonlinear dynamic systems [84]. These concepts are used to interpret instabilities and bifurcations, and thus a more complete representation of the stability problem is achieved. We shall start with the one-dimensional case, followed by the n-dimensional case.

6.3.1 The One-dimensional Case

Let us consider a system with a mathematical model of the following form

$$\frac{\mathrm{d}x}{\mathrm{d}t} = f(x) + b \cdot u, \tag{6.6}$$

with $x \in \mathcal{X} \subset \mathbb{R}$ and $u \in U \subset \mathbb{R}$, where $f(x)$ is a nonlinear, monotone, and increasing function, such that $f(0) = 0$, and $\mathbb{R}$ is the real line. Consider also the control law given by

$$u = \Phi(x), \tag{6.7}$$

where $\Phi(x)$ is a nonlinear function representing the FKBC and $\Phi(0) = 0$.

Without loss of generality, we will consider the case $b = 1$. As we noted in Section 6.2, the vector field associated with the closed-loop dynamical system (6.6) and (6.7),

$$\frac{\mathrm{d}x}{\mathrm{d}t} = f(x) + \Phi(x), \tag{6.8}$$

is composed of two parts: the *plant component* $f(x)$ and the *controller component* $\Phi(x)$. To visualize geometrically these vector fields we can rotate them counterclockwise by an angle of 90°. In this case we will obtain the curves shown in Fig. 6.12.

The equilibrium points of system (6.8) are given by the values of x for which

$$\frac{\mathrm{d}x}{\mathrm{d}t} = 0. \tag{6.9}$$

That is,

$$f(x) + \Phi(x) = 0 \quad \text{or} \quad \Phi(x) = -f(x). \tag{6.10}$$

With the graphical convention of rotating the vector fields, the equilibria will be located at values of x where the curve representing the controller component $\Phi(x)$ intersects the curve representing the plant component with the sign changed to $-f(x)$.

It is clear that system (6.8) has an equilibrium point at the origin, as $f(0) = 0$ and $\Phi(0) = 0$. For this equilibrium to be stable it is necessary that

$$f'(0) + \Phi'(0) < 0. \tag{6.11}$$

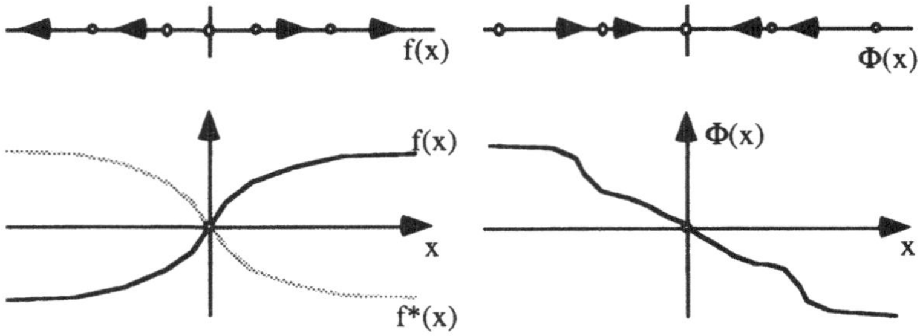

Fig. 6.12. Vector fields in the one-dimensional case.

That is, the eigenvalue at the origin has a negative real part.

Therefore, the system (6.8) is globally stable if it meets the following conditions

Condition 1: $\Phi'(0) < -f'(0)$,
Condition 2: $|\Phi(x)| < |f(x)| \qquad \forall x \neq 0$.

Condition 1 ensures the equilibrium stability at the origin, while condition 2 prevents the appearance of other equilibria (no intersection of curves $\Phi(x)$ and $-f(x)$ if this condition is satisfied).

Not meeting condition 2 for some values of x produces intersection of curves $\Phi(x)$ and $-f(x)$ at points other than the origin. This is shown in Fig. 6.13 where new equilibrium points appear.

To see how these new equilibrium points are generated, assume that some "deformation" is produced in $\Phi(x)$, and/or $f(x)$, so that an originally stable situation is changed. The question then is: under what circumstances will stability be lost? It is well known that loss of global stability may occur in at least one of the following cases:

- The equilibrium at the origin becomes unstable.
- New bifurcations are produced (intersections of $\Phi(x)$ with $-f(x)$).

As we shall see, it is possible to give a measure of how far a system is from loss of stability (how large are the "deformations" the system can be exposed to without losing stability). With this measure we will be able to define some stability indices.

To that end, let us reconsider condition 1. This condition can be rewritten

$$\Phi'(0) + f'(0) < 0 \tag{6.12}$$

or

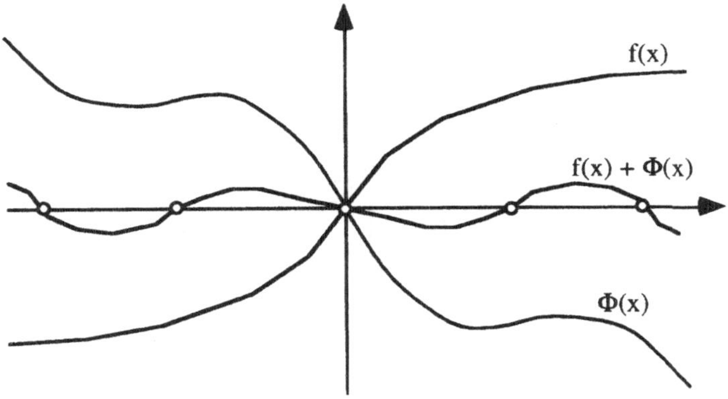

Fig. 6.13. Appearance of multiple attractors.

$$-(\Phi'(0) + f'(0)) > 0. \tag{6.13}$$

Then the quantity $-(\Phi'(0) + f'(0))$ gives a measure of the robustness of the system against loss of stability at the origin. The higher the value of $-(\Phi'(0) + f'(0))$ the further away the system is from an unstable condition. This unstable condition is reached for $-(\Phi'(0) + f'(0)) = 0$.

In a similar way, a measure can be associated with condition 2, giving the minimum distance between $\Phi(x)$ and $-f(x)$. The greater this distance, the more robust the system is against "deformations" of $\Phi(x)$ and $-f(x)$. A natural candidate for this minimum distance is given by

$$\min |\Phi(x) + f(x)|. \tag{6.14}$$

However, $|\Phi(x) + f(x)|$ will take its minimum value at the origin, where it will be 0. To avoid this, a certain region around the origin should be avoided. The boundary values of this region are β_1 and β_2. These are defined as the closest values to the origin such that $\Phi'(\beta_1) = -f'(\beta_1)$ and $\Phi'(\beta_2) = -f'(\beta_2)$. Figure 6.14 shows the geometrical interpretation of β_1 and β_2.

It is interesting to see what happens in the second case when new bifurcations are produced. Assume a stable situation (no intersections of $\Phi(x)$ with $-f(x)$). As $\Phi(x)$ and/or $-f(x)$ are "deformed" to approach each other, the "contact" between them will be realized when they acquire a common tangent. Then the contact is reduced to a single point and a bifurcation is produced. As the deformation is increased, two crossing points will appear. Under normal circumstances, one of these crossing points will give rise to a stable equilibrium point, and the other to an unstable one. This new equilibrium point will destroy the global stability at the origin, as two attraction areas appear in the state space. In other words, the origin will no longer be the only attractor of the system.

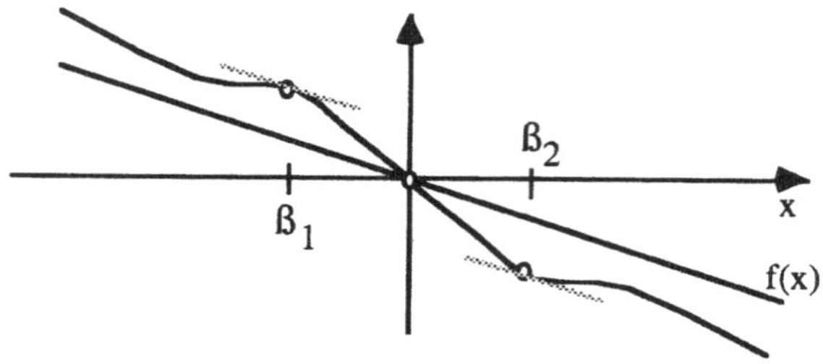

Fig. 6.14. Graphical interpretation of the region $B = (\beta_1, \beta_2)$.

Summarizing the above discussion, two indices can be defined,

$$\begin{aligned} I_1 &= -(\Phi'(0) + f'(0)), \\ I_2 &= \min_{B'} |\Phi(x) + f(x)|, \end{aligned} \tag{6.15}$$

where B' is the complement of the region around the origin $B = (\beta_1, \beta_2)$.

6.3.2 The *n*-dimensional Case

Let us start with $n = 2$. In this case the system has the form

$$\begin{aligned} \frac{\mathrm{d}\mathbf{x}}{\mathrm{d}t} &= f(\mathbf{x}) + \mathbf{b} \cdot u, \\ u &= \Phi(\mathbf{x}), \end{aligned} \tag{6.16}$$

with $\mathbf{x} \in \mathcal{X} \subset \mathbb{R}^2$ and $u \in \mathcal{U} \subset \mathbb{R}$, where $\mathbb{R}^2$ is the real plane. This equation can be written in a more detailed form as

$$\begin{aligned} \frac{\mathrm{d}x_1}{\mathrm{d}t} &= f_1(x_1, x_2) + b_1 \cdot \Phi(x_1, x_2), \\ \frac{\mathrm{d}x_2}{\mathrm{d}t} &= f_2(x_1, x_2) + b_2 \cdot \Phi(x_1, x_2), \end{aligned} \tag{6.17}$$

where $f_1(0,0) = f_2(0,0) = \Phi(0,0) = 0$ and f_1 and f_2 are monotone functions. As in the one-dimensional case $\Phi(x_1, x_2)$ stands for the FKBC. Assume that the equilibrium at the origin is stable, i.e., the two eigenvalues of the linearized system around the origin have a negative real part.

Generally, there are two ways in which the system can become unstable:

- A real eigenvalue crosses the imaginary axis and acquires a positive sign. This bifurcation from a stable node to a saddle point is also called *static bifurcation.*

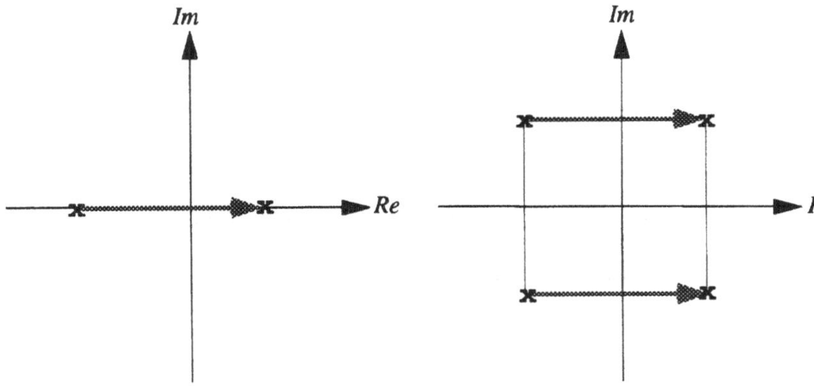

Fig. 6.15. (a) Static bifurcation; (b) Hopf bifurcation.

- A pair of complex poles crosses the imaginary axis and both poles take positive real values. This bifurcation is called *Hopf bifurcation.*

Figure 6.15 describe these two situations.

The simplest way to linearize a nonlinear system around a point x_0 in its state space is to consider the series development

$$f(x) = f(x_0) + J(x - x_0) + \dots, \tag{6.18}$$

where J is the so-called *Jacobian* matrix of $f(x)$ at x_0. This matrix is given by

$$J = \begin{bmatrix} \dfrac{\partial f_1}{\partial x_1} & \dfrac{\partial f_2}{\partial x_1} \\ \dfrac{\partial f_2}{\partial x_1} & \dfrac{\partial f_2}{\partial x_2} \end{bmatrix} = \begin{bmatrix} a_{11} & a_{12} \\ a_{21} & a_{22} \end{bmatrix}. \tag{6.19}$$

If the linearized system is stable, the Lyapunov criterion guarantees stability at the origin of the nonlinear model.

The characteristic polynomial of J is defined as

$$P(s) = \det(s \cdot I - J). \tag{6.20}$$

It is easy to see that

$$\det(s \cdot I - J) = s^2 - s \cdot (a_{11} + a_{22}) + a_{11} \cdot a_{22} - a_{12} \cdot a_{21}. \tag{6.21}$$

The characteristic polynomial J can be expressed as

$$P(s) = \det(s \cdot I - J) = s^2 + a_1 \cdot s + a_2, \tag{6.22}$$

where

$$\begin{aligned} a_1 &= -(a_{11}+a_{22}) = -\mathrm{tr}(J), \\ a_2 &= a_{11}\cdot a_{22} - a_{12}\cdot a_{21} = \det(J). \end{aligned} \tag{6.23}$$

A static bifurcation is produced when a real zero (eigenvalue) of the characteristic polynomial crosses the imaginary axis (see Fig. 6.15a), i.e., when one of the roots of $P(s)$ is $s = 0$. This only will happen in equation (6.22) when $a_2 = 0$. So this is the condition for loss of stability by a static bifurcation in a second order dynamic system. It should be noted that the larger the value of a_2 the further the system is from the bifurcation point. Therefore, the system is more robust. Then the value of a_2 gives a measure of the degree of relative stability, and it is quite natural to define a stability index such as

$$I_1 = a_2 = \det(J). \tag{6.24}$$

As we already said, a Hopf bifurcation is produced when two complex eigenvalues cross the imaginary axis and their real parts become positive (see Fig. 6.15b). In the case of our second order system, this means that the two complex eigenvalues of the system have zero real part. The requirement for this is that $a_1 = 0$ in the characteristic polynomial. Then the condition to have a Hopf bifurcation reduces to $a_1 = -\mathrm{tr}(J) = 0$. A value of a_1 far from 0 is a guarantee for not having a Hopf bifurcation. In this way we can define a second stability index as

$$I_1' = -\mathrm{tr}(J). \tag{6.25}$$

To summarize, a static bifurcation will be produced when $a_2 = 0$, and a Hopf bifurcation when $a_1 = 0$. Then the relative stability of the equilibrium point at the origin can be measured by the two indices

$$\begin{aligned} I_1 &= \det(J), \\ I_1' &= -\mathrm{tr}(J). \end{aligned} \tag{6.26}$$

Let us now generalize the index I_2. Remember that in the one-dimensional case a bifurcation occurs when the vector field of the FKBC exactly compensates the vector field of the plant, thus giving rise to a global vector field of value zero. In the two-dimensional case we are now dealing with, the vector field of the controller has, at every point of the state space, the direction given by (b_1, b_2). Therefore, the compensation of the vector components of the plant and of the controller can only occur in the region of the state space where the plant component has the direction (b_1, b_2). So we can define the *auxiliary subspace* as

$$\frac{f_1(x_1,x_2)}{b_1} = \frac{f_2(x_1,x_2)}{b_2}. \tag{6.27}$$

This expression represents the one-dimensional subspace of the state space, where the study of the occurrence of static bifurcations is to be performed. Thus the analysis is similar to the case of $n = 1$. Then an index I_2 can be defined as the minimum distance between the plant and controller components, calculated on the auxiliary subspace and excluding the region B around the origin:

$$I_2 = \min_{B'} |f(x) + b \cdot \Phi(x)|. \tag{6.28}$$

For $n > 2$, the generalization is straightforward. Let J be the Jacobian of the nonlinear system around the origin and let

$$P(s) = s^n + a_1 \cdot s^{n-1} + ... + a_{n-1} \cdot s + a_n \tag{6.29}$$

be its corresponding characteristic polynomial. The generalization of the index I_1 leads to

$$I_1' = a_n = (-1)^n \cdot \det(J). \tag{6.30}$$

The index I_1' is determined using the condition for a Hopf bifurcation. This condition is equivalent to the one for having two pure imaginary axes. The characteristic polynomial (6.29) can be rewritten in the form

$$P(s) = p_1(s) \cdot (w^2 + s^2) + b_1 \cdot s + b_2. \tag{6.31}$$

Then the condition for having two pure imaginary axes is

$$b_1 = b_2 = 0. \tag{6.32}$$

For example, consider the case $n = 3$. The corresponding characteristic polynomial is

$$P(s) = s^3 + a_1 \cdot s^2 + a_2 \cdot s + a_3, \tag{6.33}$$

which can be rewritten in the form of equation (6.31) as

$$P(s) = (s + a_1) \cdot (w^2 + s^2) + (a_2 - w^2) \cdot s + a_3 - a_1 \cdot w^2. \tag{6.34}$$

Therefore $b_1 = a_2 - w^2$ and $b_2 = a_3 - a_1 w^2$. The Hopf condition (6.32) reduces in this case to

$$\begin{aligned} a_2 - w^2 &= 0, \\ a_3 - a_1 \cdot w^2 &= 0. \end{aligned} \tag{6.35}$$

That is, $a_1 \cdot a_2 - a_3 = 0$. Then we can take I_1' as

$$I_1' = a_1 \cdot a_2 - a_3. \tag{6.36}$$

Thus, I_1' is the index for the measure of the "distance" between the complex poles having a negative real part and crossing the imaginary axis at points where instability occurs.

It can be demonstrated [153] that the Hopf condition is equivalent to

$$\det(H_{n-1}) = 0, \tag{6.37}$$

where H_{n-1} is the minor principal of order $n - 1$ of the Hurwitz matrix H. Matrices H and H_{n-1} are given as follows

$$H = \begin{bmatrix} a_1 & a_3 & a_5 & \dots \\ 1 & a_2 & a_4 & \dots \\ 0 & a_1 & a_3 & \dots \\ & \dots & & \dots \\ & \dots & & a_n \end{bmatrix}; \quad H_{n-1} = \begin{bmatrix} a_1 & a_3 & a_5 & \dots \\ 1 & a_2 & a_4 & \dots \\ 0 & a_1 & a_3 & \dots \\ & \dots & & \dots \\ & \dots & & a_{n-1} \end{bmatrix}. \tag{6.38}$$

Consequently,

$$I_1' = \det(H_{n-1}). \tag{6.39}$$

It can also be shown [153] that

$$I_1 \cdot I_1' = \det(H). \tag{6.40}$$

The generalization of the index I_2 for the case $n > 2$ is straightforward. In this case the one-dimensional auxiliary subspace will be given by

$$\frac{f_1(x_1, x_2, \dots, x_n)}{b_1} = \frac{f_2(x_1, x_2, \dots, x_n)}{b_2} = \dots = \frac{f_n(x_1, x_2, \dots, x_n)}{b_n}. \tag{6.41}$$

The remarkable fact is that for every n, the auxiliary subspace is one-dimensional (if the dimension of $\mathcal{U}$ is one).

Example 6.1 Let us consider a nonlinear system described by the following equation

$$\frac{\mathrm{d}^2 y}{(\mathrm{d}t)^2} + a_1 \cdot (1 - y^2) \cdot \frac{\mathrm{d}y}{\mathrm{d}t} + a_2 \cdot y = b_1 \cdot u. \tag{6.42}$$

This system can be represented in the state space by means of

$$\begin{aligned} \dot{x}_1 &= x_2 \\ \dot{x}_2 &= -a_2 \cdot x_1 - a_1 \cdot x_2 + a_1 \cdot x_1^2 \cdot x_2 + b_1 \cdot u, \end{aligned} \tag{6.43}$$

where $x_1 = y$, $b_1 = a_2 = 12.7388$, $a_1 = 2.2165$, and $|u| \leq 15$.

For $u = 0$, the system has an unstable limit cycle (Fig. 6.16). This corresponds to a system with a hypercritical Hopf bifurcation. Considering the saturation in u, a FKBC can be designed to increase the stability domain (limited by the unstable limit cycle) as much as possible and to improve the dynamic characteristics of this system in the stable domain. Thus a robust design is obtained.

Assume we have a set of FKBC rules provided by an expert, and given in a table form as in Fig. 6.17. Additionally, Fig. 6.18 represents the FKBC as a nonlinear function $\Phi(\mathbf{x})$.

The Jacobian of the feedback system is given by

$$J_c = \begin{bmatrix} 0 & 1 \\ -a_2 + b_1 \cdot \dfrac{\partial \Phi(x_1, x_2)}{\partial x_1} & -a_1 + b_1 \cdot \dfrac{\partial \Phi(x_1, x_2)}{\partial x_2} \end{bmatrix}. \tag{6.44}$$

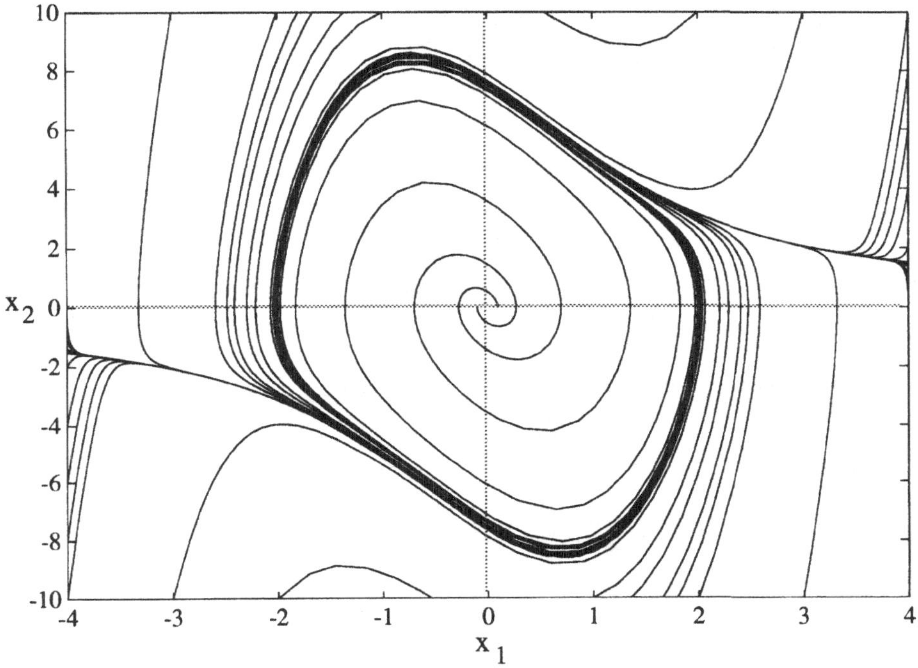

Fig. 6.16. Open system response. The system has an unstable limit cycle.

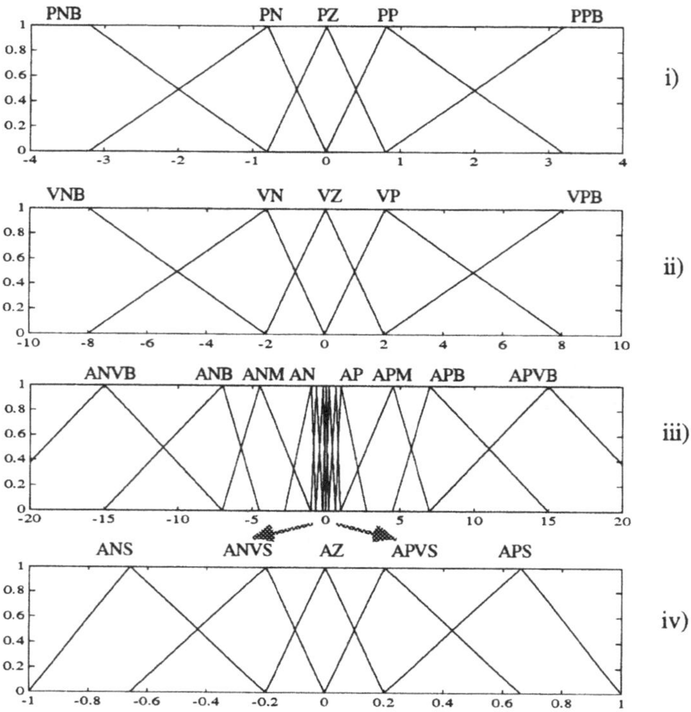

a)

x_2 \ x_1	PNB	PN	PZ	PP	PPB
VNB	1 APVB	2 APM	3 APM	4 APM	5 APVB
VN	6 APB	7 AP	8 APS	9 APS	10 APM
VZ	11 AP	12 APVS	13 AZ	14 ANVS	15 AN
VP	16 ANM	17 ANS	18 ANS	19 AN	20 ANB
VPB	21 ANVB	22 ANM	23 ANM	24 ANM	25 ANVB

b)

Fig. 6.17. (a) Fuzzy sets representing the meaning of linguistic values. (b) The rule base in a table form. The variables X_1 and X_2 can for example represent position and velocity. In (i) PNB, PN, PZ, PP and PPB represent linguistic values of X_1, such as negative big, negative, zero, positive, and positive big. In (ii) VNB, VN, VZ, VP and VPB represent linguistic values of X_2, such as negative big, negative, zero, positive, and positive big. In (iii) ANVB, ANB, AN, ANM, ANS, AZ, APS, APM, AP, APB and APVB represent linguistic values of the control output such as negative very big, negative big, negative, negative medium, negative small, zero, positive small, positive medium, positive, positive big, and positive very big. (iv) is a magnification of the control action interval $(-1, 1)$.

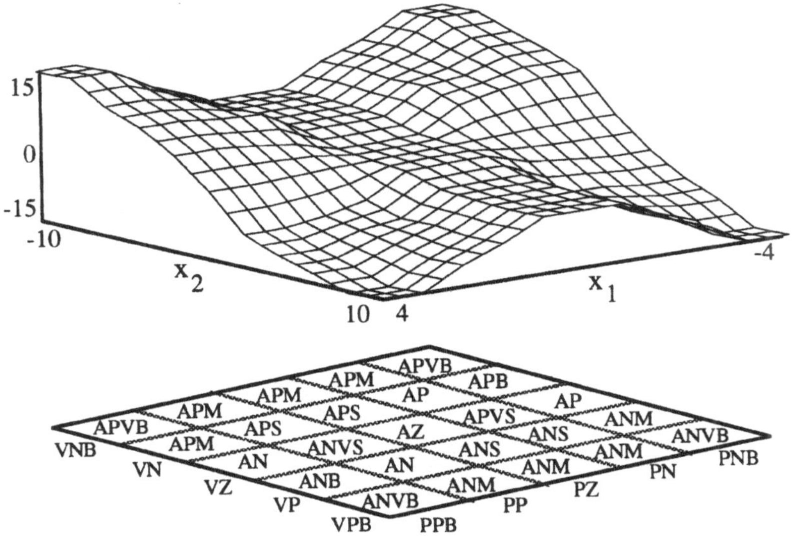

Fig. 6.18. The nonlinear function $\Phi(\mathbf{x})$ representing the FKBC.

The derivatives Φ_{x1} and Φ_{x2} at the origin can be obtained by using interpolation techniques. For the case of linear interpolation and the values of a_1, a_2 and b_1 given as $b_1 = a_2 = 12.7388$, $a_1 = 2.2165$, we obtain

$$J_c = \begin{bmatrix} 0 & 1 \\ -15.9236 & -6.4204 \end{bmatrix}. \tag{6.45}$$

Then the corresponding characteristic polynomial is given by

$$P(s) = s^2 + 6.4024 \cdot s + 15.9236. \tag{6.46}$$

Using equation (6.26) the indices can be obtained directly from the characteristic polynomial

$$I_1' = 6.4024; \qquad I_1 = 15.9236. \tag{6.47}$$

The auxiliary subspace is determined by

$$x_2 = 0. \tag{6.48}$$

Figure 6.19 presents the contributions of $f(\mathbf{x})$ and $\Phi(\mathbf{x})$ on the auxiliary subspace. As shown, the index I_2 is not defined, providing excellent robustness in the face of new attractors.

The system in the stable domain is more robust than the original one. Figure 6.20 gives a state diagram of the closed-loop system. As shown, the stability domain is considerably increased. Finally, Figs. 6.21 and 6.22 compare the transient response of the closed-loop system with that of the open-loop system. A significant improvement in system behavior is also apparent.

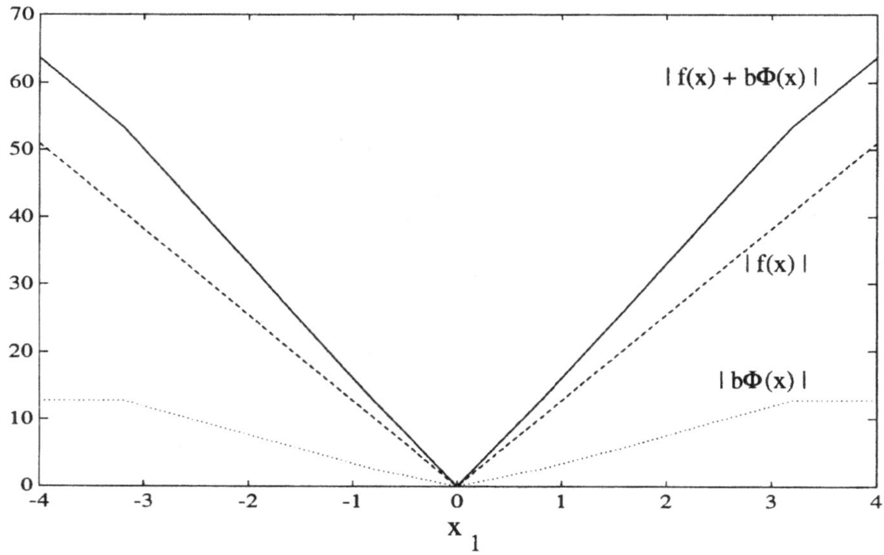

Fig. 6.19. Auxiliary subspace determined by $x_2 = 0$ The index I_2 is undefined in this case.

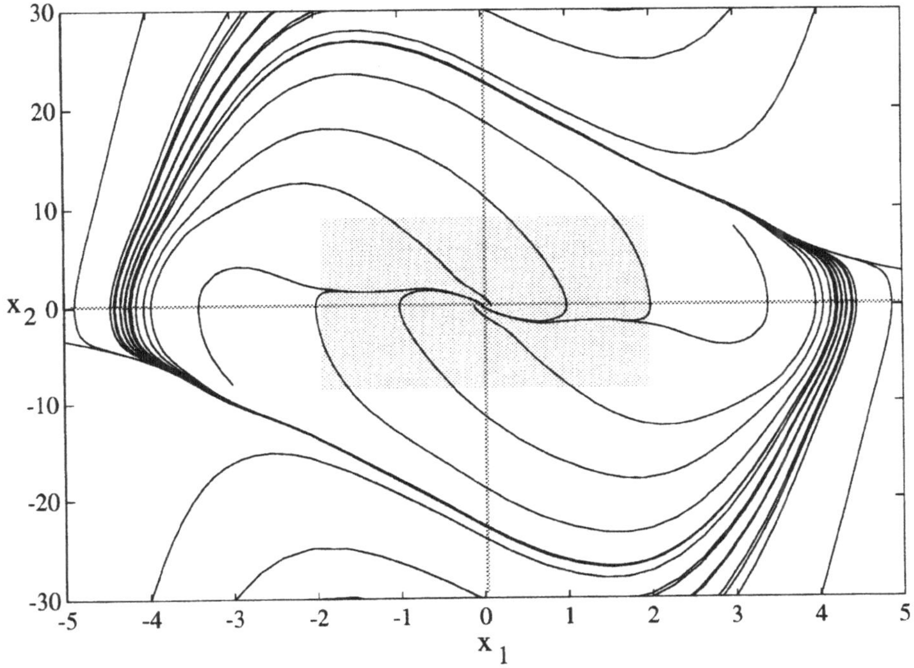

Fig. 6.20. Closed-loop response. The gray rectangle represents the limit cycle bounds of the open-loop system.

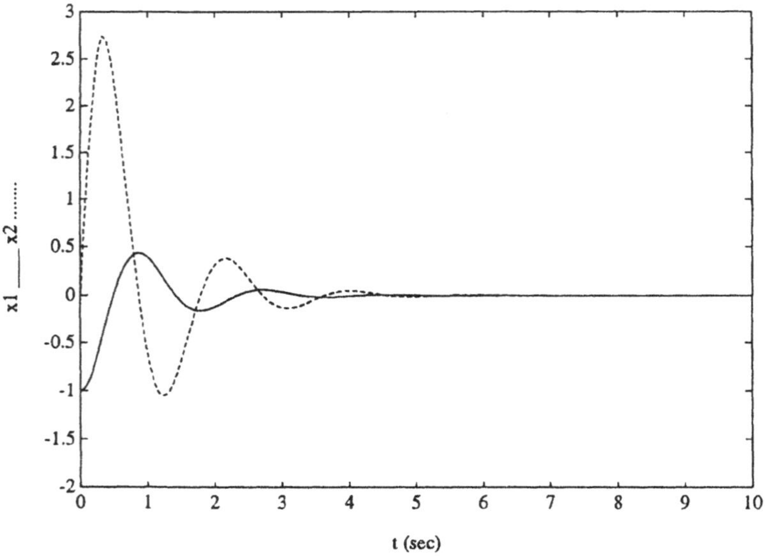

Fig. 6.21. Open-loop system response.

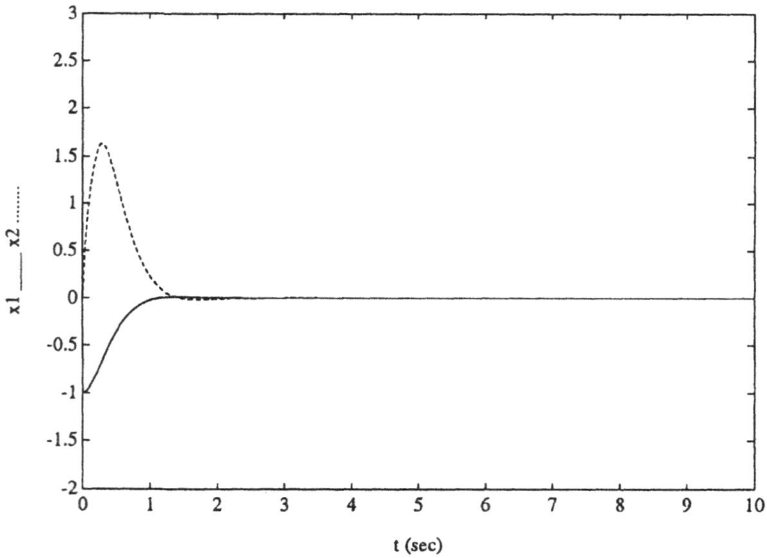

Fig. 6.22. Closed-loop system response.

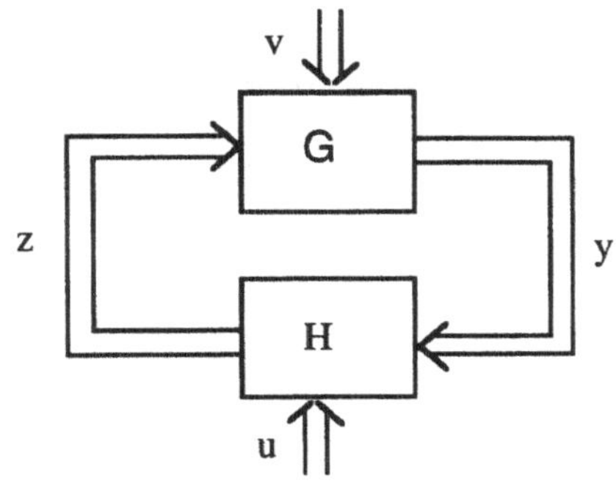

Fig. 6.23. A general feedback system.

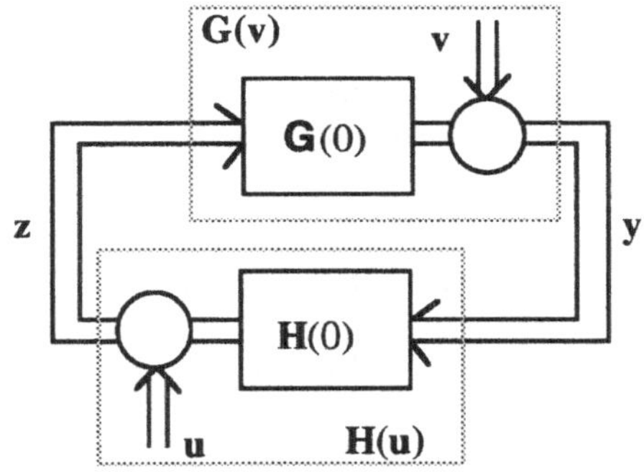

Fig. 6.24. Feedback system with additive inputs u and v.

6.4 Input-Output Stability

In this section we will present the basics of input-output stability. For a more detailed account see Safonov [181] and Vidyasagar [221].

Following Safonov [181], we deal with a feedback system as shown in Fig. 6.23. Typically, $G(v)$ represents the plant and $H(u)$ the controller. The scalar or vector signal y will be the control variable, and z the control output.

A particular case is that of Fig. 6.24, where u and v enter additively. In this case, v will be the reference variable and u may represent disturbances or initial conditions.

6.4.1 The Spaces of Signals

In the input-output stability analysis, $\mathbf{x}(t)$ is a scalar or vector function of time, taken from a normed vector space $\mathcal{X}$, i.e., a vector space with a norm on it. Two usually used norms are the 2-norm and the infinity-norm

$$| \mathbf{x}(t) |_2 = \left(\int_0^\infty | \mathbf{x}(t) |^2 \, dt \right)^{1/2} , \tag{6.49}$$

$$| \mathbf{x}(t) |_\infty = \text{ess sup} \; \{ \, | \mathbf{x}(t) | : \; t \in [0, \infty] \, \} , \tag{6.50}$$

where ess sup denotes the essential supremum.

We denote by L_2 and L_∞ the normed spaces of signals that have a finite 2-norm and infinity-norm, respectively. The space L_2 represents the set of all finite-energy signals, and the space L_∞ represents the set of all bounded signals.

The problem formulation must be able to cope with unbounded signals usual in control: (1) steps with undefined 2-norm, and (2) ramps with undefined 2- and infinity-norms. We define the *extended space* $\mathcal{X}_e$ in the following way. Consider the truncation $\mathbf{x}_T(t)$ of a signal $\mathbf{x}(t)$ after time T, defined by: $\mathbf{x}_T(t) = \mathbf{x}(t)$ for $t \leq T$, and $\mathbf{x}_T(t) = 0$ for $t > T$. The extended space $\mathcal{X}_e$ is the space formed with all signals $\mathbf{x}(t)$ such that their corresponding truncations $\mathbf{x}_T(t)$ belong to $\mathcal{X}$. In this way, the extended normed spaces contain all physically conceivable signals.

6.4.2 The Input-Output Notion of a System

A system G that takes inputs $\mathbf{x}(t)$ and yields outputs $\mathbf{y}(t)$ from the extended space $\mathcal{X}_e$ is considered as a *relation* $G \subset \mathcal{X}_e \times \mathcal{X}_e$. This relation is formed by all pairs $(\mathbf{x}(t), \mathbf{y}(t))$ such that $\mathbf{y}(t)$ is a possible output produced by the input $\mathbf{x}(t)$.

The difference with the usual notion of a system as an *operator* is that, given an input signal $\mathbf{x}(t)$, an operator produces *one and only one* output, $\mathbf{y}(t) = G\mathbf{x}(t)$, while a relation may produce *none, one, or more* outputs. The advantage of using relations in this context is that one can not only cope with responses to different conditions, but can also separate the problem of existence and uniqueness of operators from the problem of stability.

6.4.3 The Input-Output Notion of Stability

A system G is said to be *finite-gain stable* when the *gain* of G, $g(G)$, defined by

$$g(G) = \sup_{\mathbf{x}_T(t) \neq 0} \left\{ \frac{| (G\mathbf{x}(t))_T |}{| \mathbf{x}_T(t) |} \right\} < \infty . \tag{6.51}$$

The intuitive idea behind this is that we can make the system output small provided that we make its inputs small enough. The above definition for an open-loop system G is translated to the feedback system of Fig. 6.24. Let us denote by F the *closed-loop system*, with closed-loop inputs (u, v) and closed-loop outputs (y, z). The intuitive idea of the closed-loop stability of F is that we can obtain small (y, z) provided that we make (u, v) small enough. For example, y may represent the error of the plant output with respect to the reference (set-point) v, and u may represent disturbances to be avoided.

Other similar definitions of stability can be provided (for details see [181]). The input-output stability does not clarify the asymptotic approach of the control variables to zero. This property can be ensured based on the above mentioned equivalence between input-output stability and Lyapunov internal stability [221].

The main result in input-output stability, the *Small Gain Theorem* [244], states that a sufficient condition for the stability of the closed-loop system F (see Fig. 6.26) is that $g(G) \cdot g(H) < 1$. From this theorem, one can derive the circle criterion and the conicity criterion, which can be directly applied to FKBC stability. These criteria will be considered in Sections 6.5 and 6.6.

Finally, the gain of a general system is a quantity that can be difficult to obtain. For two classes of systems it is easy to derive the gain: (1) nonlinear static (memoryless) systems (*NLS*) and (2) linear time-invariant systems (*LTI*). The gain of the *NLS* system defined by $y = H(x)$ is given as

$$g(H) = \sup_{|x| \neq 0} \left\{ \frac{|H(x)|}{|x|} \right\}, \tag{6.52}$$

where the simplification is due to the fact that the time-dependence of the general formula disappears. The gain of the *LTI* system G given by its frequency response $G(j\omega)$ is given by

$$g(G) = \sup_{\omega} \overline{\sigma}\{G(j\omega)\}, \tag{6.53}$$

where $\overline{\sigma}(A)$ denotes the maximum singular value of the matrix A. The singular values of A are the square roots of the eigenvalues of $A * A$, with '$*$' denoting transpose conjugation. The two last formulas are directly applicable and extensively used in practical input-output stability analysis.

6.5 The Circle Criterion

Ray et al. [173, 174] proposed an analysis technique for FKBC stability based on the circle criterion. In the main, this study is based on considering the FKBC as a multi-level relay, as previously introduced by Kickert [115]. Even though this appears restrictive, it should not be considered as such. In practice, the concepts of limiting nonlinear functions are used for any nonlinear function.

For the FKBC to behave as a multi-level relay, two conditions on the fuzzy sets describing the meaning of the linguistic values of control output u have to hold:

1. The fuzzy sets have symmetric membership functions.
2. The Middle-of-Maxima defuzzification method is used in obtaining the crisp value of u.

Figure 6.25 shows the input-output map of the multi-level relay FKBC.

6.5.1 Stability Analysis – SISO Case

Consider a SISO (Single Input Single Output) linear system, represented by its transfer function $g(s)$, which is rational and asymptotically stable (Fig. 6.26). The nonlinear function representing the FKBC is described using $f(e)$, where e is the closed-loop error, with the constraints

$$k_1 \leq f(e) \leq k_2, \tag{6.54}$$

where k_1 and k_2 are two real numbers, which, for the sake of simplicity, we shall consider to be positive.

Under these conditions, the system is global and asymptotically stable if the Nyquist plot of $g(s)$ does not cross or go around, in a counterclockwise direction, the circle whose diameter is defined by $(-k_1^{-1}, -k_2^{-1})$, as shown in Fig. 6.27a.

6.5.2 Stability Analysis – MIMO Case

Let $G(s)$ be a transfer matrix of dimension $m \times m$ (Multi-Input Multi-Output case), which is rational and asymptotically stable. The nonlinear feedback is defined by the diagonal matrix $F(e,t)$, with components $f_i(e,t)$, and meets the restriction

$$k_1 \leq f_i(e,t) \leq k_2, \qquad \text{where } i = 1, \ldots, m. \tag{6.55}$$

If $\hat{g}_{ij}(jw)$ is the inverse of $g_{ij}(jw)$, with circles of radius $\hat{r}_i(jw)$, determined by

$$\hat{r}_i(jw) = \sum_{\substack{k=1 \\ k \neq i}} [\| \hat{g}_{ik}(jw) \|], \tag{6.56}$$

in the case of row dominance (see paragraph below), or

$$\hat{r}_i(jw) = \sum_{\substack{k=1 \\ k \neq j}} [\| \hat{g}_{kj}(jw) \|], \tag{6.57}$$

in the case of column dominance (see paragraph below), then the MIMO system is global and asymptotically stable if the Nyquist band of the inverse system crosses neither the origin nor the circle of diameter $(-k_1, -k_2)$, and encircles both of them the same number of times in a clockwise direction (see Fig. 6.27b).

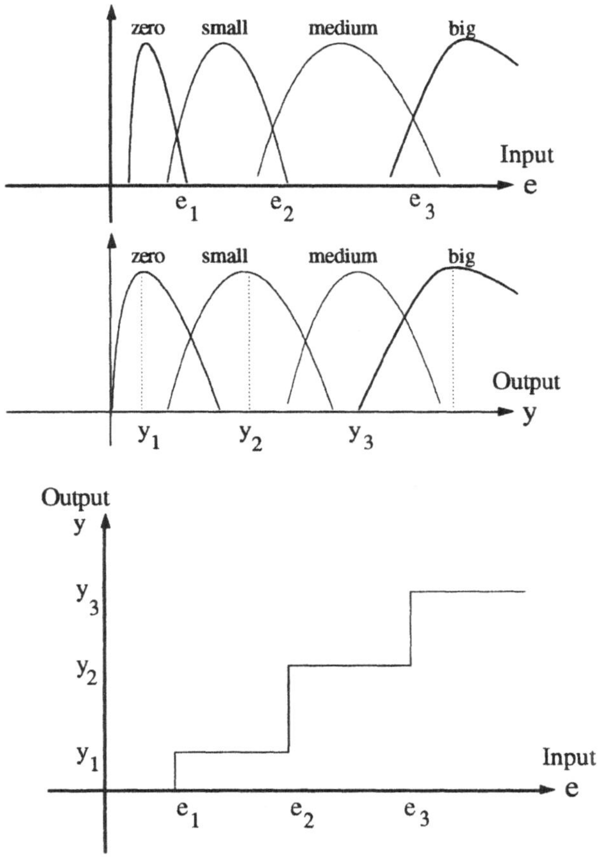

Fig. 6.25. The FKBC as a multi-level relay.

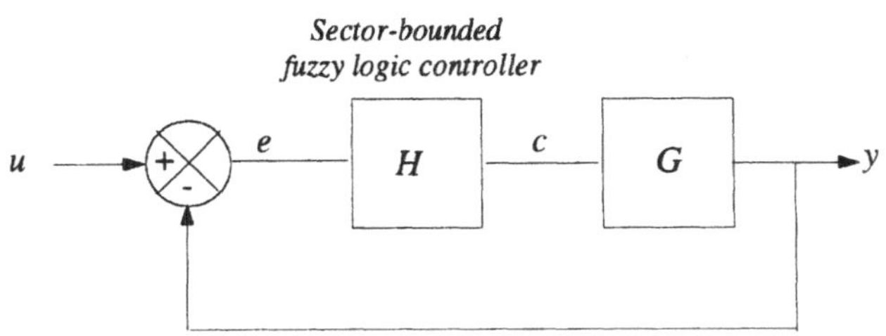

Fig. 6.26. The SISO system.

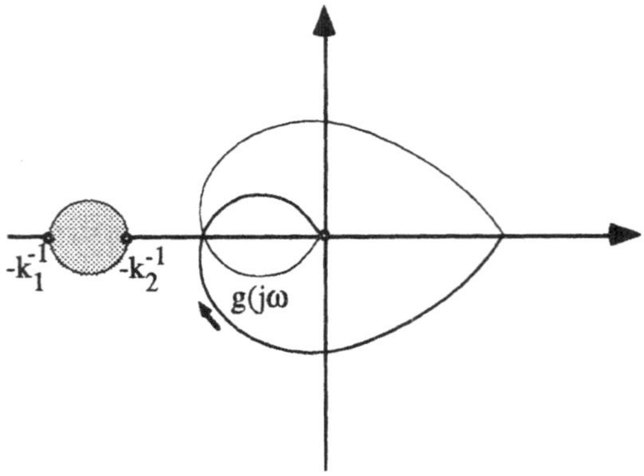

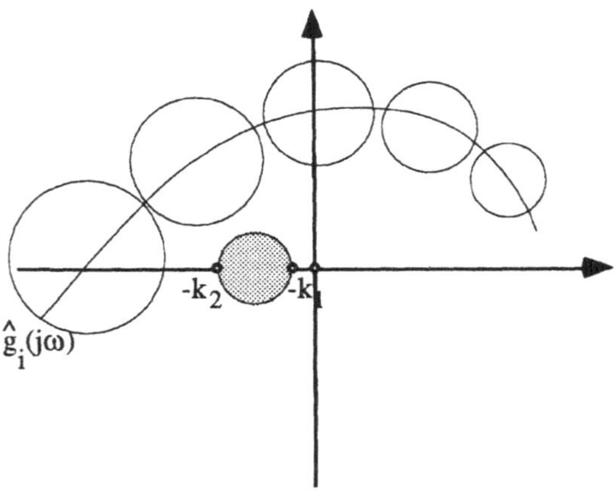

Fig. 6.27. Nyquist plot for (a) SISO case and (b) MIMO case.

Dominance Conditions for Stability Analysis
Let $G(s)$ be an $m \times m$ rational transfer function matrix, defined for all s on a subset D of the complex plane. Usually D will be the Nyquist contour consisting of the imaginary axis from $-jR$ to jR and a semicircle of large radius R in the right half plane, enclosing all poles and zeros of $G(s)$. Let us write $\hat{G}(s) = G^{-1}(s)$, if $G^{-1}(s)$ exists. Then $\hat{g}_{ii}(s)$ are the diagonal entries of $\hat{G}(s)$ and, in general, $\hat{g}_{ii}(s) \neq (g_{ii}(s))^{-1}$. The rational $m \times m$ matrix $\hat{G}(s)$ is said to be *row diagonal dominant* on D when, for all $i = 1, \ldots, m$ and for all s on D,

$$| \hat{g}_{ii}(s) | > \sum_{\substack{j=1 \\ j \neq i}}^{m} | \hat{g}_{ij}(s) | . \tag{6.58}$$

In a dual way, $\hat{G}(s)$ is said to be *column diagonal dominant* on D when

$$| \hat{g}_{ii}(s) | > \sum_{\substack{j=1 \\ j \neq i}}^{m} | \hat{g}_{ji}(s) | . \tag{6.59}$$

The diagonal dominance has a simple graphical interpretation. For each $i = 1, \ldots, m$, the $\hat{g}_{ii}(s)$ maps the imaginary axis $s = j\omega$ in the contour D onto the curve $\hat{g}_{ii}(j\omega)$ in the complex plane. Let us define the radius

$$d_i(j\omega) = \sum_{\substack{j=1 \\ j \neq i}}^{m} | \hat{g}_{ij}(j\omega) |, \tag{6.60}$$

and the closed disk $\mathcal{D}_i(\omega)$ centered at $\hat{g}_{ii}(j\omega)$ and with radius $d_i(j\omega)$. Consider the union $\mathcal{B}_i$ of all these disks, for all ω. This union is called the *inverse Nyquist band*, and looks like a band in the complex plane, centered along the curve $\hat{g}_{ii}(j\omega)$ and of variable width, depending on the $d_i(j\omega)$. Then the system $G(s)$ is row dominant when all the bands $\mathcal{B}_i$, for $i = 1, \ldots, m$, exclude the origin. A similar graphical test for column dominance can be built based on columnwise defined radius

$$d_i'(j\omega) = \sum_{\substack{j=1 \\ j \neq i}}^{m} | \hat{g}_{ji}(j\omega) | . \tag{6.61}$$

An application of diagonal dominance conditions to control systems stability was developed by Rosenbrock [176] and Cook [39], who presented multivariable versions of the circle criterion, based on frequency response bands that were called *Gershgorin* bands. A complete design method can be found in Rosenbrock [177] or in Bell et al. [16].

Additional Remarks Under normal conditions, considering the FKBC as a multi-level relay leads to $k_1 = 0$. This would cause the circle to become a semiplane $(-\infty, -k_2^{-1})$ in the SISO case, as shown in Fig. 6.28. Furthermore, in the MIMO case the circle has a diameter defined by $(0, k_2)$, as shown in Fig. 6.29.

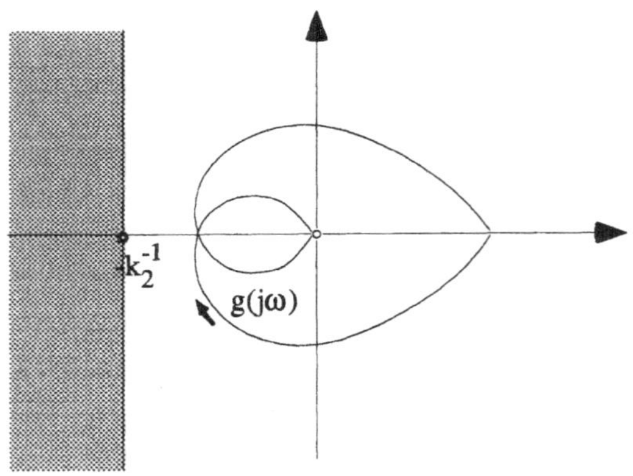

Fig. 6.28. SISO case for $k_1 = 0$.

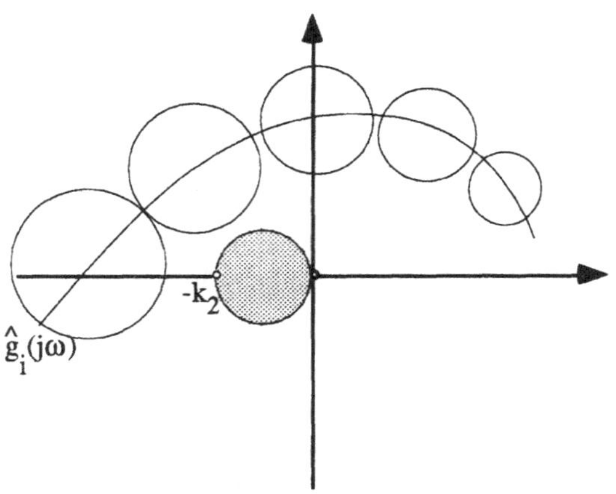

Fig. 6.29. MIMO case for $k_1 = 0$.

S \ E	NB	NM	NS	ZE	PS	PM	PB
NB	NB	NM	NM	NS	ZE	PM	PB
NM	NB	NB	NB	NM	NS	ZE	PS
NS	NB	NM	NM	NB	NB	NS	PS
ZE	NB	NM	NS	ZE	PS	PM	PB
PS	NM	NS	ZE	PS	PS	PM	PM
PM	NS	NM	NS	PM	PM	PB	PB
PB	ZE	PS	PS	PM	PB	PB	PB

E represents error
S represents error sum
ZE represents zero
PS, PM, PB represent positive small, positive medium and positive big.
NS, NM, NB represent negative small, negative medium
and negative big respectively

Table 6.1. FKBC for subsystem 1

Example 6.2 Consider the system defined by the following transfer matrix (see Ray et al. [173])

$$G(s) = \begin{bmatrix} \dfrac{1}{(s+1)^3} & \dfrac{1}{2(s+1)^3} \\ \dfrac{1}{(s+2)^3} & \dfrac{3}{2(s+2)^3} \end{bmatrix}. \tag{6.62}$$

The inverse matrix is defined by

$$\hat{G}(s) = \begin{bmatrix} \frac{2}{3}(s+1)^3 & -\frac{1}{2}(s+1)^3 \\ (s+2)^3 & (s+2)^3 \end{bmatrix}. \tag{6.63}$$

The open-loop system is column dominant. Thus, the original MIMO system can be considered as two relatively uncoupled subsystems. Tables 6.1 and 6.2 give the FKBC for each one of these subsystems. In both tables S denotes sum-of-errors and E denotes error. The controller slopes are defined by the pairs $(0,1)$ and $(0,4)$. The feedback system is stable, as shown in Fig. 6.30.

6.5.3 Application of the Circle Criterion to Design

The previously mentioned criterion requires a linear and asymptotically stable plant. Ray et al. [173] used this criterion as an aid in the design of a FKBC. If the linear plant is stable, then the definition of the circle limits that guarantee stability defines the admissible limits for the FKBC.

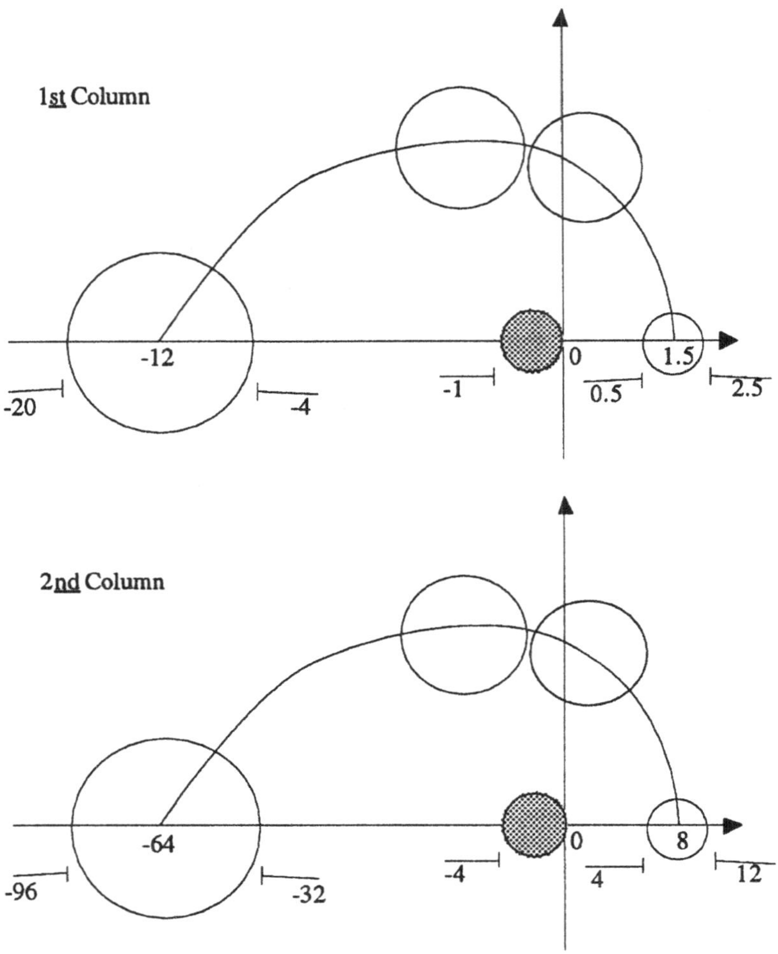

Fig. 6.30. Nyquist plot of Example 6.2.

S \ E	NB	NM	NS	ZE	PS	PM	PB
NB	NB	NM	NS	NS	ZE	PS	PS
NM	NM	NM	NM	NS	NS	ZE	PS
NS	NB	NM	NS	NM	NS	NS	ZE
ZE	NB	NM	NS	ZE	PS	PS	PM
PS	NS	NS	ZE	PS	PS	PM	PB
PM	NS	ZE	PM	PS	PM	PM	PM
PB	ZE	PM	PS	PM	PS	PM	PM

E represents error
S represents error sum
ZE represents zero
PS, PM, PB represent positive small, positive medium and positive big.
NS, NM, NB represent negative small, negative medium and negative big respectively

Table 6.2. FKBC for subsystem 2

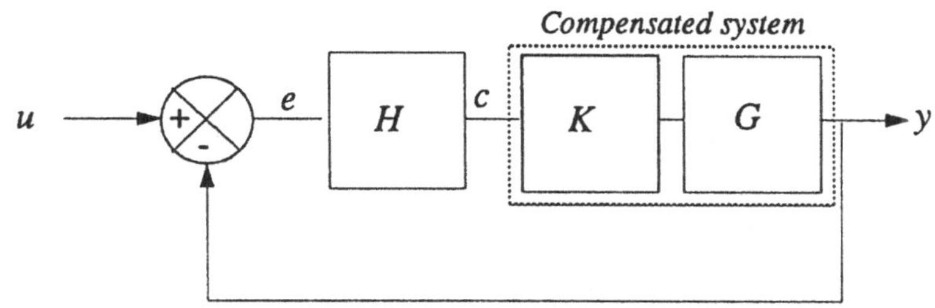

Fig. 6.31. Design of the FKBC for an unstable plant.

On the other hand, if the plant is unstable, a linear compensator is introduced in such a way that the compensated plant is stable (Fig. 6.31). The conditions are fixed on the circle in a similar way.

6.6 The Conicity Criterion

Consider the closed-loop system F of Fig. 6.32. If the small-gain condition $g(G) \cdot g(H) < 1$ holds, then we can ensure closed-loop stability. If this condition does not hold we can not conclude instability, because it is a *sufficient* condition.

One way to increase the applicability of the small-gain condition is to add and subtract from F the same block C. This gives rise to the transformed closed-loop $T(F, C)$ of Fig. 6.33. Under not very restrictive conditions, normally met

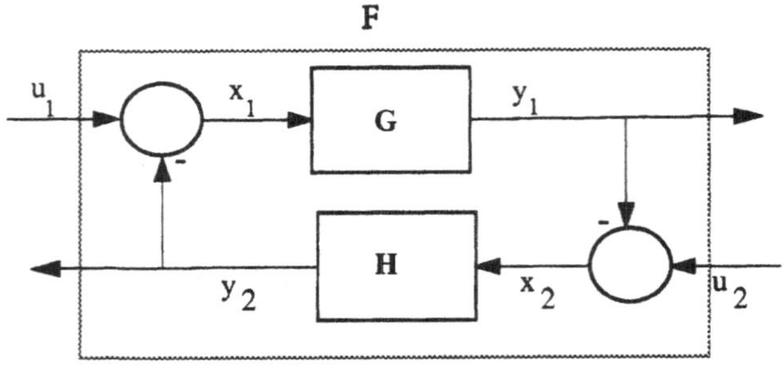

Fig. 6.32. Canonical closed-loop system.

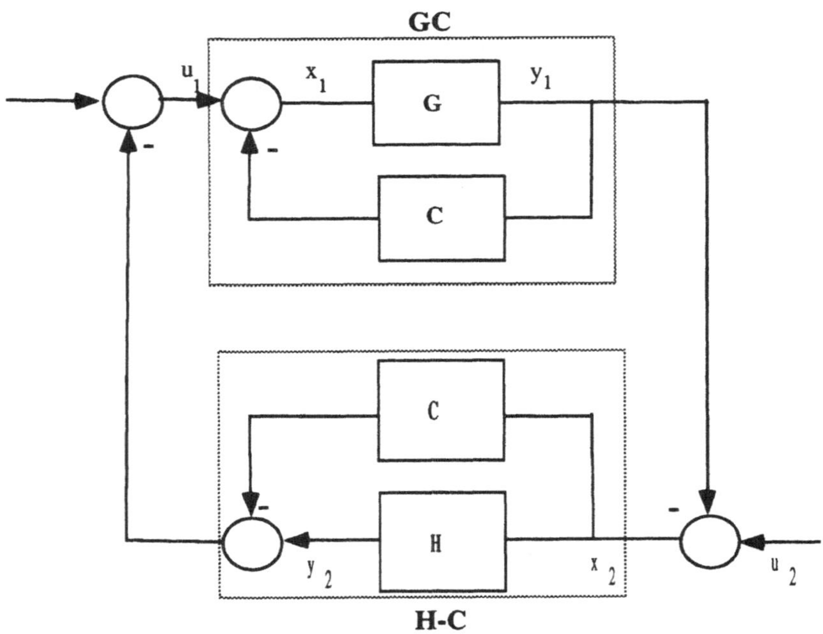

Fig. 6.33. The transformed closed-loop system.

in practice, the stability conditions of F and $\mathcal{T}(F, C)$ are equivalent, as stated in the *loop transformation theorem* [46].

If we apply the small gain condition to the transformed loop $\mathcal{T}(F, C)$, we obtain the *conicity criterion*, formulated as follows:

> The feedback system F is stable if there is a linear operator C and a positive number $r > 0$ such that
>
> **(a)** $g(H - C) < r$,
>
> **(b)** $g(\mathit{Feedb}(G, C)) \leq 1/r$,
>
> in the above $\mathit{Feedb}(G, C)$ denotes the feedback configuration of the blocks G and C.

The auxiliary elements C and r are called *center* and *radius*, respectively. An intuitive graphical interpretation of the conicity criterion can be given [244], at least in the scalar case (C scalar), in the sense that the graph of H in the plane $x - y$ must lie inside a sector of limiting slopes $C + r$ and $C - r$. Also, condition **(b)** can be related to the usual circle criterion condition defined on the scalar frequency response $G(j\omega)$ (see Section 6.5).

The conicity criterion when stated in this abstract way contains as a particular case (compare Safonov [181]) the circle criterion. To apply the conicity criterion in practice, and thus compute the gains involved using the formulas in Section 6.4 for *NLS* and *LTI* systems, the closed-loop dynamics must be organized in two blocks. One block, say G, must have a *LTI* model, and the other block, say H, must have a *NLS* model.

In this way, the center C must be chosen: (1) static, to compute $g(H - C)$ with the *NLS* formula, and (2) linear, to compute $g(\mathit{Feedb}(G, C))$ with the *LTI* formula.

This fact forces C to be a linear static (memoryless) operator, i.e., a matrix of appropriate dimensions. At this point the question is how to construct the center matrix C so that the conicity conditions hold. As no a priori information is in general available, one possibility is to explore all centers that may verify the conicity inequalities. This leads to the following definitions and criteria [12]. The *conic deviation* $d_H(C)$ of the nonlinear controller H from the center matrix C is given as

$$d_H(C) = g(H - C). \tag{6.64}$$

The *conic robustness* $r_G(C)$ of the linear plant G with the feedback C is given by

$$r_G(C) = \frac{1}{g\left(G(I + C \cdot G)^{-1}\right)}. \tag{6.65}$$

Then $d_H(C)$ gives a measure of the difference between C and H, and $r_G(C)$ gives a measure of the robustness of the closed-loop formed with the plant G with the feedback C. If C is chosen so that the closed-loop approaches instability, then the gain approaches infinity, and $r_G(C)$ converges to zero. Using this, the conicity stability conditions are that

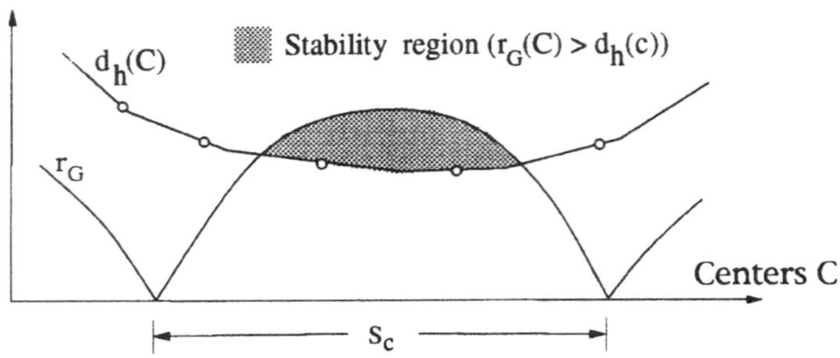

Fig. 6.34. Interpretation of the conicity criterion. S_c is a subset of stabilizing centers.

$$\exists C, r : d_H(C) < r \leq r_G(C). \tag{6.66}$$

The intuitive idea behind this is that the feedback system formed with the plant G and controller H, is stable if, for some linear law $y = Cx$, the conic robustness $r_G(C)$ of G with the feedback C (instead of H) is greater than the deviation $d_H(C)$ between H (actual controller) and C (linear controller).

These ideas are illustrated in Fig. 6.34. The x-axis is the space of linear centers C. It is of dimension 1 when C is a scalar; of dimension 2 when C is a 1×2 matrix; of dimension $n + m$ when C is a $n \times m$ matrix. Then the conic deviation $d_H(C)$, and conic robustness $r_G(C)$, as functions of C, are represented by curves, surfaces or hypersurfaces. It may happen that $r_G(C)$ is defined only on a subset S_c of stabilizing centers. The stability condition holds for the gray area in Fig. 6.34.

This formulation is a corollary of the conicity criterion, but adds implicitly the idea of exploring among all possible centers. As it may happen that the stability condition does not hold for some centers, but holds for others, we must look for an appropriate center. In Barreiro and Aracil [12], techniques to organize this search are given.

If the plant G has a transfer function $G(s)$ and frequency response $G(j\omega)$, the stability conditions are restated as follows for some r and C:

1. Nonlinear conicity ($d_H(C) < r$):

$$\forall x : \mid H(x) - Cx \mid < r \cdot \mid x \mid . \tag{6.67}$$

2. Linear conicity ($r \leq r_G(C)$), which is divided into:

   (a) Stability of the linear closed-loop

$$F(s) = G(s) \cdot (I + C \cdot G(s))^{-1}, \tag{6.68}$$

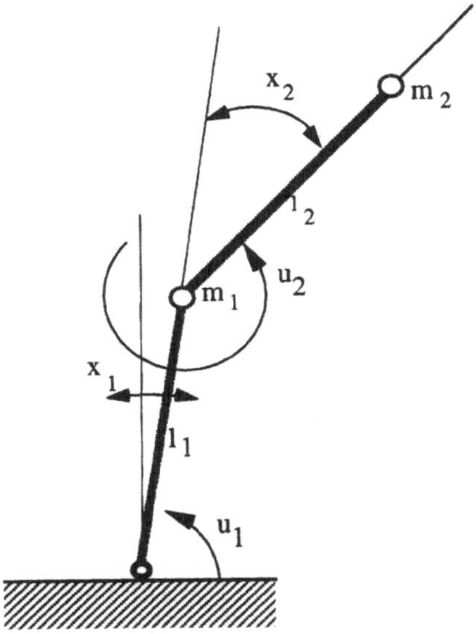

Fig. 6.35. 2-joints manipulator.

(b) Frequency response condition

$$\sup_{\omega} \overline{\sigma}\{F(j\omega)\} = \sup_{\omega} \overline{\sigma}\left\{G(j\omega) \cdot (I + G(j\omega))^{-1}\right\} \geq 1/r. \quad (6.69)$$

This reformulation of the conicity criterion can be easily applied to the analysis and design of FKBC. In the following we present the application of this method to the robust control of a two-joints manipulator.

Example 6.3 Consider the 2-joints manipulator shown in Fig. 6.35 [42]. It has 2-rotational joints with joint angles, velocities and accelerations given respectively by x_i, $\dot{x}_i$ and $\ddot{x}_i$ $(i = 1, 2)$. Assuming the masses m_1 and m_2 located at the distal ends of the links with lengths ℓ_1 and ℓ_2 (see Fig. 6.35) the dynamic model is given by

$$\mathbf{M}(\mathbf{x}) \cdot \ddot{\mathbf{x}} = \mathbf{u} + \mathbf{V}(\mathbf{x}, \dot{\mathbf{x}}) + \mathbf{G}(\mathbf{x}) + \mathbf{F}(\mathbf{x}, \dot{\mathbf{x}}) + \mathbf{z}, \quad (6.70)$$

where $\mathbf{x}$, $\dot{\mathbf{x}}$, $\ddot{\mathbf{x}}$ are two-dimensional vectors; u is a two-dimensional vector with the control torques; $\mathbf{M}(\mathbf{x})$ is a 2×2 mass matrix given as

$$\mathbf{M}(\mathbf{x}) = \begin{bmatrix} \ell_2{}^2 m_2 + 2\ell_1\ell_2 m_2 c_2 + \ell_1{}^2(m_1 + m_2) & \ell_2{}^2 m_2 + \ell_1\ell_2 m_2 c_2 \\ \ell_2{}^2 m_2 + \ell_1\ell_2 m_2 c_2 & \ell_2{}^2 m_2 \end{bmatrix}; \quad (6.71)$$

$\mathbf{V}(\mathbf{x}, \dot{\mathbf{x}})$ is a 2×1 vector of centrifugal and Coriolis terms

$$\mathbf{V}(\mathbf{x},\dot{\mathbf{x}}) = \begin{bmatrix} m_2\ell_1\ell_2 s_2\dot{x}_2^2 + 2m_2\ell_1\ell_2 s_2\dot{x}_1\dot{x}_2 \\ -m_2\ell_1\ell_2 s_2\dot{x}_1^2 \end{bmatrix}; \tag{6.72}$$

$\mathbf{G}(\mathbf{x})$ is an 2×1 vector of gravity terms

$$\mathbf{G}(\mathbf{x}) = \begin{bmatrix} -m_2\ell_2 g c_{12} - (m_1+m_2)\ell_1 g c_1 \\ -m_2\ell_2 g c_{12} \end{bmatrix}; \tag{6.73}$$

$\mathbf{F}(\mathbf{x},\dot{\mathbf{x}})$ is a model of frictions; and $\mathbf{z}$ is a 2×1 vector of unknown signals due to external disturbances and unmodelled dynamics. Furthermore, s_2, c_1, c_2, and c_{12} represent $\sin(x_2)$, $\cos(x_1)$, $\cos(x_2)$, and $\cos(x_1 + x_2)$, respectively.

Let

$$\mathbf{q}(\mathbf{x},\dot{\mathbf{x}}) = \mathbf{V}(\mathbf{x},\dot{\mathbf{x}}) + \mathbf{G}(\mathbf{x}) + \mathbf{F}(\mathbf{x},\dot{\mathbf{x}}). \tag{6.74}$$

Using the computed torque technique [42], the joint dynamics are decoupled and linearized. The control law is then given by

$$\mathbf{u} = -\hat{\mathbf{q}}(\mathbf{x},\dot{\mathbf{x}}) + \hat{\mathbf{M}}(\mathbf{x}) \cdot (\mathbf{h}(\mathbf{e},\dot{\mathbf{e}}) + \ddot{\mathbf{x}}_{\mathbf{d}}), \tag{6.75}$$

where $\mathbf{e} = \mathbf{x}_d - \mathbf{x}$, $\dot{\mathbf{e}} = \dot{\mathbf{x}}_d - \dot{\mathbf{x}}$; $\mathbf{x}_d$, $\dot{\mathbf{x}}_d$, $\ddot{\mathbf{x}}_d$ are the desired joint angles, angular velocities, and accelerations respectively; $\mathbf{h}(\mathbf{e},\dot{\mathbf{e}})$ is a feedback function. If the estimated values of parameters in $-\hat{\mathbf{q}}(\mathbf{x},\dot{\mathbf{x}})$ and $\hat{\mathbf{M}}(\mathbf{x})$ are close to the real values, and there are no external disturbances and unmodelled dynamics ($\mathbf{z} = \mathbf{0}$), then, the computed torque control law reduces the dynamics of each joint to a double integrator

$$\ddot{e} = -u, \tag{6.76}$$

$$u = h(e,\dot{e}). \tag{6.77}$$

In the above the bold-face letters are omitted, thus indicating that now we work with the scalar uncoupled dynamics of a *separate joint*. These dynamics can be regarded as a feedback configuration formed with a linear one-input two-output plant (input $-u$ and outputs $e, \dot{e}$), whose transfer function is a double integrator

$$g(s) = \begin{pmatrix} 1/s^2 \\ 1/s \end{pmatrix}. \tag{6.78}$$

The negative feedback is given by the nonlinear static law $h(e,\dot{e})$, that can be written as

$$u = h(e,\dot{e}) = k_p(e,\dot{e}) \cdot e + k_v(e,\dot{e}) \cdot \dot{e}. \tag{6.79}$$

In the above, position and velocity gains are nonlinear functions of the error e and derivative of the error $\dot{e}$. A fixed constant value for the position and speed gains $k_p(e,\dot{e})$, and $k_v(e,\dot{e})$ will suffice in the ideal case of exact model estimation. However, estimation errors, unmodelled dynamics, and other nonlinearities, mainly the actuator saturations, will deteriorate the control performance.

It will be shown how a "fuzzy design" of the gains will solve the actuator saturation problem. However, before that, we will show how the conicity criterion imposes some restrictions on this design. In this way the conicity criterion can be regarded as a complementary *stability tool* in the fuzzy design process. Let us apply the conicity conditions given above.

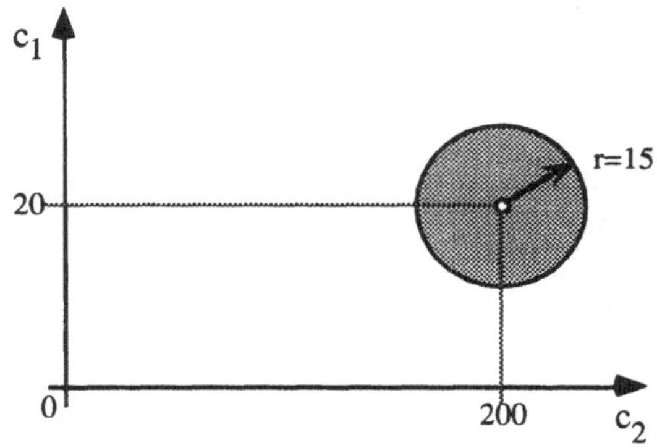

Fig. 6.36. Admissible control gains.

Stability of Linear Feedback

For some center $C = (c_1, c_2)$ the feedback system formed with $g(s)$ and C must be stable. This yields

$$c_1 > 0 \qquad \text{and} \qquad c_2 > 0. \tag{6.80}$$

Following Craig [42], we choose the central values of position and velocity gains as follows:

$$c_1 = 200; \qquad c_2 = 20. \tag{6.81}$$

Linear Conicity

For some radius $r > 0$,

$$r \leq r_g(c) = 1 \;/\; \sup_{\omega} \left(\frac{(c_1 - \omega)^2 + (c_2 \cdot \omega)^2}{1 + \omega^2} \right)^{1/2} \tag{6.82}$$

so that, for a center $c = (200, 20)$, this gives rise to the radius $r \leq r_g(c) \approx 15$.

Nonlinear Conicity

From the condition

$$\forall e, \dot{e} : |\, h(e, \dot{e}) - (c_1 \cdot e + c_2 \cdot \dot{e}) \,| < r \cdot |\, (e, \dot{e}) \,|, \tag{6.83}$$

we obtain

$$\left((k_p - c_1)^2 + (k_v - c_2)^2 \right)^{1/2} \leq r. \tag{6.84}$$

In other words, the position and speed gains $k_p(e, \dot{e})$ and $k_v(e, \dot{e})$ must always be interior to the circle of center c and radius r_g in the gain plane as shown in Fig. 6.36. Note that if the gains were fixed, then any positive value $k_p > 0$ and $k_v > 0$ will ensure linear stability. The circle restriction of Fig. 6.36 can be regarded as the price paid for the nonlinear variability of the gains.

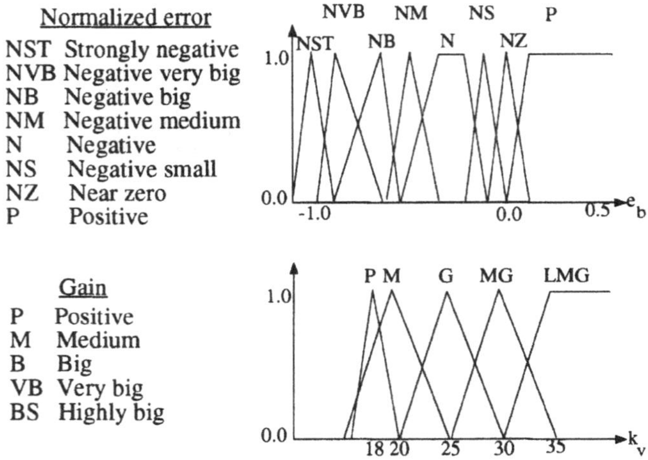

a)

If e_b is NST then k_v is P	If e_b is NVB then k_v is M
If e_b is NB then k_v is B	If e_b is NM then k_v is VB
If e_b is N then k_v is BS	If e_b is NS then k_v is VB
If e_b is NZ then k_v is M	If e_b is P then k_v is P

b)

Fig. 6.37. (a) Membership functions. (b) If-then rules, where e_b represents the normalized error defined by $e_b = e_v \cdot \text{sgn}(\dot{x}_d - \dot{x})$.

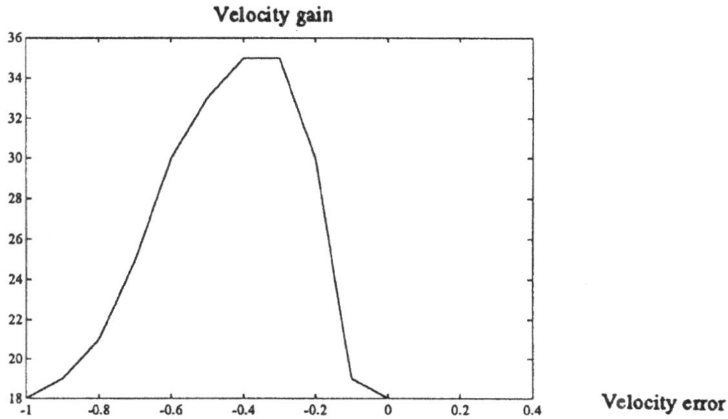

Fig. 6.38. The nonlinear function $k_v = \Phi(\dot{e})$.

Based on the conicity stability conditions, the *fuzzy design* of the error gains can be accomplished in the following way. First, assume the following working conditions for the manipulator:

- References: ramp of slope 0.5 rad/sec.
- Saturation: the saturation of the variable u is assumed to be ± 20 rad/sec^2.
- Design objectives: obtain a fast ramp response with absolute errors as small as possible and avoiding, if possible, the saturation regions.

Since the references are ramp signals, at start time, or at the times of slope changes, there will be a large speed error $\dot{e}$. Hence, the position gain has little effect and will be kept constant at the value $k_p = 200$. The velocity gain will be adjusted as a function of the velocity error, $k_v = \Phi(\dot{e})$, by means of a set of if-then rules.

It should be noted that the final conicity stability condition is

$$c_2 - r \approx 20 - 15 = 5 \ \le \ k_v = \Phi(\dot{e}) \ \le \ c_2 + r \approx 20 + 15 = 35. \qquad (6.85)$$

This condition can be guaranteed provided that the rule-consequent membership functions for k_v be in the range $[5, 35]$. Figure 6.37 shows the definitions of rules and membership functions. This formulation of the FKBC and the membership functions after fuzzification, rule firing and defuzzification results in a nonlinear function $k_v = \Phi(\dot{e})$ that verifies the stability condition. This function is shown in Fig. 6.38.

Finally, Fig. 6.39 shows the simulation results. It should be noted that the fuzzy speed gain $k_v = \Phi(\dot{e})$ gives a better global response than the linear gain settings $k_v = 5$ (with large initial errors), or $k_v = 35$ (with a slow convergence). Also, the whole model has been simulated with the two joints coupled. Simulations [72] show that even for considerable errors (50%) in the estimates $\hat{\mathbf{M}}(\mathbf{x})$

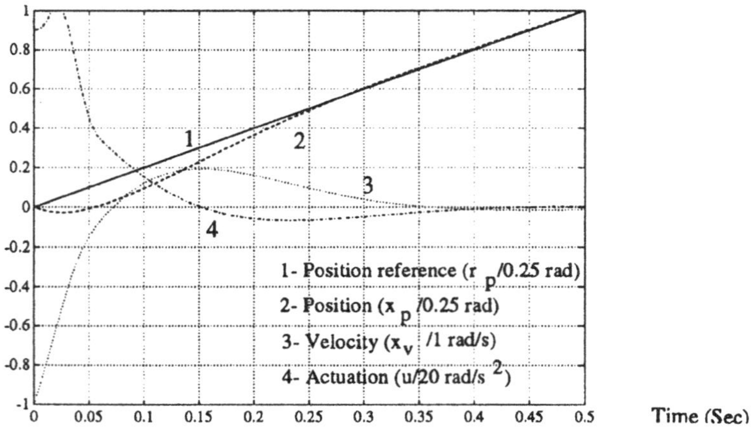

Fig. 6.39. Results of the fuzzy design.

and $\hat{\mathbf{q}}(\mathbf{x}, \dot{\mathbf{x}})$, of the actual $\mathbf{M}(\mathbf{x})$ and $\mathbf{q}(\mathbf{x}, \dot{\mathbf{x}})$, used in Craig [42], the fuzzy design of the speed gain works well, showing inherent robustness.

References

1. Allard, J.L., and Kaemmerer, W.F., "The Goal/subgoal Knowledge Representation for Real-time Process Monitoring", *Proceedings IJCAI'87*, Milano, Italy, 1987, pp. 394–398.
2. D'Ambrosio, B., *Qualitative Process Theory Using Linguistic Variables,* New York, Berlin, Springer-Verlag, 1989.
3. Aracil, J., García-Cerezo, A., and Ollero, A., "Stability Analysis of Fuzzy Control Systems: A Geometrical Approach." In: C.A. Kulikovsky, R.M. Huber, and G.A. Ferrate (eds.), *A.I. Expert Systems and Languages in Modelling and Simulation,* Amsterdam, North-Holland, 1988, pp. 323–330.
4. Aracil, J., Ollero, A., and García-Cerezo, A., "Stability Indices for the Global Analysis of Expert Control Systems," *IEEE Trans. on Systems, Man, and Cybernetics,* **19**(5)(1989)998–1007.
5. Aracil, J., García-Cerezo, A., Barreiro, A., and Ollero, A., "Design of Expert Fuzzy Controllers Based on Stability Criteria," *Proc. of the IFSA '91*, Brussels, 1991.
6. Arzen, K.-E., "Use of Expert Systems in Closed-loop Feedback Control," *Proceedings of American Control Conference*, Miami, Florida, 1986, pp. 140–145.
7. Assilian, S., and Mamdani, E.H., "Learning Control Algorithms in Real Dynamic Systems," *Proc. 4th Int. IFAC/IFIP Conf. on Digital Computer Appl. to Process Control,* Zürich, March 1974.
8. Åström, K.J., and Wittenmark, B., *Adaptive Control*, Addison-Wesley, 1989.
9. Åström, K.J., Anton, J.J., and Arzen, K.-E., "Expert Control" *Automatica*, **22**(1986)277–288
10. Baboshin, N., and Naryshkin, D., "On Identification of Multidimensional Fuzzy Systems," *Fuzzy Sets and Systems*, **35**(1990)325–331.
11. Bare, W.H., Mulholland, R.J. and Sofer, S.S., "Design of a Self-tuning Expert Controller for a Gasoline Refinery Catalytic Reformer," *Proceedings American Control Conference*, 1988, pp. 247–253.
12. Barreiro, A., and Aracil, J., "Stability of Uncertain Dynamical Systems," *Preprints of the IFAC Symposium on AI in Real-Time Control*, Delft, The Netherlands, 1992, pp. 177–182.
13. Bartolini, G., Casalino, G., Davoli, F., Mastretta, M., Minciardi, R. and Morten, E., "Development of Performance Adaptive Fuzzy Controllers with Application to Continuous Casting Plants." In: R. Trappl (ed), *Cybernetics and Systems Research*, Amsterdam, North-Holland, 1982, pp. 721-728.

14. Batur, C. and Kasparian, V., "A Real-time Fuzzy Self-tuning Control," *Proceedings of IEEE International Conference on Systems Engineering*, 1989, pp. 1810-1815.
15. Bekey, A.G., "The Human Operator in Control Systems," In: K.B. De Greene (ed.) *Systems Psychology*, New York, McGraw-Hill, 1970.
16. Bell, D.J., Cook, P.A., and Munro, N. (eds.), *Design of Modern Control Systems*, Peter Peregrinus, IEE Control Engineering Series, 1982.
17. Bellman, R.E., and Giertz, M., "On the Analytic Formalism of the Theory of Fuzzy Sets," *Information Science*, **5**(1973)149–156.
18. Bellman, R.E. and Zadeh, L.A., "Decision-making in a Fuzzy Environment," *Management Science*, **17**(1970)B141-B164.
19. Bellman, R.E., and Zadeh, L.A., "Local and Fuzzy Logics." In: J.M. Dunn, and G. Epstein (eds), *Modern Uses of Multiple-valued Logic*, Dordrecht, D. Reidel, 1977, pp. 105–165.
20. Bennett, M.E., "Real-time Continuous AI Systems", *IEE Proceedings, D*, **134**(4)(1987)272–277.
21. Berenji, H.R. et al., "An Experiment-based Comparative Study of Fuzzy Logic Control," *Proceedings American Control Conference*, Pittsburgh, PA, 1989, pp. 2752–2753.
22. Bezdek, J.C., *Pattern Recognition with Fuzzy Objective Function Algorithms*, New York, Plenum Press, 1981.
23. Black, M., "Vagueness: An Exercise in Logical Analysis," *Philos. Sci.*, **4**(1937)427–455.
24. Bocklisch, S.F., "Beratungssysteme mit unscharfen Klassifikatoren und Klassifikatorennetzen," *msr Berlin*, **30**(8)(1987)344–348.
25. Bolc, L. and Borowik, P., *Many-valued Logics. Volume 1: Theoretical Foundations*, Berlin, Springer-Verlag, 1992.
26. Bonissone, P.P., *A Pattern Recognition Approach to the Problem of Linguistic Approximation in Systems Analysis*, Memo UCB/ERL M78/57, Electronics Research Lab., Univ. of California, Berkeley.
27. Boverie, S. et al., Fuzzy Logic Control Compared with other Automatic Control Approaches, *Proceedings 30th IEEE-CDC Conf. Decision and Control*, Brighton, UK, December 11–13, 1991.
28. Box, G.E.P., and Jenkins, G.M., *Time Series Analysis: Forecasting and Control*, Holden-Day, 1989.
29. Braae, M., and Rutherford, D.A., "Selection of Parameters for a Fuzzy Logic Controller", *Fuzzy Sets and Systems*, **2**(1979)185–199.
30. Braae, M., and Rutherford, D.A., "Theoretical and Linguistic Aspects of the Fuzzy Logic Controller," *Automatica*, **15**(1979)553–577.
31. Bristol, E.H., "Pattern Recognition: An Alternative to Parameter Estimation in Adaptive Control", *Automatica*, **13**(1977), 197–202.
32. Buckley, J.J., Siler, W., and Tucker, Douglas: "A Fuzzy Expert System," *Fuzzy Sets and Systems*, **20**(1986)1–16.
33. Cao, Zhiqiang, Kandel, Abraham, and Li, Lihong, "A New Model of Fuzzy Reasoning," *Fuzzy Sets and Systems*, **36**(1990)311–325.
34. Cheeseman, P.C., "An Inquiry into Computer Understanding," *Computational Intelligence*, **4**(1)(1988)58–66.

35. Chen, Y.Y., and Tsao, T.C., "A Description of the Dynamical Behavior of Fuzzy Systems," *IEEE Trans. on Systems, Man, and Cybernetics,* **19**(4)(1989)745–755.
36. Cheng, W., Shen-yue, H., and Hua-kon, H., "Model Reference Fuzzy Adaptive Control," *Proceedings IFAC 9th Triennial World Congress*, Budapest, Hungary, 1984, pp. 889-892.
37. Chiu, S., and Togai, M., "A Fuzzy Logic Programming Environment for Real-Time Control," *Int. J. of Approximate Reasoning*, **2**(1988)163–175.
38. Clarke, D.W. "Implementation of Adaptive Controllers". In: C.J. Harris and S.A. Billings (eds.), *Self-tunung and adaptive control*, Peter Peregrinus, UK, 1981.
39. Cook, P.A., "Modified Multivariable Circle Theorems." In: D.J. Bell (ed), *Recent Mathematical Developments in Control*, Academic Press, 1972, pp. 367–372.
40. Corsberg, D., "Alarm Filtering: Practical Control Room Upgrade Busing Expert Systems Concepts", *InTech*, April 1987, pp. 39–42.
41. Cox, R.T., "On Inference and Inquiry—An Essay in Inductive Logic." In: L. Levine, and M. Tribus (eds.), *The Maximum Entropy Formalism*, Cambridge, MA, MIT Press, 1979.
42. Craig, J.J., *Adaptive Control of mechanical manipulators*, Addison-Wesley, 1985.
43. Czogala, E., and Pedrycz, W., "On Identification in Fuzzy Systems and its Applications in Control Problems," *Fuzzy Sets and Systems*, **6**(1981)73-83.
44. Da, R., *A Critical Study of Widely Used Fuzzy Implication Operators and their Influence on the Inference Rules*, PhD thesis, State University of Ghent, 1990.
45. Daugherity, W.C. et al., "Performance Evaluation of a Self-tuning Fuzzy Controller, *Proceedings IEEE International Conference Fuzzy systems*, San Diego, March 8–12, 1992, pp. 389–397.
46. Desoer, C.R., and Vidyasagar, M., *Feedback Systems. Input-Output Properties,* Academic Press, 1975.
47. Dijkman, J.G., Haeringen, H. van, and Lange, S.J. de, "Fuzzy Numbers," *Journal of Mathematical Analysis and Applications*, **92**(2)(1983)301–341.
48. Dijkman, J.G., and Haeringen, H. van, *Fuzzy Sets,* lecture notes Delft University of Technology, Delft.
49. Dombi, J., "A General Class of Fuzzy Operators and Fuzziness Measures Induced by Fuzzy Operators," *Fuzzy Sets and Systems*, **8**(1982)149–163.
50. Driankov, D., "Inference with a Single Fuzzy Conditional Proposition," *Fuzzy Sets and Systems,* **24**(1987)51–63.
51. Driankov, D., *Towards a Many-Valued Logic of Quantified Belief*, Thesis 192, Department of Computer and Information Science, Linköping.
52. Driankov, D., *Time for Some Fuzzy Thinking*, Internal Report Siemens A.G., Munich, 1990.
53. Dubois, D., and Prade, H., "Fuzzy Real Algebra: Some Results," *Fuzzy Sets and Systems*, **2**(1979)327–348.
54. Dubois, D., and Prade, H., *Fuzzy Sets and Systems: Theory and Applications,* Orlando, FL, Academic Press, 1980.
55. Dubois, D., and Prade, H., "A Class of Fuzzy Measures Based on Triangular Norms," *Int. J. of General Systems*, **8**(1982),43–61.
56. Dubois, D., and Prade, H., "Fuzzy Logics and the Generalized Modus Ponens Revisited," *Cybernetics and Systems*, **15**(1984)293–331.

57. Dubois, D., and Prade, H., "The Generalized Modus Ponens under Sup-min Composition: a theoretical study." In: M.M. Gupta, A. Kandel, W. Bandler, and J.B. Kiszka (eds), *Approximate Reasoning in Expert Systems*, Amsterdam, North-Holland, 1985, pp. 217–232.
58. Dubois, D., and Prade, H., "Twofold Fuzzy Sets and Rough Sets—Some Issues in Knowledge Representation," *Fuzzy Sets and Systems,* **23**(1987)3–18.
59. Dubois, D., and Prade, H., *Théorie des Possibilités—Applications à la Représentation des Connaissances en Informatique*, Paris, Mason, 1987 (2ème édition 1988). Translated into English: *Possibility Theory—An Approach to Computerized Processing of Uncertainty*, New York, Plenum Press, 1988.
60. Dubois, D., and Prade, H., "On the Combination of Uncertain and Imprecise Pieces of Information in Rule-Based Systems—A Discussion in the Framework of Possibility Theory," *Int. J. of Approximate Reasoning*, **2**(1988)65–87.
61. Dubois, D., Martin-Clouaire, R., and Prade, H., "Practical Computing in Fuzzy Logic." In: M.M. Gupta, and T. Yamakawa (eds.), *Fuzzy Computing—Theory, Hardware and Applications,* Amsterdam, North-Holland, 1988, pp. 11–34.
62. Dubois, D., and Prade, H., "Comments on 'An Inquiry into Computer Understanding'," *Computational Intelligence,* **4**(1)(1988)73–76.
63. Dubois, D., and Prade, H., "Basic Issues on Fuzzy Rules and their Application to Fuzzy Control," *Proceedings of the IJCAI-91 Workshop on Fuzzy Control,* Sydney, 1991, pp. 5–17.
64. Efstathiou, J., "Rule-Based Process Control Using Fuzzy Logic." In: E. Sanchez and L.A. Zadeh (eds.), *Proceedings of Approximate Reasoning in Intelligent systems,* Oxford, UK, 1987, pp. 145–158.
65. Eshrag, F., and Mamdani, E.H., "A General Approach to Linguistic Approximation." In: Mamdani, E.H., Gaines, B.R. (eds), *Fuzzy Reasoning and Its Applications,* London, Academic Press, 1981, pp. 169–187.
66. Eykhoff, P., *System Identification, Parameter and State Estimation,* London, J. Wiley, 1974.
67. Forsman, K., and Strömberg, J-E., *A Short Introduction to Fuzzy Control for Control Theorists; With Annotated Bibliography*, Research Report LiTH-ISY-I-1339, University of Linköping, Linköping, Sweden, 1992.
68. Fox, M.S., Allen, B.P., and Strohm, G.A., "Job-shop Scheduling: An Investigation in Constraint-based Reasoning", *2nd National Conference on AI*, 1982, pp. 155–158.
69. Fox, M.S., *Constraint-directed Search: A Case Study of Job-shop Scheduling*, Ph.D. thesis, Carnegie-Mellon University, 1983.
70. Fujiwara, R. et al., "An Intelligent Load-flow Engine for Power Systems Planning", *Proceedings 1985 Power Industry Computer Applications Conference*, 1985, pp. 236–241.
71. García-Cerezo, A., *Aplicaciones del Razonamiento Aproximado en el Control y Supervisión de Procesos,* Ph.D. thesis, University of Santiago, Spain, 1987.
72. García-Cerezo, A., Barreiro, A., Aracil, J., and Ollero, A., "Design of Robust Intelligent Control of Manipulators," *Proceedings of the IEEE Int. Conference on Systems Engineering,* Dayton, Ohio, USA, 1991, pp. 225–228.
73. García-Cerezo, A., Ollero, A., and Aracil J., "Stability of Fuzzy Control Systems by using Nonlinear Systems Theory," *Preprints of the IFAC Symposium on AI in Real-Time Control,* Delft, The Netherlands, 1992, pp. 171–176.

74. Giles, R., "Lukasiewicz Logic and Fuzzy Set Theory." In: E.H. Mamdani, and B.R. Gaines (eds.), *Fuzzy Reasoning and Its Applications,* London, Academic Press, 1981, pp. 117–131.
75. Glas, M. de, "Theory of Fuzzy Systems," *Fuzzy Sets and Systems,* **21**(1983)65–77.
76. Glas, M. de, "Invariance and Stability of Fuzzy System," *J. Math. Analysis Appl.,* **199**(1984)299–319.
77. Glorennec, P.Y., "Adaptive Fuzzy Control," *Proc. of the IFSA '91*, 1991, pp. 33-36.
78. Goguen, J.A., "On Fuzzy Robot Planning." In: Zadeh, L.A. et al. (eds), *Fuzzy Sets and Their Application to Cognitive and Decision Processes,* New York, Academic Press, 1975, pp. 429–447.
79. Gorzałgzany, M.B., "A Method of Inference in Approximate Reasoning Based on Interval-Valued Fuzzy Sets," *Fuzzy Sets and Systems,* **21**(1987)1–17.
80. Gottwald, S., and Pedrycz, W., "Problems of the Design of Fuzzy Controllers." In: M.M. Gupta, A. Kandel, W. Bandler, and J.B. Kiszka (eds), *Approximate Reasoning in Expert Systems*, Amsterdam, North-Holland, 1985, pp. 393–405.
81. Graham, B., *Fuzzy Identification and Control*, Ph.D. thesis, University of Queensland, Australia, 1986.
82. Graham, B., and Newell, R., "Fuzzy Identification and Control of a Liquid Level Rig," *Fuzzy Sets and Systems*, **26**(1988)255–273.
83. Graham, B., and Newell, R., "Fuzzy Adaptive Control of a First-order Process," *Fuzzy Sets and Systems*, **31**(1989)47–65.
84. Guckemheimer, J., and Holmes, P., *Nonlinear Oscillations, Dynamical Systems and Bifurcations of Vector Fields,* New York, Springer-Verlag, 1983.
85. Hall, L., and Kandel, A., *Designing Fuzzy Expert Systems,* Cologne, Verlag TÜV Rheinland, 1986.
86. Hamacher, H., "Über logische Verknüpfungen unscharfer Aussagen und deren zugehörigen Bewertungsfunktionen." In: R. Trappl and F. de P. Hanika, *Progress in Cybernetics and Systems Research*, Vol. II, Hemishere Publ. Corp., New York, 1975, pp. 276–287.
87. Handelman, D.A., and Stengel, R.F., "An Architecture for Real-time Rule-based Control," *Proceedings of the American Control Conference*, Minneapolis, 1987.
88. Hansen, P.D., and Kraus, T.W., "Expert Systems and Model Based Self-tuning Controllers," In: D.M. Considine and G.D. Considine (eds.) *Standard Handbook of Industrial Automation*, New York, Chapmann and Hall, 1986.
89. Hayashi, S., "Auto-tuning Fuzzy PI Controller," *Proc. of the IFSA '91*, 1991, pp. 41–44.
90. Hellendoorn, H., "Closure Properties of the Compositional Rule of Inference," *Fuzzy Sets and Systems,* **35**(1990)163–183.
91. Hellendoorn, H., "The Generalized Modus Ponens Confronted with Some Criteria," *Proceedings of the TIMS/ORSA-conference*, Las Vegas, 1990.
92. Hellendoorn, H., *Reasoning with Fuzzy Logic*, Ph.D. thesis, Delft University of Technology, Delft, 1990, 141p.
93. Hellendoorn, H., "The Generalized Modus Ponens Considered as a Fuzzy Relation," *Fuzzy Sets and Systems*, **46**(1)(1992)29–48.

94. Hellendoorn, H., "Fuzzy Logic and Fuzzy Control." In: R. van der Vleuten et al. (eds.), *Clear Applications of Fuzzy Logic*, (Proceedings IEEE-Symposium Delft, 17 Oct. 1991) Delft, pp. 57–82.
95. Hellendoorn, H. & Thomas, C., *On Defuzzification in Fuzzy Controllers*, Internal Report SICSL-92/3 Siemens AG, München, 1992.
96. Hendy, R.J., *Supervisory Control using Fuzzy Set Theory*, Ph.D. thesis, University of Queensland, 1980.
97. Heshmaty, B., and Kandel, A., "Fuzzy Linear Repression and its Application to Forecasting in Uncertain Environment," **15**(1985)159–191.
98. Higashi, M., and Klir, G., "Identification in Fuzzy Relation Systems," *IEEE Trans. Systems, Man, and Cybernetics*, **14**(1984)349–355.
99. Huang, C.Y., and Stengel, R.F., "Failure Mode Determination in a Knowledge Based Control System," *Proceedings of the American Control Conference*, Minneapolis, 1987.
100. Huang, L.-J., and Tomizuka, M., "A Self-Paced Fuzzy Tracking Controller for Two-Dimensional Motion Control," *IEEE Trans. on Systems, Man, and Cybernetics*, **20**(5)(1990)1115–1124.
101. Isaka, S. et al., "On the Design and Performance Evaluation of Adaptive Fuzzy Controllers," *Proc. of the 27th Conf. on Decision and Control*, Austin, TX, 1988, pp. 1068-1069.
102. Isaka, S., and Sebald, A.V., "An Optimization Approach for Fuzzy Controller Design," *Proceedings American Control Conference*, Pittsburgh, PA, 1989, pp. 1485–1490.
103. Isermann, R., "Parameter Adaptive Control Algorithms: A Tutorial," *Automatica*, **18**(1982)513–528
104. Johnson, B.W., *Design and Analysis of Fault Tolerant Digital Systems*, Reading, MA, Addison-Wesley, 1989.
105. Junhong, N., "The EA FCS: a New and Efficient Adaptive Control Scheme," *Proceedings of IEEE International Conference on Systems Engineering*, 1989.
106. Jones, A.H. and Shenassa, M.H., "Real-time Expert Tuners for PI Controllers Incorporating Closed-loop Step-response Identifiers," *3rd IEEE International Symposium on Intelligent Control*, Arlington, VA, August 24–26, 1988.
107. Jones, A.H. and Porter, B., "Design of Adaptive Digital Set-point Tracking PID Controllers Incorporating Recursive Step-response Matrix Identifiers for Multivariable Plants," *Proceedings 24th IEEE Conference on Decision and Control*, Fort Lauderdale, December 1985.
108. Jurke, M., "Optimization and Decision-making under Multicriterial Conditions by Means of Fuzzy-LP-models," *Syst. Anal. Model. Simul.*, **7**(9)(1990)721–724.
109. Kandel, A., and Lee, S.C., *Fuzzy Switching and Automata: Theory and Applications*, New York, Crane Russak, 1979.
110. Kandel, A., *Fuzzy Mathematical Techniques with Applications*, Reading, MA, Addison-Wesley, 1986.
111. Karwowski, W., Mulholland, N.O., Ward, T.L., and Jagannathan, V., "A Fuzzy Knowledge Base of an Expert System for Analysis of Manual Lifting Tasks," *Fuzzy Sets and Systems*, (1987).
112. Kaufman, A., *Introduction to the Theory of Fuzzy Subsets, Vol. 1*, New York, Academic Press, 1975.

113. Kaufman, A., and Gupta, M.M., *Fuzzy Mathematical Models in Engineering and Management Science,* Amsterdam, North-Holland, 1988.
114. Kawaji, S., and Matsunaga, N., "Fuzzy Control of VSS Type and its Robustness," *Proc. of the IFSA '91,* Vol. "engineering," Brussels, July 1991, pp. 81–88.
115. Kickert, W.M., and Mamdani, E.H., "Analysis of Fuzzy Logic Controllers," *Fuzzy Sets and Systems,* **1**(1)(1978)29–44.
116. Kickert, W.M., and Mamdani, E.H., *Analysis of Fuzzy Logic Controllers,* Queen Mary College, Research Report 2, 1985.
117. Kiszka, J.B., Gupta, M.M., and Nikiforuk, M.N., "Energetistic Stability of Fuzzy Dynamic Systems," *IEEE Trans. on Systems, Man, and Cybernetics,* **15**(1985)783–792.
118. Klir, G.J., and Folger, T.A., *Fuzzy Sets, Uncertainty and Information,* Englewood Cliffs, NJ, Prentice Hall, 1988.
119. Kochen, M., "Applications of Fuzzy Sets in Psychology." In: L.A. Zadeh, K.S. Fu, K. Tanaka, and M. Shimura, (eds), *Fuzzy Sets and their Applications to Cognitive and Decision Processes,* New York, Academic Press, 1975, pp. 395–408.
120. Kosko, B., *Neural Networks and Fuzzy Systems,* Englewood Cliffs, NJ, Prentice Hall, 1992.
121. Kraus, T.W., and Myron, T.J., "Self-tuning PID Controller uses Pattern Recognition Approach", *Control Engineering,* (6)(1984)106–111.
122. Kuhn, T.S., *The Structure of Scientific Revolutions,* University of Chicago Press, 1970, 2nd edition.
123. Lamb, D.E, Dhurjati, P., and Chester, D.L., "Development of an Expert System for Fault Identification in a Commercial Scale Chemical Process," *Proceedings 6th International Workshop on Expert Systems,* **2**, 1371–1382.
124. Ledley, R.S., *Digital Computer and Control Engineering,* New York, McGraw-Hill, 1960.
125. Lee, C.C., "Fuzzy Logic in Control Systems: Fuzzy Logic Controller—Part I and II," *IEEE Trans. on Systems, Man, and Cybernetics,* **20**(2)(1990)404–418 and **20**(2)(1990)419–435.
126. Liu, K., and Gertler, J., "An On-line Stabilization of Adaptive Controllers by De-tuning in a Supervisory Framework," *Proceedings of the American Control Conference,* Minneapolis, 1987, pp. 194–200.
127. Locke, M., "Fuzzy Modelling in Expert Systems for Process Control," *Syst. Anal. Model. Simul.,* **7**(9)(1990)715–719.
128. Maeda, A., Someya, R., and Funabashi, M., "A Self-Tuning Algorithm for Fuzzy Membership Functions using a Computational Flow Network," *Proc. of the IFSA '91,* 1991.
129. Mamdani, E.H., "Application of Fuzzy Algorithm for Control of Simple Dynamic Plant," *Proc. IEEE,* **121**(12)(1974)1585–1888.
130. Mamdani, E.H., and Baaklini, N., "Prescriptive Method for Deriving Control Policy in a Fuzzy Logic Controller," *Electronic Letters,* **11**(1975)625-626.
131. Mamdani, E.H., "Advances in the Linguistic Synthesis of Fuzzy Controllers," *Int. J. Man-Mach. Stud.,* **8**(1976)669–678.
132. Mamdani, E.H., Procyk, T., and Baaklini, N., "Application of Fuzzy Logic to Controller Design Based on Linguistic Protocol." In: E.H. Mamdani and B.R.

Gaines (eds), *Discrete Systems and Fuzzy Reasoning*, Queen Mary College, University of London, 1976, pp. 125–149.

133. Mamdani, E.H., "Advances in the Linguistic Synthesis of Fuzzy Controllers." In: E.H. Mamdani, and B.R. Gaines (eds.), *Fuzzy Reasoning and Its Applications,* London, Academic Press, 1981, pp. 325–334.
134. Mamdani, E.H., and Assilian, S., "An Experiment in Linguistic Synthesis with a Fuzzy Logic Controller." In: E.H. Mamdani, and B.R. Gaines (eds.), *Fuzzy Reasoning and Its Applications,* London, Academic Press, 1981, pp. 311–323.
135. Mamdani, E.H., "Application of Fuzzy Algorithms for Control of Simple Dynamic Plant," *Proc. IEE,* **121**(12)(1984).
136. Manzoul, M. A., "Faults in Fuzzy Logic Systolic Arrays," *Cybernetics and Systems: An Int. Journal,* **21**(1990)513–524.
137. Masuoka, R. et al., "Neurofuzzy System - Fuzzy Inference using a Structured Neural Network," *Proc. of the Int. Conf. on Fuzzy Logic and Neural Networks,* Iizuka, Fukuoka, Japan, July 22-24, 1990, pp. 173-178.
138. McCahon, C.S., and Lee, E.S., "Comparing Fuzzy Numbers: The Proportion of the Optimum Method," *Int. J. of Approximate Reasoning,* 4(1990)159–181.
139. Miller, W.T., Sutton, R.S., and Werbos, P.J. (eds), *Neural Networks for Control,* Cambridge, MA, MIT Press, 1990.
140. Milne, R., AI for On-line Diagnosis", *IEE Proceedings, D,* **134**(4)(1987)238–244.
141. Milliken, K.R. et al., YES/MVS and the Automation of Operations for Large Computer Complexes, *IBM Systems Journal,* **25**(2)159–180.
142. Mizumoto, M., Fukami, S., and Tanaka, K., "Some Methods of Fuzzy Reasoning." In: Gupta, M.M., Ragade, R.K., and Yager, R.R. (eds), *Advances in Fuzzy Set Theory and Applications,* Amsterdam, North-Holland, 1979, pp. 117–136.
143. Mizumoto, M., Fukami, S., and Tanaka, K., "Some Properties of Fuzzy Numbers," In: Gupta, M.M., Ragade, R.K., and Yager, R.R. (eds), *Advances in Fuzzy Set Theory and Applications,* Amsterdam, North-Holland, 1979, pp. 153–164.
144. Mizumoto, M., and Zimmermann, H.-J., "Comparison of Fuzzy Reasoning Methods," *Fuzzy Sets and Systems,* **8**(1982)253–283.
145. Mizumoto, M., "Improvement Methods of Fuzzy Controls," *Proc. of the IFSA '89,* Seattle, WA, pp. 60–62.
146. Murphy, G.I., *Basic Automatic Control Theory,* Princeton, NJ, D. Van Nostrand, 1962.
147. Nauta Lemke, H.R. van, *Vage Verzamelingen 2, 3,* Syllabus Control and System Theory (ℓ88A), Delft University of Technology, Delft.
148. Negoita, C.V., and Ralescu, D.A., *Applications of Fuzzy Sets to System Analysis,* Basel, Birkhauser Verlag, 1975.
149. Nola, A. di, Sessa, S., Pedrycz, W., and Sanchez, E., *Fuzzy Relation Equations and their Applications to Knowledge Engineering,* Dordrecht, Kluwer, 1989.
150. Nomura, H., Hayashi, I., and Wakami, N., "A Self-Tuning Method of Fuzzy Control by Descent Method," *Proc. of the IFSA '91,* Brussels, 1991, pp. 155–158.
151. Norton, J.P., *An Introduction to Identification,* London, Academic Press, 1986.
152. Ollero, A., and García-Cerezo, "Direct Digital Control, Autotuning and Supervision using Fuzzy Logic," *Fuzzy Sets and Systems,* **30**(1989)135–153.
153. Ollero, A., García-Cerezo, A., and Aracil, J., *Design of Fuzzy Control Systems,* Dpto. Ing. Sist., University Malaga, Research Report, 1992.

154. Osborne, R.L., Gonzalez, A.J., and Weeks, C.A., "First Years Experience with On-line Generator Diagnostics," *Proceedings American Control Conference*, Chicago, IL, 1986.
155. Palm, R., "Fuzzy Controller for a Sensor Guided Robot Manipulator," *Fuzzy Sets and Systems,* **29**(1)(1989)133–149.
156. Palm, R., "Sliding Mode Fuzzy Control," *Proceedings IEEE International Conf. on Fuzzy Systems FUZZ-IEEE'92,* March 8–12, 1992, San Diego, pp. 519-526.
157. Pappis, C.P., "Application of a Measure of Proximity to Fuzzy Control Algorithms," *Proc. of the IFSA '89*, Seattle, pp. 90–93.
158. Pawlak, Z., "Rough Sets and Fuzzy Sets," *Fuzzy Sets and Systems,* **17**(1985)99–102.
159. Pedrycz, W., "An Identification Algorithm in Fuzzy Relational Systems," *Fuzzy Sets and Systems*, **13**(1984)153-167.
160. Pedrycz, W., "Identification in Fuzzy Systems," *IEEE Trans. Systems, Man, and Cybernetics*, **14**(1984)361-366.
161. Pedrycz, W., "Approximate Solutions of Fuzzy Relational Equations," *Fuzzy Sets and Systems*, **28**(1988)183-202.
162. Pedrycz, W., "Processing in Relational Structures: Fuzzy Relational Equations," *Fuzzy Sets and Systems,* **40**(1990)77–106.
163. Pedrycz, W., "Fuzzy Modelling: Fundamentals, Construction and Evaluation," *Fuzzy Sets and Systems,* **41**(1991)1–15.
164. Pedrycz, W., *Fuzzy Control and Fuzzy Systems,* 2nd revised edition, Research Studies Publ., 1992.
165. Peng, X.T., Kandel, A., and Wang, P., "Concepts, Rules, and Fuzzy Reasoning: A Factor Space Approach," *IEEE Trans. on Systems, Man, and Cybernetics,* **21**(1)(1991)194–205.
166. Pétieau, A.M., Moreau, A., and Willaeys, D., "Tools for Approximate Reasoning in Expert Systems," *Proceedings of the TIMS/ORSA conference*, Las Vegas, NV, 1990.
167. Pfluger, N., Yen, J. & Langari, R., "A Defuzzification Strategy for a Fuzzy Logic Controller Employing Prohibitive Information in Command Formulation," *Proceedings of the IEEE International Conference on Fuzzy Systems*, San Diego, CA, 1992, pp. 717–723.
168. Porter, B., Jones, A.H., and MCKeown, C.B., "Real-time Expert Tuners for PI Controllers," *IEE Proceedings, D* **134**(4)(1987).
169. Procyk, T.J., and Mamdani, E.H., "A Linguistic Self-Organizing Process Controller," *Automatica,* **15**(1)(1979)15–30.
170. Puccia, Ch.J., and Levins, R., *Qualitative Modeling of Complex Systems,* Cambridge, MA, Harvard University Press, 1985.
171. Ragade, R.K., and Gupta, M.M., "Fuzzy Set Theory: Introduction." In: M.M. Gupta (ed), *Fuzzy Automata and Decision Processes,* New York, North-Holland, 1977, pp. 105–131.
172. Rahman, S. et al., "An Expert System Based Algorithm for Short-term Load Forecast," *Proceedings PES Winter Power Meeting*, 1987.
173. Ray, K.S., and Majumder, D.D., "Application of the Circle Criteria for Stability Analysis of Linear SISO And MIMO Systems Associated with Fuzzy Logic Controller," *IEEE Trans. Systems, Man, and Cybernetics*, **14**(2)(1984)345-349.

174. Ray, K.S., Ghosh, A., and Majumder, D.D., "L2-Stability and the Related Design Concept for SISO Linear System Associated with Fuzzy Logic Controllers," *IEEE Trans. Systems, Man, and Cybernetics*, **14**(6)(1984)932-939.
175. Rescher, N., *Many-Valued Logic*, New York, McGraw-Hill, 1969.
176. Rosenbrock, H.H., "Multivariable Circle Theorems." In: D.J. Bell (ed), *Recent Mathematical Developments in Control*, London, Academic Press, 1972, pp. 345–365.
177. Rosenbrock, H.H., *Computer Aided Control System Design*, London, Academic Press, 1974.
178. Rumelhart, D. E. and McClelland, J.L., *Parallel Distributed Processing: Explorations in the Microstructure of Cognition, Vols. 1 and 2*, Cambridge, MA, MIT Press, 1986.
179. Rytoft, C., Peterson, M., Rosenberg, B., and Erzen, K.-E., *Knowledge-Based Real-Time Control Systems*, Feasibility Study, IT-4. ABB Corporate Research, Satt Control, TeleLogic, and Automatic Control, Lund Institute of Technology, Internal project report, 1988.
180. Sadler, D.R., *Numerical Methods for Nonlinear Regression*, St. Lucia, University of Queensland Press, 1975.
181. Safonov, M.G., *Stability and Robustness of Multivariable Feedback Systems*, Cambridge, MA, MIT Press, 1980.
182. Sakagushi, T. et al, "Prospects of Expert Systems in Power Systems Operation", *Proceedings 9th Power Systems Computation Conference*, 1987, Cascais, Portugal.
183. Sanoff, S.P., and Wellstead, P.E., "Expert Identification and Control," *Technical Report PB85-155562*, National Technical Information Service, Springfield, VA, 1984
184. Saridis, G., "Towards the Realization of Intelligent Controls," *Proceedings of the IEEE*, **67**(1979)1115–1133.
185. Savic, D.A., and Pedrycz, W., "Evaluation of Fuzzy Linear Regression Models," *Fuzzy Sets and Systems*, **39**(1991)51–63.
186. Scharf, E.M., and Mandic N.J., "The Application of a Fuzzy Controller to the Control of Multi-degree-freedom Robot Arm." In: M. Sugeno (ed), *Industrial Applications of Fuzzy Control*, Amsterdam, North-Holland, 1985, pp. 41–62.
187. Schweitzer, B., and Sklar, A., "Associative Functions and Statistical Triangle Inequalities," *Publicationes Mathematicae Debrecen*, **8**(1961)169–186.
188. Schweitzer, B., and Sklar, A., "Associative Functions and Abstract Semigroups," *Publicationes Mathematicae Debrecen*, **10**(1963)69–81.
189. Shafer, G., *A Mathematical Theory of Evidence*, Princeton, NJ, Princeton University Press, 1976.
190. Shao, S., "Fuzzy Self-Organizing Controller and its Application for Dynamic Processes," *Fuzzy Sets and Systems*, **26**(1988)151–164.
191. Shin, K.G., and Cui, X., "Design of a Knowledge-Based Controller for Intelligent Control Systems," *Proceedings American Control Conference*, Pittsburgh, PA, 1989, pp. 1461–1466.
192. Siler, W., Buckley, J., and Tucker, D., "Functional Requirements for a Fuzzy Expert System Shell." In: E. Sanchez, and L.A. Zadeh (eds), *Approximate Reasoning in Intelligent Systems, Decision and Control*, Oxford, Pergamon Press, 1987, pp. 21–31.

193. Siler, W., Tucker, D., and Buckley, J., "A Parallel Rule Firing Fuzzy Production System With Resolution of Memory Conflicts by Weak Fuzzy Monotonicity," *Int. J. Man-Machine Studies,* **26**(1987)321–332.

194. Siler, W., and Ying, H., "Fuzzy Control Theory: The Linear Case," *Fuzzy Sets and Systems,* **33**(1989)275–290.

195. Slotine, J.E., "The Robust Control of Robot Manipulators," *Intern. J. Robotics Research (MIT),* **4**(2)(1985)49–64.

196. Smith, S.F., Fox, M.S., and Ow, P.S., "Construction and Maintaining Detailed Production Plans", *AI Magazine,* **7**(4)(1986)45–61.

197. Sripada, N.R., Fischer, D.G., and Morris, A.J., "AI Application for Process Regulation and Servo Control," *IEE Proceedings, D,* **134**(4)(1987)251–259.

198. Stephanopoulos, G., *Chemical Process Control: An Introduction to Theory and Practice,* Englewood Cliffs, NJ, Prentice Hall, 1984.

199. Sugeno, M., and Murakami K., "Fuzzy Parking Control of Model Car," *Proceedings 23rd IEEE Conf. on Decision and Control,* Las Vegas, 1984.

200. Sugeno, M., "An Introductory Survey of Fuzzy Control," *Information Sciences,* **36**(1985)59–83.

201. Sugeno, M. (ed.), *Industrial Applications of Fuzzy Control,* Amsterdam, North-Holland, 1985.

202. Sugeno, M., and Kang, G.T., "Fuzzy Modelling and Control of Multilayer Incinerator," *Fuzzy Sets and Systems,* **18**(1986)329-346.

203. Sugeno, M., and Kang, G.T., "Structure Identification of Fuzzy Model," *Fuzzy Sets and Systems,* **28**(1988)15–33.

204. Takagi, T., and Sugeno, M., "Fuzzy Identification of Systems and Its Applications to Modeling and Control," *IEEE Trans. on Systems, Man, and Cybernetics,* **15**(1)(1985)116–132.

205. Takagi, H., *Fusion Technology of Fuzzy Theory and Neural Networks—Survey and Future Directions,* Report of the Central Research Laboratories, Matsushita Electrical Industrial Co. Ltd., 1990.

206. Tanaka, H., Uejima, S., and Asai, K., "Linear Repression Analysis with Fuzzy Model." *IEEE Trans. Systems, Man, and Cybernetics,* **6**(1982)903–907.

207. Tanaka, H., "Fuzzy Data Analysis by Possibilistic Linear Models." *Fuzzy Sets and Systems,* **24**(1987)363–375.

208. Tanaka, K., and Sano, M., "A New Tuning Method of Fuzzy Controllers," *Proc. of the IFSA '91,* Brussels, 1991, pp. 207–210.

209. Tanaka, K., Sugeno, M., "Stability Analysis and Design of Fuzzy Control Systems" *Fuzzy Sets and Systems,* **45**(1992)135–156.

210. Tang, K.L., and Mulholland, R.J., "Comparing Fuzzy Logic with Classical Controllers Designs," *IEEE Trans. on Systems, Man, and Cybernetics,* **17**(6)(1987)1085–1087.

211. Tanscheit, R., and Scharf, E.M., "Experiments with the Use of a Rule-based Self-organising Controller for Robotic Applications," *Fuzzy Sets and Systems,* **26**(1988)195–214.

212. Taunton, C., "Expert Systems in Process Control: An Overview," *ISA'89 Advanced Control conference,* NEC, Birmingham, 1989, pp. 5.1–5.5.

213. Thole, U., Zimmermann, H.-J., and Zysno, P., "On the Suitability of Minimum and Product Operators for the Intersection of Fuzzy Sets," *Fuzzy Sets and Systems,* **2**(1979)167–180.

214. Togai, M., "A Fuzzy Inverse Relation Based on Gödelian Logic and its Applications," *Fuzzy Sets and Systems,* **17**(1985)211–219.
215. Tong, R.M., "Analysis of Fuzzy Control Algorithms using the Relation Matrix," *Int. J. Man-Machine Studies*, **8**(1976)679–686.
216. Tong, R.M., "Synthesis of Fuzzy Models for Industrial Processes—Some Recent Results," *Int. J. General Systems*, **4**(1978)143–162.
217. Tong, R.M., "The Evaluation of Fuzzy Models Derived from Experimental Data," *Fuzzy Sets and Systems*, **4**(1980)1–12.
218. Toth, H., "From Fuzzy-Set Theory to Fuzzy Set-Theory: Some Critical Remarks on Existing Concepts," *Fuzzy Sets and Systems,* **23**(1987)219–237.
219. Turksen, I.B., "Measurement of Membership and their Acquisition." *Fuzzy Sets and Systems*, **40**(1991)5–38.
220. Utkin, V.I., "Variable Structure Systems with Sliding Mode: A Survey," *IEEE Trans. Automatic Contr.,* **22**(1977)212–222.
221. Vidyasagar, M., "New Directions of Research in Nonlinear System Theory," *Proceeding of the IEEE,* **77**(8)(1986)1060–1090.
222. Vachtsevanos, G., and Hexmoor, H., "A Fuzzy Logic Approach to Robotic Path Planning with Obstacle Avoidance," *Proceedings of the 25th Conference on Decision and Control,* Athens, Greece, 1986, pp. 1262–1264.
223. Wakileh, B.A.M., and Gill, K.F., "Use of Fuzzy Logic in Robotics," *Computers in Industry,* **10**(1988)35–46.
224. Wang, B.H., and Vachtsevanos, G., "Learning Fuzzy Logic Control: An Indirect Control Approach", *Proceedings IEEE International Conference Fuzzy Systems,* March 8–12, San Diego, 1992, pp. 297–303.
225. Weber, S., "A General Concept of Fuzzy Connectives, Negations and Implications Based on t-norms and t-co-norms," *Fuzzy Sets and Systems*, **11**(1983)115–134.
226. Xu, C.-W., and Lu, Y.-Z., "Fuzzy Model Identification and Self Learning for Dynamic Systems," *IEEE Trans. on Systems, Man, and Cybernetics,* **17**(4)(1987)683–689.
227. Yager, R.R., "Robot Planning with Fuzzy Sets," *Robotica*, **1**(1983)41–50.
228. Yamakawa, T., "Intrinsic Fuzzy Electronic Circuits for Sixth Generation Computer." In: M.M. Gupta, and T. Yamakawa (eds.), *Fuzzy Computing*, Amsterdam, North-Holland, 1988, pp. 157–171.
229. Yamakawa, T., "Stabilization of an Inverted Pendulum by a High-Speed Fuzzy Logic Controller Hardware System," *Fuzzy Sets and Systems*, **32**(1989)161–180.
230. Yamashita, Y., Matsumoto, S., and Suzuki, M., "Start-up of a Catalytic Reactor by Fuzzy Controller," *J. Chemical Engineering of Japan*, **21**(1988)277-281.
231. Yamazaki, T., and Mamdani, E.H., "On the Performance of a Rule-based Self-organising Controller," *Proc. of IEEE Conf. on Applications of Adaptive and Multivariable Control*, Hull, England, 1982, pp. 50-55.
232. Ying, H., Siler, W., and Buckley, J.J., "Fuzzy Control Theory: A Nonlinear Case," *Automatica,* **26**(3)(1990)513–520.
233. Zadeh, L.A., "Fuzzy Sets," *Information and Control,* **8**(1965)338–353.
234. Zadeh, L.A., "Probability Measures of Fuzzy Events," *J. of Mathematical Analysis and Applications*, **23**(1968)421–427.
235. Zadeh, L.A., "Quantitative Fuzzy Semantics," *Information Sciences,* **3**(1971)159–176.

236. Zadeh, L.A., "A Fuzzy-Set-Theoretic Interpretation of Linguistic Hedges," *Journal of Cybernetics,* **2**(1972)(3)4–34.
237. Zadeh, L.A., "Outline of a New Approach to the Analysis of Complex Systems and Decision Processes," *IEEE Trans. on Systems, Man and Cybernetics,* **3**(1)(1973)28–44.
238. Zadeh, L.A., "A Rationale for Fuzzy Control," *Trans. ASME, Ser. G,* **94**(1974)3–4.
239. Zadeh, L.A., "Calculus of Fuzzy Restrictions." In: Zadeh, L.A. et al. (eds), *Fuzzy Sets and Their Applications to Cognitive and Decision Processes,* New York, Academic Press, 1975, pp. 1–39.
240. Zadeh, L.A., "The Concept of a Linguistic Variable and its Application to Approximate Reasoning," *Information Science,* Part I: **8**(1975)199–249, Part II: **8**(1975)301–357. Part III: **9**(1975)43–80.
241. Zadeh, L.A., "PRUF—A Meaning Representation Language for Natural Languages," *Int. J. of Man-Machine Studies,* **10**(1978)395–460. (Also in: Mamdani, E.H., and Gaines, B.R. (eds), *Fuzzy Reasoning and Its Applications,* London, Academic Press, 1981, pp. 1–66.)
242. Zadeh, L.A., "Fuzzy Sets as a Basis for a Theory of Possibility," *Fuzzy Sets and Systems,* **1**(1978)3–28.
243. Zadeh, L.A., "The Role of Fuzzy Logic in the Management of Uncertainty in Expert Systems," *Fuzzy Sets and Systems,* **11**(1983)199–227.
244. Zames, G., "On the I-O stability of Time Varying Nonlinear Feedback Systems," *IEEE Trans. on Automatic Control,* **11**(1966)228–238.
245. Zimmermann, H.-J., "Testability and Meaning of Mathematical Models in Social Sciences," *Mathematical Modelling,* **1**(1980)123–129.
246. Zimmermann, H.-J., and Zysno, P., "Latent Connectives in Human Decision Making," *Fuzzy Sets and Systems,* **4**(1980)37–51.
247. Zimmermann, H.-J., *Fuzzy Set Theory and its Applications,* Dordrecht, Kluwer, 1985.
248. Zimmermann, H.-J., *Fuzzy Sets, Decision Making, and Expert Systems*, Boston, Kluwer, 1987.

Index

Lightning Source UK Ltd.
Milton Keynes UK
UKOW07f1852031214

242609UK00002B/167/P

9 783642 082344